W0091732

Molecular Beacons

METHODS IN MOLECULAR BIOLOGY

METHODS IN MOLECULAR BIOLOGY™

John M. Walker, SERIES EDITOR

METHODS IN MOLECULAR BIOLOGY™

Molecular Beacons: Signalling Nucleic Acid Probes, Methods, and Protocols

Edited by

Andreas Marx

Department of Chemistry, University of Konstanz, Konstanz, Germany

and

Oliver Seitz

Institute of Chemistry, Humboldt University Berlin, Berlin, Germany

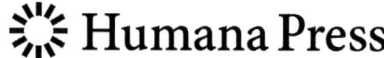 Humana Press

© 2008 Humana Press Inc.
999 Riverview Drive, Suite 208
Totowa, New Jersey 07512
www.humanapress.com

All rights reserved. No part of this book may be reproduced, stored in a retrieval system, or transmitted in any form or by any means, electronic, mechanical, photocopying, microfilming, recording, or otherwise without written permission from the Publisher.

The content and opinions expressed in this book are the sole work of the authors and editors, who have warranted due diligence in the creation and issuance of their work. The publisher, editors, and authors are not responsible for errors or omissions or for any consequences arising from the information or opinions presented in this book and make no warranty, express or implied, with respect to its contents.

Due diligence has been taken by the publishers, editors, and authors of this book to assure the accuracy of the information published and to describe generally accepted practices. The contributors herein have carefully checked to ensure that the drug selections and dosages set forth in this text are accurate and in accord with the standards accepted at the time of publication. Notwithstanding, since new research, changes in government regulations, and knowledge from clinical experience relating to drug therapy and drug reactions constantly occur, the reader is advised to check the product information provided by the manufacturer of each drug for any change in dosages or for additional warnings and contraindications. This is of utmost importance when the recommended drug herein is a new or infrequently used drug. It is the responsibility of the treating physician to determine dosages and treatment strategies for individual patients. Further, it is the responsibility of the health care provider to ascertain the Food and Drug Administration status of each drug or device used in their clinical practice. The publishers, editors, and authors are not responsible for errors or omissions or for any consequences from the application of the information presented in this book and make no warranty, express or implied, with respect to the contents in this publication.

This publication is printed on acid-free paper. ⬭∞⬭

ANSI Z39.48-1984 (American National Standards Institute) Permanence of Paper for Printed Library Materials.

Production Editor: Christina Thomas

Cover illustration: Figure 2, Chapter 10, by Ulf B. Christensen, "EasyBeacons™ for the Detection of Methylation Status of Single CpG Duplets."

Cover design by Nancy Fallatt

For additional copies, pricing for bulk purchases, and/or information about other Humana titles,contact Humana at the above address or at any of the following numbers: Tel: 973-256-1699; Fax: 973-256-8341; E-mail: orders@ humanapr.com or visit our website at http://humanapress.com

Photocopy Authorization Policy:
Authorization to photocopy items for internal or personal use, or the internal or personal use of specific clients, is granted by Humana Press Inc., provided that the base fee of US $30.00 per copy is paid directly to the Copyright Clearance Center at 222 Rosewood Drive, Danvers, MA 01923. For those organizations that have been granted a photocopy license from the CCC, a separate system of payment has been arranged and is acceptable to Humana Press Inc. The fee code for users of the Transactional Reporting Service is: [978-1-58229-700-6/08 $30].

Printed in the United States of America.   10   9   8   7   6   5   4   3   2   1

eISBN: 978-1-60327-040-3

Library of Congress Control Number: 2007938799

Preface

Significant DNA sequence data have been made available by several important genome projects. We clearly live in an exciting era, in which the allocation of functions to a mere string of the four letters in the code of life plays a major if not the predominant role of ongoing and future research. In functional genomics, techniques are required that allow to examine the occurrence of gene products (be it RNA or protein material) and their interplay with other gene products, if possible, in real time. Again, once an important function has been identified, there will be demands to trace or screen that function in in vitro and in vivo systems. For example, it may be necessary to detect dissimilarities in RNA expression between individuals or it may be required to detect pathogens as rapid as possible. Progress in functional genomics as well as molecular diagnostics requires the availability of methods that allow quantitative measurements in almost any environment, if desired even within living cells, with easy to use and inexpensive equipment.

Recently, we have witnessed the emergence of powerful nucleic acids-based reporter probes that are widely employed in studies ranging from genome diagnostics to basic enzymology. These versatile probes have been designed to signal binding of the probe to the target in homogeneous solution by changes of optical properties, most commonly fluorescence since fluorescence is a quantity that can be measured most conveniently and with high sensitivity by both experienced and first time users. It was perhaps the advent of polymerase chain reaction (PCR), enabling the extraction and enrichment of nucleic acid targets from virtually any source, that provided an important stimulus to probe development. It was soon recognized that probes can report the presence of target nucleic acids, for example, using hybridization as a means to induce changes of the distance between two communicating chromophores. In the mid-1990s, a seminal paper introduced new structured probe molecules and a new term, Molecular Beacons, which laid emphasis on one of the preferred properties that is positive signaling wherein formation of the probe—target complex results in increases of fluorescence rather than decreases. The conceptual elegance served as invaluable inspiration to chemists, physicists, and biologists in the field. While originally being used for fluorogenic nucleic acid probes with a stem-loop structure, molecular beacon now refers to a whole family of fluorescence-up probes.

v

Positive signaling probes find numerous applications in modern life sciences, e.g., in rapid nucleic acids diagnostics, gene expression analysis, basic enzymology up to the development of setups for drug-screening purposes. This book provides no comprehensive overview. It rather intends to show a diverse set of instructive examples, from probe design to applications in clinical settings, guided by experts in the field who provide easy-to-follow experimentals. The book commences with an introduction to the basic principles of fluorescence written by Larry Morrison. It is continued with a section describing applications of fluorogenic probes in real-time PCR, which currently is the gold standard for quantitative DNA and RNA analyses. Five studies, the multiplex detection of mutations in a bacterial gene that confer resistance to antibiotics (David Perlin), the detection of Salmonella pathogenes in foodstuff (Arvind Bhagwat), the identification of transgenic mice (Gunter Schmidtke), the gene expression analysis in carcinoma samples (Sebastian Lukasiak), and the absolute quantitation of viral RNA in cats (Regina Hofmann-Lehmann) are testimony to the usefulness of fluorogenic probe technologies. Thomas Schmittgen extends the potential of real-time PCR by showing high-throughput profiling of the expression of microRNA precursors as a more accurate alternative to microarray-based profiling. Two contributions advocate simplifications of the probe technologies. David Whitcombe shows that the primer of a PCR and the probe can be combined in one, the Scorpion primer. Thomas Froehlich's and Oliver Geulen's work indicates that single-labeled probes can be used as an alternative to traditional dual-labeled probes. This theme is carried to the section on innovative probe technologies. Ulf Christensen presents a new simplified molecular beacon design, EasyBeacons, and demonstrates the utility in DNA methylation profiling. The chemical reaction between two probe molecules, termed Quenched Autoligation probes by Eric Kool, is the basis of a RNA-detection method that functions even within living cells. Two contributions (HyBeacons presented by David French and FIT-Probes introduced by Oliver Seitz) suggest the incorporation of single smart chromophores to omit the need for two interacting fluorescence labels and to increase the responsiveness towards single-base mutations in real-time PCR analysis. An article by Carl Wittwer shows that further simplification and improvement of the cost-effectiveness is feasible by using unlabeled probes and high performance DNA stains in asymmetric real-time PCR. The final section of the book is dedicated to fluorogenic nucleic acid probes that can be used for the detection and characterization of non-nucleic acid targets. Weihong Tan demonstrates that molecular beacons are not only limited to the detection of nucleic acids, but also be used to report on DNA—protein interactions. Daniel Summerer shows how molecular beacon technology can be fashioned into a new tool that facilitates the molecular enzymology of DNA-polymerases. Increasing the sensitivity by signal amplification is the aim of Najafi-Shoushtari and Michael Famulok, who show that nucleic acid targets can control the conversion of fluorogenic substrates by ribozymes. This principle also allows the fashioning of catalytic RNA-based reporters of RNA—protein interactions as described by Jörg Hartig and Michael Famulok.

Andreas Marx

Oliver Seitz

Contents

Contributors

HIROSHI ABE • *Nano Medical Engineering Laboratory Discovery Research Institute, Riken, Wako, Japan*

SERGEY BALASHOV, PhD • *Public Health Research Institute, UMDNJ-New Jersey Medical School, International Center for Public Health, Newark, NJ*

ARVIND A. BHAGWAT, PhD • *Produce Quality and Safety Laboratory, Henry A. Wallace Beltsville Agricultural Research Center, USDA, Beltsville, MD*

AUDREY CHAN • *Biotechnology Program, School of Life Sciences & Chemical Technology, Ngee Ann Polytechnic, Singapore*

ULF BECH CHRISTENSEN, MSc., Chem., Cell Biol. and Mol. Biol. • *PentaBase, Soendersoe, Denmark*

TRINA CHUA • *Biotechnology Program, School of Life Sciences & Chemical Technology, Ngee Ann Polytechnic, Singapore*

SAÚL RUIZ CRUZ, PhD • *Horicultural Products and Cereal Technology Lab, Center of Research for Food and Development (CIAD A.C.), Hermosillo, Sonora, Mexico*

MARIA ERALI • *Advanced Technology Group, Institute for Clinical and Experimental Pathology, ARUP Laboratories, Salt Lake City, UT*

MICHAEL FAMULOK, DR. • *LIMES Institute, Program Unit Chemical Biology & Medicinal Chemistry University of Bonn, Bonn, Germany*

DAVID FRENCH • *LGC, Middlesex,UK*

THOMAS FROHLICH, DR. • *Roche Diagnostics GmbH, Roche Applied Science, Penzberg, Germany*

OLIVER GEULEN, DR. • *Roche Diagnostics GmbH, Roche Applied Science, Penzberg, Germany*

GUSTAVO A. GONZÁLES, PhD • *Horicultural Products and Cereal Technology Lab, Center of Research for Food and Development (CIAD A.C.), Hermosillo, Sonora, Mexico*

JÖRG S. HARTIG, DR. • *Department of Chemistry, University of Konstanz, Konstanz, Germany*

REGINA HOFMANN-LEHMANN, • *Clinical Laboratory, University of Zeurich, Winterthurerstrasse, Zurich, Switzerland*

INMAI JIANG • *College of Pharmacy, Ohio State University, Columbus, OH*

ERIC T. KOOL, PhD • *George and Hilda Daubert Professor of Chemistry, Department of Chemistry, Stanford University, Stanford, CA*

EUN JOO LEE, MS • *College of Pharmacy, Ohio State University, Columbus, OH*

SEBASTIAN LUKASIAK • *Chair of Immunology, Department of Biology, University of Konstanz Universitätsstr, Konstanz, Germany*

ANDREAS MARX, PhD • *Department of Chemistry, University of Konstanz, Konstanz, Germany*

LARRY E. MORRISON, PhD • *Abbott Molecular, Des Plaines, IL*

HANI SEYED NAJAFI-SHOUSHTARI, PhD • *Kekulé-Institute, University of Bonn, Bonn, Germany*

ROBERT PALAIS, PhD • *Department of Mathematics, University of Utah, Salt Lake City, UT*

STEVEN PARK • *Public Health Research Institute, UMDNJ-New Jersey Medical School, International Center for Public Health, Newark, NJ*

JITU PATEL, PhD • *Food Technology and Safety Laboratory, Henry A. Wallace Beltsville Agricultural Research Center, USDA, Beltsville, MD*

DAVID S. PERLIN, PhD • *Public Health Research Institute, UMDNJ-New Jersey Medical School, International Center for Public Health, Newark, NJ*

GUNTER SCHMIDKTE • *Department of Biology, University of Konstanz Konstanz, Germany*

THOMAS SCHMITTGEN, PhD • *College of Pharmacy, Ohio State University, Columbus, OH*

OLIVER SEITZ, DR. • *Institute of Chemistry, Humboldt University Berlin, Berlin, Germany*

ADAM P. SILVERMAN, PhD • *Department of Chemistry, Stanford University, Stanford, CA*

ELK SOCHER, Dipl.-Chem • *Institute of Chemistry, Humboldt-University of Berlin, Berlin, Germany*

DANIEL SUMMERER, DR. • *The Scripps Research Institute, La Jolla, CA*

WEIHONG TAN • *Department of Chemistry, University of Florida, Gainesville, FL*

DAVID WHITCOMBE • *DXS Ltd, Manchester, UK*

CARL WITTWER, MD, PhD • *Department of Pathology, University of Utah School of Medicine, Salt Lake City, UT*

I

INTRODUCTION

1

Basic Principles of Fluorescence and Energy Transfer

Larry E. Morrison

Abstract

Fluorescence is highly sensitive to environment, and the distance separating fluorophores and quencher molecules can provide the basis for effective homogeneous nucleic acid hybridization assays. Molecular interactions leading to fluorescence quenching include collisions, ground state and excited state complex formation, and long-range dipole-coupled energy transfer. These processes are well understood and equations are provided for estimating the effects of each process on fluorescence intensity. Estimates for the fluorescein-tetramethylrhodamine donor–acceptor pair reveal the relative contributions of dipole-coupled energy transfer, collisional quenching, and static quenching in several common assay formats, and illustrate that the degree of quenching is dependent upon the hybridization complex formed and the manner of label attachment.

Key Words: Fluorescence; energy transfer; collisional quenching; fluorescence lifetime; static quenching; quantum yield; homogeneous assays; DNA-binding assays.

1. Introduction

Fluorescence has been used to great advantage in the design of homogeneous nucleic acid hybridization assays, and is particularly useful in detecting specific DNA sequences generated by the polymerase chain reaction (PCR) in real time (RT-PCR). The success of fluorescence in real-time assays arises from the exceptional sensitivity of fluorescence to the environment of the fluorophore. This environmental sensitivity arises from the fact that fluorescence lifetimes are in the range of about 0.1 to 100 ns, a range that makes many intermolecular processes competitive with fluorescence. Considerably shorter, and fluorescence would be unresponsive to many intermolecular phenomena; considerably longer, and fluorescence would be rarely observed in solution. It is the purpose of this chapter to describe the basic tenants of fluorescence, and the intermolecular processes that modulate fluorescence, in order to better understand why

From: *Methods in Molecular Biology, Vol. 429: Molecular Beacons: Signalling Nucleic Acid Probes, Methods and Protocols* Edited by: A. Marx and O. Seitz © Humana Press Inc., Totowa, NJ

fluorescence has been so advantageous to the implementation of homogeneous hybridization assays, and to aid in the further development and improvement of these assay formats. First, concepts basic to fluorescence and its environmental sensitivity are presented. This is followed by application to RT-PCR assay formats, taking into consideration the probe geometries and possible label interactions. Since fluorescence intensity is the most commonly and easily measured fluorescence property, measurement of fluorescence lifetimes and fluorescence polarization will not be discussed. Only the most basic elements of fluorescence can be provided in a single chapter, and full texts on the subject should be consulted for greater detail *(1,2)*. Review articles are also available for fluorescence-based homogeneous DNA-assay formats *(3,4)*.

2. Basic Fluorescence

2.1. Excitation and Emission

In a typical fluorescence experiment, the fluorescent label, or fluorophore, is exposed to light that is absorbed by the fluorophore, thereby promoting an electron in the ground state of the fluorophore to an electronically excited state. Return to the ground state is accompanied by emission of light—the fluorescence. For efficient absorption of the excitation light, the wavelength must correspond to the energy difference between the ground state and an excited state. The process is represented schematically in **Fig. 1A** for a hypothetical molecule. In **Fig. 1**, horizontal lines represent electronic and vibrational energy levels, and upward arrows represent absorption of a photon of light. Electronic levels are represented by heavier lines than the vibrational levels. The vertical axis represents the energy of various levels so that longer vertical lines represent higher energy transitions than shorter lines plotted in **Fig. 1B** is the fluorescence excitation spectrum of the hypothetical molecule in which the horizontal axis corresponds to wavelength and the vertical axis corresponds to the intensity of light emitted when excitation is at the corresponding wavelength. Vertical lines are placed in the spectrum at wavelengths corresponding to the energy of the absorption transition aligned directly above in part A. Wavelength (λ) is related to energy (E) and frequency (v) by the expressions:

$$E = hv = hc / \lambda \qquad \text{Eq. (1)}$$

where h is Plank's constant and c is the speed of light.

In the gas phase, the lines of absorbance tend to be narrow and transitions corresponding to promotion from the lowest vibrational level (0 level) of the lowest electronic state (S_0 or ground state) to different vibrational levels (levels 0, 1, 2, …) of electronically excited states (S_1, S_2, …) can be resolved. In solution the energy transitions are broadened as energy levels change slightly with time due to changes in the orientations and collisions with neighboring solvent

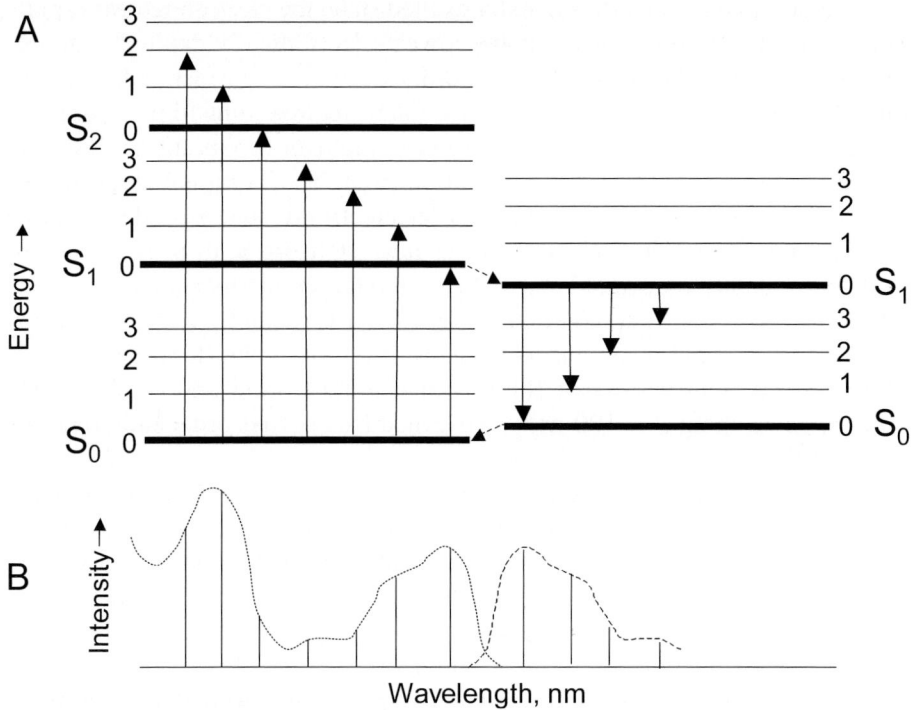

Fig. 1. (A) Diagram showing electronic (heavy horizontal lines) and vibrational (thin horizontal lines) energy levels and absorbance (upward arrows) and emission transitions (downward arrows) between energy levels for a hypothetical fluorophore. The vertical axis is energy. **(B)** Diagram showing excitation and emission spectra corresponding to transitions indicated in **(A)**. Vertical lines approximate gas phase spectra and broader curves approximate excitation and emission spectra in solution.

molecules, as represented by the stippled line in **Fig. 1B**. At room temperature absorbance originates predominately from the lowest vibrational level of the lowest excited state as predicted by the Boltzmann distribution.

The magnitude of the absorbance is determined by the probability of the corresponding transition. The Frank–Condon principle states that transitions within a molecule are more likely between energy levels for which the interatomic distances are unchanged. This is because changes in electron distributions are more rapid, on the order of 10^{-15} s, than changes in the more massive nuclei that occur within the time frame of molecular vibrations (10^{-12} s). Excited states can have higher mean interatomic spacing due to promotion of the electron from bonding to nonbonding orbitals, such that transitions from the lowest vibrational level of S_0 may preferentially occur to higher vibrational levels of S_1 that can achieve matching interatomic distances.

Once promoted to an electronically excited state, the electron returns rapidly to the lowest vibrational level of the lowest electronically excited state, S_1. The time for this transition is in the order of 10^{-12} s, corresponding to the timescale of intermolecular vibrations. The electron transitions from one vibrational level to the next, eventually crossing over to vibrational levels of the next lower excited state. The transition from the vibrational levels of one electronic state to those of the next lower electronic state is called internal conversion (IC) and the energy is lost as heat. IC from S_1 to S_0 is not as fast as IC between higher excited states due to the larger energy spacing between vibrational levels of S_1 and S_0. The lower rate of IC from S_1 permits emission of a photon of light to compete as a means of returning the fluorophore to S_0. This de-excitation process of fluorescence is first order and occurs in the time range of about 0.1–100 ns, as governed by the first-order radiative rate constant, k_r.

Figure 1A includes de-excitation of S_1 by fluorescence, represented by the downward arrows. Depending upon the energy of the photon, the excited electron can return to a number of different vibrational levels of S_0. Often fluorophores have similar spacing of the vibrational levels in both S_0 and S_1, and considering the Frank—Condon principle, a fluorescence emission spectrum often shows a "mirror image" relationship to the absorbance spectrum. This is demonstrated in **Fig. 1B** in which vertical lines have been placed at wavelengths corresponding to the energy of the transitions represented by downward arrows aligned directly above in **Fig. 1A**. Interactions with solvent molecules can significantly broaden the transition energies or completely mask the vibrational structure, as indicated by the dashed line representing a solution-phase fluorescence emission spectrum.

The longest wavelength absorbance transition, corresponding to excitation from the lowest vibrational level of S_0 to the lowest vibrational level of S_1, and the shortest wavelength emission transition, corresponding to de-excitation between the same two levels, should occur at the same wavelength. However, reorientation of neighboring solvent molecules within the lifetime of the excited state can lower the energy of the excited state and raise that of the ground state thereby lowering the energy of the emitted photon. This is indicated in **Fig. 1A** by the dashed arrows leading to a lower S_1 and higher S_0 state. The difference between the longest wavelength absorbance maximum and the shortest wavelength emission maximum is called the Stoke's shift. Stoke's shifts are relatively large in polar solvents when the dipole moment of S_1 exceeds that of the S_0. Upon excitation the dipole moment increases rapidly with the redistribution of electrons, before solvent molecules can reorient. Over the ensuing nanoseconds, reorientation of the solvent molecules, called solvent relaxation, opposes the new dipole and lowers the overall energy. The fluorophore then reverts to its original dipole upon photon emission while being surrounded by the newly oriented solvent molecules which must reverse their orientation to return the molecule completely to its original energy state. The ability of neighboring molecules to oppose dipole changes within the fluorophore is one of the many causes of the environmental sensitivity of fluorescence.

2.2. Competition with Fluorescence—Intrinsic Factors

As described earlier, IC competes with fluorescence and only molecules with smaller rate constants for IC, k_{IC}, from S_1 to upper vibrational levels of S_0, will have substantial fluorescence. Other processes also compete with fluorescence, including intersystem crossing (IS). An IS is the transfer of the excited electron between electronic levels of different multiplicity, for example from a singlet state to a triplet state. These transitions are "forbidden" and are considerably slower than transitions between states of the same multiplicity, and typically do not compete well with fluorescence. However, spin—orbit coupling can blur the distinctions between states making the singlet-to-triplet conversion more likely in some molecules. Once in the triple state the excited electron can return to S_0 with emission of a photon, called phosphorescence, with a characteristically longer emissive lifetime, k_p, owing again to the "forbidden" nature of the transition. Phosphorescence lifetimes are typically in the range of microseconds to seconds, and as such phosphorescence is usually very weak, succumbing to more rapid quenching processes.

The magnitudes of k_r, k_{IC}, and k_{IS} are intrinsic to the fluorophore and determine the lifetime of the excited state as well as the efficiency of fluorescence, measured as the fluorescence quantum yield, Φ_F. Equation (2) describes the time dependence of the fluorescence intensity, I, following an infinitely short pulse of excitation light:

$$I \equiv d(h\nu) / dt = -\Phi_F d(*F) / dt = \Phi_F (k_r + k_{IC} + k_{IS})(*F) \qquad \text{Eq. (2)}$$

where $*F$ is the number of electronically excited fluorophores at time t and $h\nu$ is a photon of emitted light. Integration of Equation (2), leads to the first-order rate equation:

$$*F = (*F_0)e^{-(k_F t)} = (*F_0)e^{-(t/\tau_F)} \qquad \text{Eq. (3)}$$

where $*F_0$ is the number of fluorophores absorbing a photon of excitation light to form the initially excited fluorophore population at $t = 0$, k_F is the rate constant for conversion of S_1 to S_0 by both radiative and nonradiative processes ($k_F = k_r + k_{IC} + k_{IS}$), and τ_F is the lifetime of the excited state, equal to $1/k_F$, and therefore equal to the fluorescence lifetime. The fluorescence quantum yield, Φ_F, defined as the number of photons emitted per photon absorbed by the fluorophore, is determined by the relative rates of radiative and nonradiative processes, according to Equation (4):

$$\Phi_F = k_r / (k_r + k_{IC} + k_{IS}) = k_r / k_F = k_r \tau_F \qquad \text{Eq. (4)}$$

such that higher rates of IC or IS serve to lower the fluorescence efficiency as IC or IS compete with fluorescence to de-excite S_1.

2.3. Competition with Fluorescence—Extrinsic Factors

Environmental factors extrinsic to the fluorophore also compete to de-excite the fluorophore, and further quench fluorescence. As the name implies, collisional quenching occurs when the fluorophore, while in its excited state, collides with a quencher molecule, Q. The rate of collisional quenching is governed by a bimolecular rate constant, k_q, and the concentration of the quencher molecule, $[Q]$. Adding this to IC and IS further reduces the fluorescence lifetime and quantum yield according to Equations (5) and (6):

$$\tau_Q = 1 / (k_F + k_q[Q]) \qquad \text{Eq. (5)}$$

$$\Phi_Q = k_r / (k_F + k_q[Q]) \qquad \text{Eq. (6)}$$

where the subscript Q indicates the presence of collisional quencher. The dependence of fluorescence on $[Q]$ is given by the Stern–Volmer equation:

$$\Phi_F / \Phi_Q = I_0 / I = 1 + k_q \tau_F [Q] \qquad \text{Eq. (7)}$$

where I_0 is the fluorescence intensity of the solution in the absence of Q.

Collision with the quencher can cause distortion of energy levels and momentary coupling between levels such that IC and IS become more favorable, thereby causing rapid de-excitation at the expense of fluorescence. Heavy atom substituents, such as iodide and bromide, are especially capable of inducing spin—orbit coupling upon collision.

Collisions can also result in formation of a complex between *F and Q that may be nonfluorescent or may have highly altered fluorescence characteristics relative to *F alone. If a dimer is formed between the same kind of molecules, *F–F, the complex is called an excimer. An example of a molecule forming excimers is pyrene, for which the excimer fluorescence is shifted to higher wavelengths. An exiplex is a complex formed between different kinds of molecules. Fluorescence changes due to excimer and exiplex formation can be the basis of molecular assays *(5)* or they can be a nuisance to be avoided *(6)*. Collisions can also result in chemical reactions; two of the more common being electron transfer and proton transfer, either one of which can drastically alter electronic structure and lead to nonfluorescent products.

Another type of quenching, static quenching, occurs by complex formation between F and Q in their ground states to form $F–Q$. $F–Q$ may be nonfluorescent, or alternatively, F may be nonfluorescent while $F–Q$ may be fluorescent. An example of the latter is ethidium bromide which has a low fluorescence quantum yield when free in solution, and becomes highly fluorescent when bound to DNA. Increases in fluorescence can result from shielding of F from endogenous quenchers (e.g., dissolved oxygen) when bound with Q *(7)*. Complex formation is governed by a stability constant, K_s, for the equilibrium:

$$K_s = [F-Q]/([F][Q])$$ Eq. (8)

Given that the concentration of fluorophore in the absence of Q is $[F_0] = [F] + [F–Q]$, fluorescence intensity is reduced according to the following relationship upon introduction of Q (assuming F–Q is nonfluorescent in this example).

$$I_0/I = [F_0]/[F] = 1 + K_s[Q]$$ Eq. (9)

F still has the same fluorescence lifetime and quantum yield since static quenching is not competing with de-excitation of *F. The "quenching" in this case results from the fact that $[F]$ is reduced by conversion of F to $F–Q$.

3. Nonradiative Dipole-Coupled Energy Transfer

There is an additional form of quenching that does not involve contact between labels, and can occur over relatively large separation distances, e.g., 20–100 Å. Excitation energy is transferred from *F to Q via coupling between the emission dipole of *F and an absorption dipole of Q for which the energies of the associated transitions are matched. The electron distribution of a particular electronic state creates a dipole which can be viewed as oscillating when energized. A coulombic coupling between this dipole on F and a dipole on Q can result in transfer of the oscillator energy without the formation of an intermediate photon, thereby making this a nonradiative process that can occur over long range on a molecular scale. The rate constant for this long-range energy transfer process, k_T, was described by Förster *(8)*:

$$k_T = 8.79 \times 10^{-25} \kappa^2 \Phi_F k_F \eta^{-4} R^{-6} \int_0^\infty E_F(\lambda)\varepsilon_Q(\lambda)\lambda^4 d\lambda$$ Eq. (10)

where κ is a factor describing the orientation between the emission dipole of *F and the absorption dipole of Q, η is the refractive index of the medium, R is the distance separating *F and Q, $E_F(\lambda)$ is the emission spectrum of *F as a function of wavelength normalized to an integrated value of 1, and $\varepsilon_Q(\lambda)$ is the spectrum of the molar extinction coefficients for light absorption by Q as a function of wavelength. The integral establishes that there must be absorption transitions in Q matching in energy to the emission transitions in *F, and κ establishes that the corresponding dipoles must be aligned for optimal transfer efficiency. A characteristic distance, R_0, is defined as the distance at which the rate of energy transfer is equal to the rate of S_1 de-excitation in the absence of energy transfer ($k_F = k_T$):

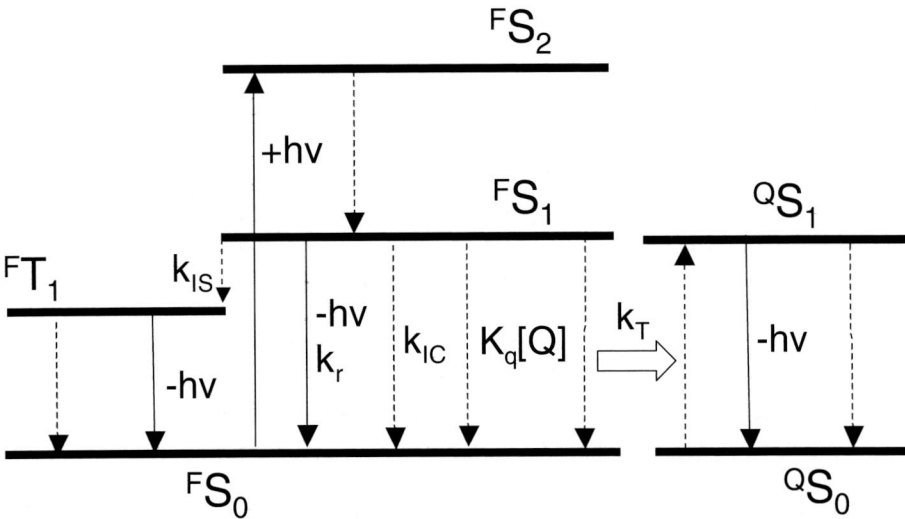

Fig. 2. Diagram showing electronic energy levels (horizontal lines) and transitions (vertical arrows) between energy levels for a hypothetical fluorophore, F (vibrational levels not shown). Energy levels for a second molecule, Q, are shown that is capable of receiving energy from $*F$ via long-range energy transfer. Solid vertical arrows correspond to transitions involving absorption or emission of a photon of light, and dashed arrows correspond to radiationless transitions.

$$(R_0)^6 = 8.79 \times 10^{-25} \kappa^2 \Phi_F \eta^{-4} \int_0^\infty E_F(\lambda) \varepsilon_Q(\lambda) \lambda^4 d\lambda \qquad \text{Eq. (11)}$$

The efficiency of energy transfer, E_T, is calculated according to Equation (12).

$$E_T = k_T / (k_F + k_T) = (R_0)^6 / [(R_0)^6 + R^6] \qquad \text{Eq. (12)}$$

The various processes which compete with fluorescence are summarized in **Fig. 2**, which includes the intrinsic processes of IC and IS, as well as the external factors of collisional quenching and energy transfer. Static quenching is not shown since its effect is not to compete with de-excitation, but to reduce the concentration of the free fluorophore. With the addition of energy transfer, lifetime and quantum yield of $*F$ become:

$$\tau_{qT} = 1 / (k_F + k_q[Q] + k_T) \qquad \text{Eq. (13)}$$

$$\Phi_{qT} = k_r / (k_F + k_q[Q] + k_T) \qquad \text{Eq. (14)}$$

where the subscript qT denotes the potential presence of both collisional quenching and energy transfer.

4. Intermolecular Distances and Quenching in DNA Probe Complexes

In homogenous nucleic acid assays, hybridization causes a dramatic change in the environment of a fluorescent label, resulting in a decrease or an increase in that label's fluorescence or a dramatic shift in another fluorescence property such as spectral distribution. Often the change in environment is a decrease or an increase in the distance separating a fluorescent label from a quenching species. In order to illustrate the sensitivity of fluorescence to formation of hybridization complexes that occurs by design in homogeneous DNA hybridization assays, the percentage of DNA quenching due to dipole-coupled energy transfer, collisional quenching, or static quenching will be estimated for various assay formats. Since these examples are intended only to illustrate the concepts, a number of assumptions are made to simplify calculations and only rough estimates of distances are employed.

4.1. Calculations of Quenching Percentages

Several of the most common probe configurations used in homogeneous hybridization assays are represented schematically in **Fig. 3** and are explained in detail below. The fluorophore and quencher labels are attached to either the same or separate probe strands depending upon the assay format. Introduction of the complementary analyte nucleic acids (also referred to as the targets) contained within a sample serves to either bring F and Q closer together or force them further apart.

In the examples to follow, two common fluorescent labels will be employed, F = fluorescein and Q = tetramethylrhodamine, attached via either 3-or 6-atom linkers to phosphate groups on the DNA backbone. In all instances, molecular distances were estimated by measurements made on molecular models, and $\tau_F = 4$ ns ($k_F = 2.5 \times 10^8$ s^{-1}) was assumed for fluorescein. For calculations of quenching due to energy transfer, $\Phi_F = 0.8$ for fluorescein attached to DNA was used, together with $\kappa^2 = 2/3$ (assumes randomization of donor and acceptor orientations within the lifetime of *F) and $\eta = 1.33$ for water (589 nm, 20°C). The emission spectrum for fluorescein and the absorbance spectrum for tetramethylrhodamine (approximated by the fluorescence excitation spectrum) are plotted in **Fig. 4**. Also plotted is the overlap spectrum, obtained by multiplying the absorbance spectrum by the normalized emission spectrum and λ^4 at each wavelength. Substituting into Equation (10) the values of Φ_F, κ^2, η, and the area under the overlap spectrum (integral in Equation [10]) provided $R_0 = 57$ Å. The percentage of fluorescence quenching due to energy transfer was calculated as $100(E_T)$ by substituting R_0 and a "representative" label separation distance into Equation (12). A proper calculation for the extent of quenching by energy transfer would use the distribution of possible separation distances within the hybridization complex, but given other approximations and assumptions,

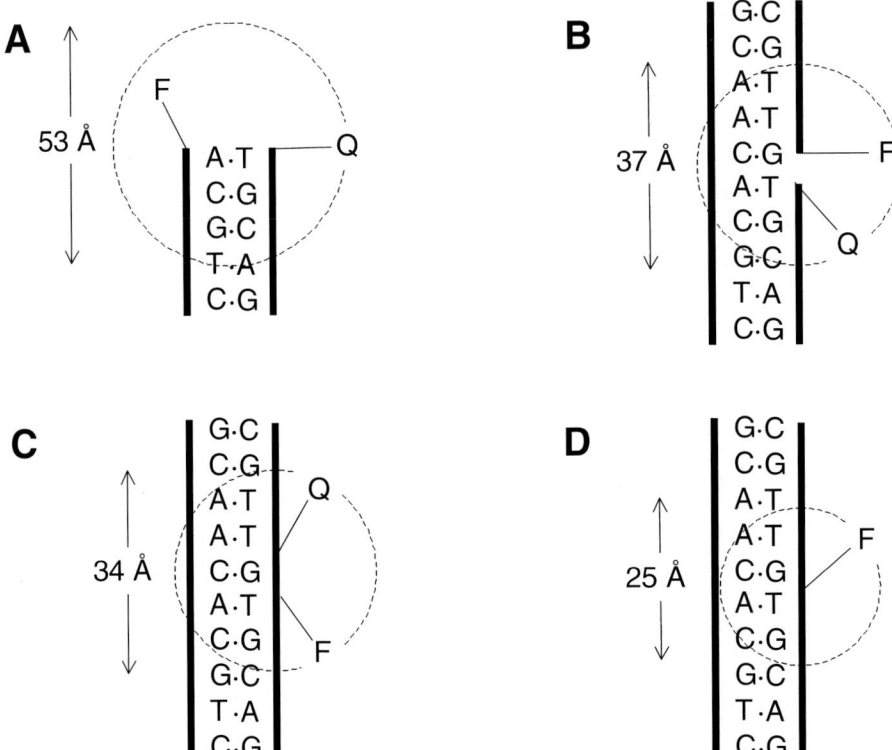

Fig. 3. Schemes for hybridization complexes formed in four different homogeneous DNA-assay formats. Circles are drawn with radii equal to half the maximum separation distance of F and Q, assuming F and Q are tethered to phosphates groups of the DNA backbones by 6-atom linkages. The indicated distances are not drawn to scale with respect to base pairs.

a "representative" separation distance equaling half the maximum separation distance will be employed in place of a distribution.

For collisional quenching, [Q] in a hybridization complex was approximated as the concentration of 1 molecule of Q in the volume defined as a sphere with diameter equal to the maximum separation between fully extended F and Q. A circle is drawn in each of the four schemes in **Fig. 3** to represent the circumference of the sphere when labels are attached via 6-atom linkers. Of course, this only roughly estimates the concentration since Q cannot access the entire volume and is excluded from a sizeable fraction of the volume by nucleic acid. For Scheme D, in which there is only one label, the distance between the extended label and the center of the helix is used as the maximum separation

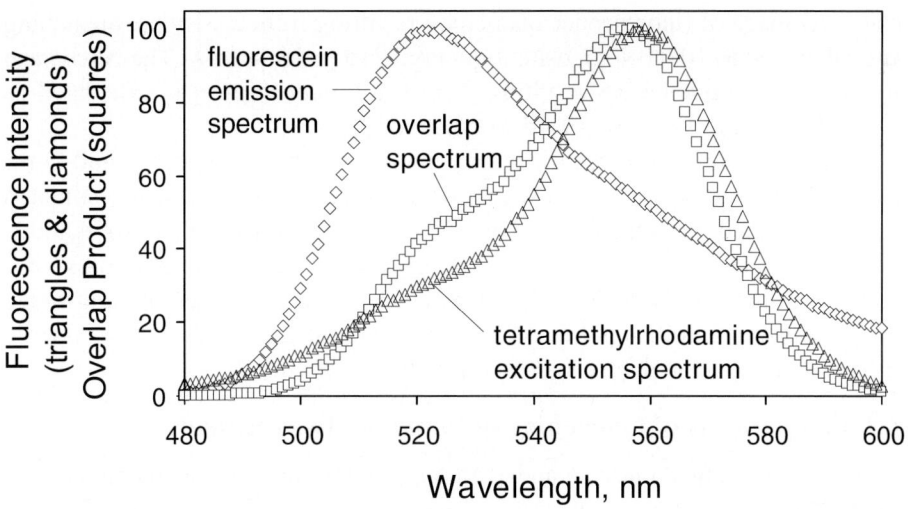

Fig. 4. Fluorescence excitation spectrum for fluorescein attached to DNA probe, emission spectrum for tetramethylrhodamine attached to DNA probe, and overlap spectrum used to calculate the long-range energy transfer rate constant. The fluorescein excitation spectrum was used to approximate the spectrum of extinction coefficients of absorbance. All three spectra are scaled to values of 100 at their respective maxima for plotting.

distance since the label in this format often interacts by intercalation between the base pairs, or by binding within the major or minor groove of the helix. The collisional rate constant for quenching, k_q, equal to the product of the diffusion controlled rate constant, k_0, and the collisional quenching efficiency, γ, was calculated using the Smoluchowski equation:

$$k_q = \gamma k_0 = 7.57 \times 10^{21} \gamma (R_F + R_Q)(D_F + D_Q) \qquad \text{Eq. (15)}$$

where R_F and R_Q are the radii of F and Q, respectively, and D_F and D_Q are the respective diffusion coefficients. The diffusion coefficients were calculated according to the Stokes–Einstein equation:

$$D = kT / (6\pi\zeta R) \qquad \text{Eq. (16)}$$

where k is the Boltzmann constant, T is the temperature on the Kelvin scale, ζ is the solvent viscosity at T, and R is either R_F or R_Q. Assuming each collision of F and Q results in quenching of F ($\gamma = 1$), using 65°C as a reasonable temperature for RT-PCR fluorescence measurements, and substituting 0.4335 cp for the viscosity of water, and 4.8 and 5.9 Å for the radii of fluorescein and tetramethylrhodamine, respectively, k_q is found to equal 4.37×10^9 $M^{-1}s^{-1}$.

The percentage of fluorescence quenching resulting from collisional quenching was calculated as $100(1-I/I_0)$, using Equation (7) to calculate I/I_0. The percentage of fluorescence quenching resulting from static quenching was calculated as $100(1-I/I_0)$, using Equation (9) to calculate I/I_0.

Table 1 lists the estimates for representative distances between F and Q for each of the hybridization complexes shown in **Fig. 3**. The distances are estimated for both 3- and 6-atom linkers to show the effect of linker length. Also included are the calculated values for $[Q]$, k_T, and $k_q[Q]$, and the percentage of fluorescence quenching that would result if Q were acting as an acceptor of long-range energy transfer, a collisional quencher, or a static quencher. The level of static quenching assumes a relatively low value of 100 M^{-1} for K_s.

4.2. Comparisons of Quenching in Common Assay Formats

Scheme A in **Fig. 3** places F and Q on opposing termini of two complementary probe sequences. The complementary sequences may be on separate strands *(9),* or may comprise the stem portion of a hairpin probe *(10)*. In the absence of the analyte DNA, the complementary probe regions hybridize to each other as shown in Scheme A, bringing F and Q near one another. With a 6-atom linker, distances between F and Q range from contact to a maximum of about 53 Å, restricting the labels to a volume of about 7.8×10^{-23} L, and resulting in $[Q]$ ~0.02 M (first row in **Table 1**). Based on a representative separation distance of 27 Å (half the maximum separation), k_T is predicted to be 100 times greater than k_F, resulting in nearly complete (99%) quenching of F. Collisional quenching is also competitive with fluorescence, although much less than energy transfer, resulting in 27% quenching of F by collisions with Q. Formation of nonfluorescent ground-state complexes at the expense of free F would result in a 68% reduction in fluorescence due to static quenching for $K_s = 100$ M^{-1}. As can be seen in the second row of **Table 1**, decreasing the linker to three atoms has little effect on quenching due to energy transfer but noticeably increases collisional and static quenching.

In the presence of the analyte nucleic acid, the complementary probe segments are hybridized to the target strands instead of each other, resulting in separation of the labels. For a complementary probe pair, this translates into a label separation distance in the order of 1000 Å in a typical RT-PCR assay in which probes would be free in solution at concentrations of one to several micromolar (third row, **Table 1**). At such distances energy transfer would be negligible (<0.001%). At $[Q] = 1$ μM, collisional quenching of F by Q would also be negligible (0.002%), as would static quenching (0.01%). For a hairpin probe, the labeled termini are separated by the length of the probe when hybridized with the complementary analyte DNA (3.4 Å/bp). For a 28 base hairpin probe that would be about 95 Å, at which quenching via energy transfer is small

Table 1
Label Separation Distances and Resulting Quenching Characteristics Estimated for Hybridization Complexes Formed in Different Homogeneous DNA Assay Formats

Scheme	Target DNA[a]	Label attachment position	Linker length, atoms	Label separation distance (Å)[b]	$[Q]$ (M)	k_T (s^{-1})	$k_q[Q]$ (s^{-1})	Quenching of F by Q (%) Energy transfer	Collisional quenching	Static quenching $K_s = 100\ M^{-1}$
A	−		6	27	0.021	2.4×10^{10}	9.3×10^7	99.0	27	68
A	−		3	23	0.033	5.5×10^{10}	1.4×10^8	99.5	36	77
A probe pair	+	Neighboring 3′- and 5′-PO$_4$	6 or 3	1000	1×10^{-6}	8.2	4.4×10^3	<0.001	0.002	0.01
A hairpin	+		6 or 3	95	1×10^{-6}	1.1×10^7	4.4×10^3	4.3	0.002	0.01
B	+		6	19	0.063	2.0×10^{11}	2.7×10^8	99.9	52	86
B	+	Neighboring 3′- and 5′-PO$_4$	3	15	0.117	7.2×10^{11}	5.1×10^8	100	67	92
B	−		6 or 3	1000	1×10^{-6}	8.2	4.4×10^3	<0.001	0.002	0.01
C nuclease	−		6	17	0.081	3.4×10^{11}	3.5×10^8	99.9	59	89
C nuclease	−	Adjacent PO$_4$'s	3	14	0.144	1.1×10^{12}	6.3×10^8	100	72	94
C nuclease	+		6 or 3	1000	1×10^{-6}	8.2	4.4×10^3	<0.001	0.002	0.01
C helix-coil	−	Terminal positions	6 or 3	34	0.010	5.3×10^9	4.4×10^7	95.5	15	50
C helix-coil	+		6 or 3	68	0.001	8.3×10^7	5.5×10^6	24.8	2.2	11.2
D	+		6	13	0.20	2.1×10^{12}	8.9×10^8		78	95
D	+	PO$_4$	3	11	0.30	4.6×10^{12}	1.3×10^9		84	97
D	−		6 or 3	1000	1×10^{-6}	8.2	4.4×10^3		0.002	0.01

[a]The presence of excess target and the absence of target are indicated by plus (+) and minus (−) signs, respectively.

[b]Representative label separation distances equal to half the estimated maximum label separation distances (see **Fig. 3**), except for labels separated by rigid double helices (i.e., Scheme C helix—coil probes hybridized to target and Scheme A hairpin probes hybridized to target) for which separation is estimated from the length of the rigid helix, and probes or probe fragments free in solution and not bound to other probe species (i.e., Scheme A probe pair in the presence of target, Scheme B in the presence of target, Scheme C after nuclease digestion of probe hybridized to target, and Scheme D in the absence of target), for which the separation distance is estimated from the probe concentration.

but measurable at 4% (*see* **Note 1**). Collisional and static quenching would be comparable to that for complementary pair probes in the presence of excess analyte DNA since the rigid helix of the hairpin probe hybridized to the target DNA should prevent contact between labels on the same hairpin molecule and quenching would be due to contact between labels on different probe molecules (fourth row, **Table 1**).

Scheme B in **Fig. 3** utilizes labels attached to probe strands hybridizing adjacent to one another on the analyte nucleic acid *(11,12)*. Relevant distances and predicted quenching for this scheme are included in **Table 1**, fifth through seventh rows, for F and Q attached to neighboring 3′- and 5′-terminal phosphate groups of the adjacent probes. The drawings in **Fig. 3** indicate that interlabel distances may be closer in this configuration relative to Scheme A, increasing the probability of energy transfer, collisional quenching, and static quenching. In contrast to Scheme A, the hybridization complex that brings F and Q close to one another occurs in the presence of the analyte DNA. In the absence of target, the labeled probes are free in solution, and for micromolar probe concentrations quenching would be the same as that predicted in Scheme A for probe pairs in the presence of excess analyte DNA.

Scheme C in **Fig. 3** places both labels on a single probe strand. In this configuration, the labels can be separated by nuclease activity, such as the 5′-to-3′ exonuclease activity of DNA polymerase utilized in the TaqMan® RT-PCR assays when the probe is hybridized to its target *(13)*. The entries in **Table 1** (eight and ninth rows) suggest that high levels of quenching can be obtained for the intact probe. Separation of labels by several nucleotides can improve enzymatic digestion while still maintaining high levels of energy transfer, and percentage quenching with intervening nucleotides can be calculated using appropriately larger values for R. After nuclease digestion, labels are free in solution at the original probe concentration, and quenching should be negligible (10th row, **Table 1**, assuming 1 μM probe).

Scheme D in **Fig. 3** utilizes a single label on a single-probe strand. The single label is affected by the different environment created when the probe hybridizes to the analyte nucleic acid *(14)*. Quenching could result from interactions with opposing nucleotides, for example by electron transfer, or by intercalation or groove-binding. Intercalation and groove-binding can also provide a protected environment. Therefore, complexes formed between F and helical structures can be either more fluorescent or less fluorescent than uncomplexed F. For compatibility with the other calculations, entries in **Table 1** (last three rows) for Scheme D assume that F bound to the helix interior is nonfluorescent, and predict that even a low affinity constant of 100 M^{-1} could result in 95% or greater quenching of F. Note that quantitative quenching is not an assay requirement, and the 68% static quenching predicted in Scheme A for labels attached by six atom linkers would form the basis of an effective assay. Static quenching is highly dependent upon K_s, and increasing K_s to a moderate value of 1000 M^{-1} would increase the static quenching in Scheme A from 68% to 96%.

In the above examples, distances between labels can be controlled by changing the positions of label attachment or altering the flexibility of the linkers, in addition to changing linker length. When both labels are attached to the same probe strand, as in Scheme C, labels can be attached to positions with intervening nucleotides in order to improve other assay characteristics, such as accessibility to nuclease. When labels are attached to different probe strands, as in Schemes A and B, probe sequences and attachment points can be selected such that probe termini are separated by intervening nonhybridized nucleotides. If fluorescence intensity of F is the only measurement in a hybridization assay, then it may not matter if one or several quenching processes contribute, however, increasing label separation can be useful if nonradiative energy transfer is desired over collisional or static quenching. For example, if the energy acceptor label Q is also fluorescent, then measuring acceptor emission confirms that donor quenching is due to energy transfer, thereby avoiding a falsely positive assay result due to a sample impurity capable of quenching F.

It is noteworthy that the single-strand probe of Scheme C can be used in a homogeneous assay format that does not rely upon enzymatic digestion to separate the two labels in response to the analyte DNA. This variation requires attachment of the F and Q to more distant positions, such as the probe termini, and relies upon the random coil of the single strand in the absence of target to allow labels to approach one another *(15)*. When hybridized to target the labels are separated by the length of the rigid double-stranded structure, the same as the hairpin probe described in Scheme A. Calculated levels of quenching are included in **Table 1** for probe hybridized to the target DNA (row 12) and in the absence of target (row 11). For these calculations, a 20 nucleotide long probe is assumed with the length of the rigid double helix (68 Å) used as the separation distance in the presence of target. In the absence of target, a representative distance of half the rigid helical length is substituted. Again, a distance distribution should be used that takes into account the flexibility of the single-stranded probe, but that level of complexity is not warranted given other approximations, and the representative distances serve to demonstrate the concept. As can be seen in the table entries, for each quenching process the "background" of unquenched fluorescence in the absence of target is higher (lower percentage of quenching) than for other formats, however, the difference in quenching in the presence and absence of target is easily measured.

5. Conclusions

The above models and discussion show that the sensitivity of fluorescence to environment provides a highly effective means to detect specific nucleic acid sequences in RT-PCR or other homogeneous assays. Tools were provided to estimate the scale of fluorescence quenching afforded by different probe geometries and different quenching mechanisms. Nonradiative energy transfer between donor and acceptor fluorophores with its strong dependence on separation distance is ideal for homogeneous assay formats in which labels become confined to separations on the order of R_0. Energy transfer rates at least two orders of magnitude greater than the rate of intrinsic fluorescence were predicted in each example using the common donor–acceptor label pair of fluorescein and tetramethylrhodamine. Collisional quenching rates were considerably lower than rates of energy transfer, but are still sufficient to form the basis of effective assay formats. Finally, binding between fluorophore and quencher, even with relatively low affinity, can produce dramatic quenching or enhancement effects when labels are confined by hybridization.

6. Notes

1. This separation distance assumes the stem regions are included in the rigid helix. However, since the stem portions are not complementary to the target they would have some random coil nature and a somewhat smaller separation distance would result.

References

1. Lakowicz, J. R. (1983) *Principles of Fluorescence Spectroscopy*. Plenum Press, New York.
2. Valeur, B. (2002) *Molecular Fluorescence: Principles and Applications*. Wiley-VCH, Weinheim.
3. Morrison, L. (1999) Homogeneous detection of specific DNA sequences by fluorescence quenching and energy transfer. *J. Fluores.* **9,** 187–196.
4. Morrison, L. E. (2003) Fluorescence in nucleic acid hybridization assays. In *Topics in Fluorescence Spectroscopy, vol 7* (Lakowicz, J. R., ed.), pp. 69–97, Kluwer Academic Publications, New York.
5. Ebata, K., Masuko, M., Ohtani, H., and Kashiwasake-Jibu, M. (1995) Nucleic acid hybridization accompanied with excimer formation from two pyrene-labeled probes. *Photochem. Photobiol.* **62,** 836–839.
6. Morrison, L. E. (1988) Time-resolved detection of energy transfer: Theory and application to immunoassays. *Anal. Biochem.* **174,** 101–120.
7. Lakowicz, J. R. and Weber, G. (1973) Quenching of fluorescence by oxygen. A probe for structural fluctuation in macromolecules. *Biochemistry* **12,** 4161–4170.
8. Forster, T. (1959) Transfer mechanisms of electronic excitation. *Disc. Faraday Soc.* **27,** 7–17.

9. Morrison, L. E., Halder, T. C., and Stols, L. M. (1989) Solution-phase detection of polynucleotides using interacting fluorescent labels and competitive hybridization. *Anal. Biochem.* **183,** 231–244.

10. Tyagi, S. and Kramer, F. R. (1996) Molecular beacons: Probes that fluoresce upon hybridization. *Nat. Biotechnol.* **14,** 303–308.

11. Heller, M. J. and Morrison, L. E. (1985) Chemiluminescent and fluorescent probes for DNA hybridization systems. In *Rapid Detection and Identification of Infectious Agents* (Kingsbury, D. T. and Falkow, S., eds.), pp. 245–256, Academic Press, Orlando.

12. Wittwer, C. T., Herrmann, M. G., Moss, A. A., and Rasmussen, R. P. (1997) Continuous fluorescence monitoring of rapid cycle DNA amplification. *Biotechniques* **22,** 130–138.

13. Lee, L. G., Connell, C. R., and Bloch, W. (1993) Allelic discrimination by nick-translation pcr with fluorogenic probes. *Nucleic Acids Res.* **21,** 3761–3766.

14. Ishiguro, T., Saitoh, J., Yawata, H., Otsuka, M., Inoue, T., and Sugiura, Y. (1996) Fluorescence detection of specific sequence of nucleic acids by oxazole yellow-linked oligonucleotides. Homogeneous quantitative monitoring of in vitro transcription. *Nucleic Acids Res.* **24,** 4992–4997.

15. Abravaya, K., Huff, J., Marshall, R., et al. (2003) Molecular beacons as diagnsotic tools: technology and applications. *Clin. Chem. Lab. Med.* **41,** 468–474.

II

APPLICATIONS

2

Multiplex Detection of Mutations

David S. Perlin, Sergey Balashov, and Steven Park

Abstract

Rapid and reliable detection of mutations at the genetic level is an integral part of modern molecular diagnostics. These mutations can range from dominant single nucleotide polymorphisms within specific loci to codominant heterozygotic insertions and they present considerable challenges to investigators in developing rapid nucleic acid-based amplification assays that can distinguish wild-type from mutant alleles. The recent improvements of real-time polymerase chain reaction (PCR) using self-reporting fluorescence probes have given researchers a powerful tool in developing assays for mutation detection that can be multiplexed for high-throughput screening of multiple mutations and cost effectiveness. Here we describe application of a multiplexed real-time PCR assay using Molecular Beacon probes for the detection of mutations in codon 54 of the *CYP51A* gene in *Aspergillus fumigatus* conferring triazole resistance.

Key Words: Real-time PCR; molecular beacons; *Aspergillus fumigatus*; triazole resistance; multiplex.

1. Introduction

The ability to detect multiple targets in a single assay represents one of the most powerful applications of real-time self-reporting probes. Multiplex assays increase throughput and reduce assay costs, which provides important benefits for diagnostic laboratories and research applications *(1)*. Most current instrumentation is limited to the simultaneous discrimination of four separate fluorophores per assay reaction, although newer instruments can handle six-color detection. Thus, in most assays only four defined targets can be distinguished if each target is labeled with a separate probe. Other limitations can include interference from the inclusion of multiple primer sets for amplification of separate targets within the assay reaction. However, depending on the requirement of the output data many more targets can be distinguished. For example, if the assay

From: *Methods in Molecular Biology, Vol. 429: Molecular Beacons: Signalling Nucleic Acid Probes, Methods and Protocols* Edited by: A. Marx and O. Seitz © Humana Press Inc., Totowa, NJ

is intended to distinguish between a wild-type sequence and multiple mutant sequences within the probe domain, then labeling the wild-type probe with one fluorophore and all others with a different fluorophore allows simultaneous detection of numerous targets with two color discrimination that can be detected with most real-time polymerase chain reaction (PCR) instruments. Such an approach is particularly amenable to the detection of drug resistance alleles, which can be confined to a single residue or to short sequence stretches *(2–4)*. In such cases, allele discrimination within a common target region is critical. Molecular beacons are ideally suited for both allele discrimination and multiplex applications *(5–8)*. The development of a robust multiplex assay for distinguishing multiple targets is shown for codon 54 of AfCYP51A, which encodes the sterol biosynthetic enzyme lanosterol demethylase *(4)*. Mutations at this locus confer resistance to the triazole drug itraconazole in the invasive mould *Aspergillus fumigatus (9,10)*, which is a major cause of death in severely immunosuppressed patients. Depending on the data output requirement, a multiplex assay can be established with either four (or six depending on instrument) outputs (four color) corresponding to each specific allele or with either two outputs (two color) representing the wild-type target sequence and any mutant allele.

2. Materials

2.1. Probe Development and Validation

1. Beacon Designer 2.12 software (PREMIER Biosoft Int., Palo Alto, CA).
2. Oligo, version 4.04, software (Molecular Biology Insights, Inc., Cascade, CO).
3. Molecular beacons were purchased from Biosearch Technologies, Inc. (Novato, CA). The probes were labeled at the 5′-end with fluorophores 5-carboxyfluorescein (FAM) and 6-carboxy-2′,4,4′,5′,7,7′-hexachlorofluorescein (HEX), carboxy-X-rhodamine (ROX), or Quasar 670 (Q670) and dabcyl (4-((-4-(dimethylamino)-phenyl)-azo)-benzoic acid) or BHQ-1 at the 3′-end stored at −20°C until used.
4. Beckman Coulter System Gold high-pressure liquid chromatography (HPLC) (Beckman Coulter, Fullerton, CA).
5. C-18 reverse-phase column (Waters, Milford, MA).
6. 20% to 70% acetonitrile in 0.1 M triethylammonium acetate (pH 6.5) filtered and degassed.
7. Ethanol, 200 proof (Sigma Aldrich, St Louis, MO) stored at −20°C until used.
8. 0.3 M sodium acetate (Sigma Aldrich) stored at 4°C until used.
9. Eppendorf Vacufuge concentrator (Brinkmann, Westbury, NY).
10. DNA grade water (Fisher Scientific, Fairlawn, NJ), stored at room temperature until used.
11. Hybridization reaction mixture contained Stratagene Core PCR buffer (Stratagene, La Jolla, CA), 4 mM MgCl$_2$, 100 pmol individual target oligonucleotide and 5 pmol molecular beacon. Each reagent was stored at −20°C until used.
12. Stratagene Mx4000 Multiplex Quantitative PCR System (Stratagene).

2.2. Real-Time PCR

1. Stratagene Core PCR buffer stored at −20°C until used.

2.3. DNA Extraction and PCR Amplification of Template

1. Sabouraud's dextrose medium (SAB): 4% dextrose, 1% bactopeptone (pH 5.7) stored at room temperature until used.
2. FastDNA® kit (Qbiogene, Inc., Carlsbad, CA).
3. FastPrep® instrument (Qbiogene).
4. PCR amplification mix contained 25 pmol of each of Af210F (GTCTCTCATTCG-TCCTTGTCCT) and Af709R (CGTTGAGAATAAACTCGTTCCC) primer, 2.5 U iTaq DNA polymerase (Bio-Rad Laboratories, Hercules, CA), 0.5 mM dNTPs, 50 mM KCl, 4 mM MgCl$_2$, 20 mM Tris–HCl (pH 8.4), and about 100 ng *A. fumigatus* chromosomal DNA.
5. iCycler thermal cycler (Bio-Rad Laboratories).
6. The cycling conditions were 1 cycle of 3 min at 95°C, 35 cycles of 30 s at 95°C, 30 s at 55°C, 1 min at 72°C 1 cycle of 3 min at 72°C.
7. Montage PCR purification kit (Millipore, Bedford, MA).

3. Methods

3.1. DNA Extraction and PCR Amplification of Template

1. The sequence of *A. fumigatus* gene CYP51A (accession number AF338659) was used for primer design for PCR amplification of a 500 nt product containing codon 54 to be used as a template for real-time PCR assay development.
2. *A. fumigatus* chromosomal DNA is isolated using the FastDNA® kit (Qbiogene, Inc., Carlsbad, CA). Cells grown 24 h in SAB are first homogenized for a 30-s interval at 6000 m/s using a FastPrep® instrument (Qbiogene).
3. Amplification of a 500 bp fragment of *cyp51A* is performed on an iCycler thermal cycler (Bio-Rad Laboratories). Each 100 µL PCR contain 25 pmol of each of Af210F (GTCTCTCATTCGTCCTTGTCCT) and Af709R (CGTTGAGAATAAACTCGT-TCCC) primer, 2.5 U iTaq DNA polymerase (Bio-Rad Laboratories), 0.5 mM dNTPs, 50 mM KCl, 4 mM MgCl$_2$, 20 mM Tris–HCl (pH 8.4), and about 100 ng *A. fumigatus* chromosomal DNA. The cycling conditions were 1 cycle of 3 min at 95°C, 35 cycles of 30 s at 95°C, 30 s at 55°C, 1 min at 72°C 1 cycle of 3 min at 72°C. PCR products were purified using the Montage PCR purification kit (Millipore).

3.2. Design Considerations for Multiplexing Allele Discriminating Molecular Beacons

1. Molecular beacons covering the locus of diversity corresponding to codon 54 of CYP51A were designed using Beacon Designer 2.12 software utilizing default software parameters (*see* **Note 1**). The following sequence was designed for the wild-type CYP51A codon 54 region including the six nucleotide complementary molecular beacon stems; CGCGATCATCAGTTAC**GGG**ATTGATCCATCGCG (codon 54 in bold).

2. Analysis of the CYP51A region surrounding the Gly54 codon for secondary structures was performed using Zuker DNA—RNA folding server (http://www. bioinfo.rpi.edu/~zukerm/) (*see* **Note 2**).

3. PCR primers were designed with Beacon Designer, version 2.12, software and Oligo, version 4.04, software and were purchased from Sigma-Genosys (Woodlands, TX).

4. A molecular beacon complementary to the test locus (wild-type allele) was synthesized with a 21-nucleotide probe target sequence (hairpin loop) domain (CATCAGT-TAC- **GGG**ATTGATCC; codon 54 in bold) and a six-nucleotide stem domains (CGCGAT) with the 5′-end labeled with FAM as the fluorophore and the six-nucleotide 3′-end (ATCGCG) modified with a dabcyl quencher (*see* **Note 3**).

5. Molecular beacons designed with probe sequences recognizing mutant alleles were otherwise identical to the wild-type beacon, and they were labeled with either HEX alone (two color multiplex) or HEX, ROX, FAM, and Quasar 670 (four color multiplex) at the 5′-end and with dabcyl at the 3′-end as described: http://www. molecular-beacons.org/PA_protocol.html#cap2 (*see* **Note 4**).

6. Molecular beacons were purchased from Biosearch Technologies, Inc. as either fully- or semi-purified products (*see* Steps 7 and 8), which were subjected to HPLC purification for the latter.

7. Purification of fluorophore- and/or quencher-coupled oligonucleotides was achieved by HPLC on a C-18 reverse-phase column (Waters), utilizing a linear elution gradient of 20% to 70% acetonitrile in 0.1 M triethylammonium acetate (pH 6.5), which was filtered and degassed and run for 25 min at a flow rate of 1 ml/min. The elution of nucleic acid peaks were monitored at 260 and 495 nm (FAM), 535 nm (HEX), 575 nm (ROX), or 649 nm (Quasar 670), as described: http://www.molecular-beacons.org/PA_protocol.html#cap2. The final product was precipitated with ethanol at 2.5:1 (v/v) and 0.3 M sodium acetate, washed with 70% ethanol, dried with a Vacufuge (Eppendorf) concentrator, and final resuspension in DNA grade nuclease-free water.

3.3. Validating Molecular Beacons

1. Molecular beacon—target hybridization was investigated using the Stratagene Mx4000 Multiplex Quantitative PCR System (Stratagene). The "Molecular Beacon Melting Curve" option was chosen in the Mx4000 software for data monitoring and analysis.

2. Each hybridization reaction mixture contained 1x Stratagene Core PCR buffer, 3 or 5 mM MgCl$_2$, 100 pmol individual oligonucleotides, and 5 pmol molecular beacons.

3. The test reactions were subjected to heating at 95°C for 3 min and cooling to 80°C with subsequent cooling down to 25°C using 112 30-s steps with a temperature gradient −0.5°C. Fluorescence output was measured at the end of each step.

4. The final data of the "Molecular Beacon Melting Curve" experiment were converted to a "SYBR Green (with Dissociation Curve)" output. The values for melting temperatures (T_m) were calculated by Mx4000 software as a temperature points corresponding to maximal values of the first derivative of the fluorescence output ($-Rn'(T)$).

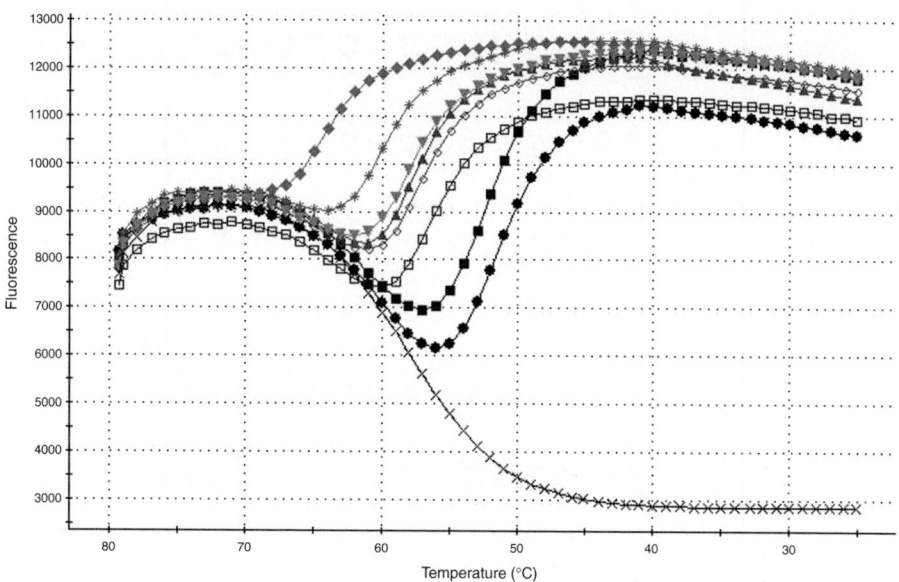

Fig. 1. Molecular beacon CYP51A codon 54 melting curves. This summarizes the results of nine separate experiments on hybridization of GGG-FAM (wild-type) molecular beacon with complement target GGG-T (♦), seven mutant allele-specific targets GTG-T (∗), AGG-T (▼), GAG-T (▲), TGG-T (◇), CGG-T (□), AAG-T (■), GAA-T (●), and no-template control (✕).

5. Molecular beacon must show appropriate thermal behavior (**Fig. 1**). In the absence of target, molecular beacons should show self-annealing of the stem sequence as temperature decreases resulting in fluorescence quenching (*see* **Note 5**).

6. Hybridization with a perfect complementary target should show maximum fluorescence response at temperatures below predicted T_m (**Fig. 1**).

7. The efficiency for annealing to different targets varies depending on target nucleotide content with experimentally derived T_m values as illustrated for each of the eight beacon-target hybrids as shown in **Table 1**. In this case, the stability of intermolecular hybrids of GGG-FAM molecular beacons and artificial oligonucleotide targets decreased in the order of GGG > GTG > AGG > GAG > TGG > CGG > AAG > GAA. Hybrids with double mismatches are expected to possess the lowest stability (*see* **Note 6**).

8. The temperature interval between Tm of the most stable mismatched beacon—target hybrid and Tm of the complement beacon—target hybrid represented the condition allowing specific allele-discriminative binding of the molecular beacon or window of discrimination. Accordingly, the window of discrimination for GGG-FAM molecular beacon was within the temperature ranges of 61.7–65.2°C under conditions of 3 mM Mg^{2+} and 63.7–67.2°C at 5 mM Mg^{2+} (**Table 1**) (*see* Note 7). Under these

Table 1
Melting Temperatures (T_m) for CYP51A Codon 54 Molecular Beacons

	Mg concentration[a]															
	3 mM for target								5 mM for target							
Beacon	GGG-T	GAA-T	AAG-T	GAG-T	AGG-T	GTG-T	TGG-T	CGG-T	GGG-T	GAA-T	AAG-T	GAG-T	AGG-T	GTG-T	TGG-T	CGG-T
AAG-FAM	48.7	44.7	**62.2**	52.7	53.2	48.2	50.7	49.7	49.7	45.2	**63.7**	54.2	54.7	54.2	52.2	52.2
GGG-FAM	**65.2**	53.7	54.2	58.7	59.7	61.7	58.2	55.7	**67.2**	55.2	55.7	61.2	61.7	63.7	61.2	59.7
GAA-HEX	45.7	**61.2**	41.7	50.2	35.2	45.2	41.7	36.2	46.7	**63.7**	47.7	52.7	41.2	46.7	42.2	41.7
AAG-HEX	46.2	39.7	**61.2**	52.2	51.7	49.7	48.2	47.2	48.7	44.7	**63.2**	53.2	53.2	51.2	50.7	49.7
GAG-HEX	55.7	55.7	56.7	**63.2**	50.2	57.7	50.7	50.7	57.2	56.2	57.2	**64.2**	51.7	58.7	52.7	52.2
AGG-HEX	54.7	42.2	55.2	49.7	**61.7**	49.7	56.2	58.2	56.7	44.7	57.2	51.7	**63.7**	52.2	58.2	59.7
GTG-HEX	54.7	47.7	48.7	54.7	49.7	**63.7**	51.7	49.7	56.2	48.7	51.2	57.2	50.7	**65.2**	53.7	52.2
TGG-HEX	56.2	47.2	53.7	52.7	58.7	53.7	**63.7**	60.2	57.7	48.2	54.7	53.2	59.7	55.2	**65.2**	61.7
CGG-HEX	54.7	40.7	47.7	48.2	54.2	50.2	54.7	**65.2**	55.2	42.7	49.2	50.2	55.2	51.7	57.2	**66.2**
TGG-ROX	52.7	39.2	49.7	46.2	56.2	48.2	**62.7**	58.2	54.2	39.7	51.2	49.2	57.7	50.2	**63.7**	59.7
CGG-ROX	52.7	40.7	47.7	46.7	53.7	49.7	53.2	**64.2**	54.2	42.2	48.7	49.7	54.7	51.2	55.2	**65.7**
GAA-Q670	42.2	**58.2**	42.7	48.7	39.7	42.7	39.2	41.2	46.7	**61.7**	42.7	51.2	40.2	48.2	44.7	43.2
GTG-Q670	44.7	46.2	47.7	55.2	42.7	**61.2**	49.7	48.2	49.7	48.2	50.2	56.2	49.2	**64.2**	51.7	50.2

[a] T_ms of complement molecular beacon–target hybrids are in boldface.

conditions, all molecular beacons (mutant and wild type) could be run in a real-time PCR experiment at the same annealing temperature of 61°C and showed excellent discrimination against both wild-type and mutant alleles.

3.4. Multiplex Panels and Real-Time PCR

1. Once target-specific hybridization with complement DNA target was validated for each beacon, the probes can be combined in a simple multiplex real-time PCR format. The advantage of this format is that, in a single assay, mutant (resistant) and wild-type (susceptible) strains can be easily distinguished. However, this format cannot be used to identify specific mutations of Gly54 because each mutant beacon was labeled with the same fluorophore (*see* Section 3.4.2.).

2. A single-reaction real-time PCR assay with eight different molecular beacons (one wild-type beacon labeled with FAM and seven mutant beacons labeled with HEX) and a 0.5-kb template corresponding to each *cyp51A* allele can be performed. In this format, the fluorescence signal indicated the presence of either wild-type or mutant sequences corresponding to itraconazole susceptibility or resistance, respectively.

3. To expand the applicability of the multiplex assay to distinguish separate alleles, the molecular beacons were labeled with different fluorophores: FAM, HEX, ROX, or the CY5 analog Q670 (**Table 1**). This configuration required two PCRs to assess all seven mutations, since only four colors could be distinguished in a single reaction. The first PCR mixture contained molecular beacons with high T_ms against complement DNA targets: GGG-FAM, GAG-HEX, GTG-Q670, and CGG-ROX. The second PCR mixture contained molecular beacons with somewhat lower T_ms: AAG-FAM, AGG-HEX, GAA-Q670, and TGG-ROX. The system specificity was further optimized by adjusting the Mg^{2+} concentration up to 3 mM for the first reaction and to 5 mM for the second reaction. A uniform annealing temperature of 57°C was used for both reactions.

4. The assay system was tested by adding individual 0.5-kb *cyp51A* templates to each of two reaction mixtures. Under these conditions, specific hybridization with complement *cyp51A* alleles was observed for all molecular beacons except the GGG-FAM wild-type beacon, which exhibited some level of nonspecific hybridization with the GTG allele. To avoid any possible false-positive results coming from nonspecific hybridization of the wild-type GGG-FAM beacon, the threshold fluorescence level was adjusted to values close to those obtained for complement beacon DNA polymerase, 0.4 mM dNTPs, 3 or 5 mM MgCl$_2$, and 100 ng chromosomal DNA.

5. PCRs were performed using the parameters, as follows: 1 cycle of 10 min at 95°C, 45 cycles of 30 s at 95°C, 30 s at 61°C and 30 s at 72°C. Annealing temperatures of 55°C and 57°C are used when PCR experiments are performed in multiplex format. The filter gain set of the Mx4000 System is changed to FAM-940 HEX-720 CY5-700 ROX- with the aim of equalization of the fluorescence signals from different molecular beacons.

6. The fluorescence is measured during the annealing step. This multiplex format allowed all seven specific *cyp51A* alleles bearing mutations in the Gly54 codon, along with the wild-type allele, to be distinguished in a real-time assay.

7. Real-time PCR experiments were performed on a Stratagene Mx4000 Multiplex Quantitative PCR System. The "Quantitative PCR (Multiple Standards)" setting was used for all real-time PCR experiments. Reagents from Brilliant® QPCR Core Reagent Kit were used for all reactions. Each 50-µl PCR contained 1x Stratagene Core PCR buffer, 20 pmol molecular beacon, 25 pmol of each of the CYP51AS and CYP51AA primers, 2.5 U Stratagene SureStart® *Taq*.

3.5. Data Processing

Fluorescence signals coming from Mx4000 System during PCR amplification were monitored using Mx4000 software in real time. At the end of each run, the amplification plots data were converted to graphic format and stored as image files or exported into Microsoft Office Excel and stored as spreadsheet files. In the case of multiplex beacons PCRs, the final results of PCR amplifications were converted from "Quantitative PCR (Multiple Standards)" type of experiment to the "Quantitative Plate Read" type of experiment.

4. Notes

1. Amplicons <100 bp are preferred for multiplex reactions.
2. Ideally, target regions should be devoid or have minimal secondary structure for reliable probe—target hybridizations.
3. The stem length and GC content can be varied to influence the temperature at which a molecular beacon will open upon hybridization with target.
4. Quasar 670 has spectral properties equivalent to Cy5 and ROX has spectral properties equivalent to Texas Red.
5. The molecular beacons should be fully quenched at temperatures at or below the probe-target hybrid T_m.
6. Codon GGG, corresponding to wild-type Gly54 target, occupied nucleotide positions 11—13 of the probe domain. However, allele discrimination can occur with any nucleotide within the 21 nt probe domain (*4,10*).
7. The strength of probe—target hybrids can be manipulated by varying divalent cation concentrations and provides an additional source for experimental window discrimination.

References

1. Mackay, I. M. (2004) Real-time PCR in the microbiology laboratory. *Clin. Microbiol. Infect.* **10**, 190–212.
2. Piatek, A. S., Tyagi, S., Pol, A. C., et al. (1998) Molecular beacon sequence analysis for detecting drug resistance in *Mycobacterium tuberculosis*. *Nat. Biotechnol.* **16**, 359–363.
3. Kostrikis, L. G., Tyagi, S., Mhlanga, M. M., Ho, D. D., and Kramer, F. R. (1998) Spectral genotyping of human alleles. *Science* **279**, 1228–1229.
4. Balashov, S. V., Gardiner, R., Park, S., and Perlin, D. S. (2005) Rapid, high-throughput, multiplex, real-time PCR for identification of mutations in the *cyp51A*

gene of *Aspergillus fumigatus* that confer resistance to itraconazole. *J. Clin. Microbiol.* **43,** 214–222.

5. Bonnet, G., Tyagi, S., Libchaber, A., and Kramer, F. R. (1999) Thermodynamic basis of the enhanced specificity of structured DNA probes. *Proc. Natl Acad. Sci. USA.* **96,** 6171–6176.

6. Tyagi, S., Bratu, D. P., and Kramer, F. R. (1998) Multicolor molecular beacons for allele discrimination. *Nat. Biotechnol.* **16,** 49–53.

7. Marras, S. A., Kramer, F. R., and Tyagi, S. (1999) Multiplex detection of single-nucleotide variations using molecular beacons. *Genet Anal.* **14,** 151–156.

8. Marras, S. A., Kramer, F. R., and Tyagi, S. (2003) Genotyping SNPs with molecular beacons. *Methods Mol. Biol.* **212,** 111–128.

9. Nascimento, A. M., Goldman, G. H., Park, S., et al. (2003) Multiple resistance mechanisms among *Aspergillus fumigatus* mutants with high-level resistance to itraconazole. *Antimicrob. Agents Chemother.* **47,** 1719–1726.

10. Diaz-Guerra, T. M., Mellado, E., Cuenca-Estrella, M., and Rodriguez-Tudela, J. L. (2003) A point mutation in the 14alpha-sterol demethylase gene cyp51A contributes to itraconazole resistance in *Aspergillus fumigatus*. *Antimicrob. Agents Chemother.* **47,** 1120–1124.

3

Detection of *Salmonella* Species in Foodstuffs

Arvind A. Bhagwat, Jitu Patel, Trina Chua, Audrey Chan, Saúl Ruiz Cruz, and Gustavo A. González Aguilar

Abstract

Conventional methods to detect *Salmonella* spp. in foodstuffs may take up to 1 wk. Methods for pathogen detection are required. Real-time detection of *Salmonella* spp. will broaden our ability to screen large number of samples in a short time. This chapter describes a step-by-step procedure using an oligonucleotide probe that becomes fluorescent upon hybridization to the target DNA (Molecular Beacon; MB) in a real-time polymerase chain reaction (PCR) assay. The capability of the assay to detect *Salmonella* species from artificially inoculated fresh- and fresh-cut produce as well as ready-to-eat meats is demonstrated. The method uses internal positive and negative controls which enable researchers to detect false-negative PCR results. The procedure uses the buffered peptone water for the enrichment of *Salmonella* spp. and successfully detects the pathogen at low level of contamination (2–4 cells/25 g) in <24 h.

Key Words: Microbial food safety; fresh produce; ready-to-eat meat; enteric pathogens; salmonelosis; food surveillance.

1. Introduction

The detection of *Salmonella* species by regulatory agencies is still primarily based on traditional microbiological culture methods which may take up to 5 d to confirm the results *(1)*. When a foodborne outbreak is suspected, faster the source of a pathogen can be identified, the sooner the public can regain confidence in the food supply *(2)*. A rapid pathogen detection method helps identify source of pathogen during outbreaks investigation, reestablishing public confidence in the food supply. However, only 2 out of the 27 outbreak investigations on fresh produce clearly identified a point of contamination, which underscores the importance and need for rapid and accurate pathogen identification methods *(3)*. Advances in biotechnology have permitted more rapid microbial identification

From: *Methods in Molecular Biology, Vol. 429: Molecular Beacons: Signalling Nucleic Acid Probes, Methods and Protocols* Edited by: A. Marx and O. Seitz © Humana Press Inc., Totowa, NJ

and surveillance *(4,5)*, and polymerase chain reaction (PCR)-based detection methods have become valuable tools for investigating foodborne outbreaks and identifying the responsible etiological agents *(6–8)*. The recently introduced real-time PCR-based format utilizes an internal fluorogenic probe that is specific to the target gene *(9–12)*. During the PCR assay, the target gene is amplified and simultaneously recognized and monitored by the fluorescent probe moiety *(13)*.

There are two types of fluorogenic PCR-based detection methods. One is based on a linear fluorogenic probe and requires the 5′–3′-nuclease activity of the DNA polymerase *(9,14,15)* (also known as TaqMan assays), while the other utilizes a fluorogenic probe which has flanking GC-rich arm sequences complementary to one another *(12,13,16)* (also known as Molecular Beacon, MB). In both types of real-time PCR probes, a fluorescent moiety is conjugated to one end of the sequence, and a quencher moiety is attached to the other end of the sequence. In the absence of target DNA sequences, the MB assumes a hairpin conformation with the two arms hybridizing to each other thus bringing the quencher into close proximity to the fluorophore (which results in no or low background fluorescence). When the target DNA is present the sequence in the loop region hybridizes, the hairpin of the MB opens, and the fluorophore and the quencher separate. In the open conformation, the fluorophore of the MB emits a detectable signal that is directly correlated with the quantity of the target template present in the PCR assay *(13,17)*. The TaqMan assay differs from the MB method in that the generation of the fluorophore signal is dependent upon 5′–3′-nuclease activity to cleave the reporter dye from the linear probe *(14,15,18)*.

Irrespective of the reporter technology employed in the PCR assay, its successful application to food samples, particularly to fresh produce, has been hindered by the lack of a convenient and relatively simple method for preparation of PCR-amplifiable DNA *(6,19–22)*. We and others have reported the presence of inhibitory compounds of plant origin that interfered with PCR biochemistry resulting in false-negative data *(5,19,22)*. Here we describe a simple, commercially available MB-based detection protocol for *Salmonella* spp. *(10,23)*. The protocol enables detection of wide-range of *Salmonella* serovars in several food matrices such as fresh and fresh-cut produce, meat, poultry, and various ready-to-eat foods.

2. Materials

All buffers and double-distilled (or reverse osmosis)-purified water used in the protocol must be sterilized by autoclaving at 15 lb pressure for 15 min.

2.1. Bacterial Strains and Culture Media

1. *Salmonella enterica* serovar Typhimurium ATCC 14028s and other serovar strains [Agona—SARB 1, Anatum—SARB 2, Dublin—SARB 12, Haifa—SARB 21,

Choleraesuis—SARB 4, Pullorum electrophoretic types Pu3 (SARB 51) and Pu4 (SARB 52) and Paratyphi A (SARB 42)] were obtained from the *Salmonella* Genetic Stock Center (Calgary, Alberta, Canada) and have been described previously *(24)*. Luria—Bertani (LB) broth and agar media (tryptone 10 g, yeast extract 5 g, NaCl 10 g, water 1 L (pH 7.2), add Difco granular agar 17 g L^{-1} for solid media), saline (0.9% NaCl in water (pH 7.0)), and buffered peptone water media (10 g bactopeptone, 5 g NaCl, 9 g Na$_2$HPO$_4$·12 H$_2$O, and 1.5 g KH$_2$PO$_4$ in 1 L of water, dissolve on magnetic stirrer, and heat if necessary. Final pH should be 7.2).
Orbital shaker incubator, 125-mL Erlenmeyer flasks, spectrophotometer to read optical density (A_{600}) of cell cultures.

2.2. DNA Extraction

1. Bench top microcentrifuge (10,000–12,000g, for 1.5 mL tubes, e.g., 5410, Eppendorf, Koln, Germany or similar), 1.5 mL screw-cap centrifuge tubes, micro-pippets and tips with incorporated cotton plugs, dry heat block or a boiling water bath (100°C), magnetic stir plate, vortex apparatus, and powder-free gloves (*see* **Note 1**).
TE buffer: 1 mM EDTA, 10 mM Tris–HCl (pH 8.0); lysis reagent A (or Instagel matrix, *see* **Note 2**) (Bio-Rad Laboratories, Hercules, CA).

2.3. Processing of Food Samples

1. Stomacher Lab Blender 400 (Seaward Medical, London, UK).
2. Stomacher model 400 bags with incorporated filter (cat. no. 6041/STR, Seaward Medical).
3. Incubator.
4. Buffered peptone water (as described above).

2.4. Real-Time PCR

1. Thermal cycler with real-time data acquisition capability for FAM and Texas Red.
2. 96-well PCR plates with optically clear sealing tape or 8-strip 200 µL PCR tubes with optically clear flat caps.
3. iQ-check *Salmonella* MB-PCR detection kit (Bio-Rad Laboratories, cat. no. 357 8100) containing reagent B (dual-labeled oligonucleotide probe based on *iagA* gene with FAM at 5′-end and DABSYL at 3′-end); reagent C (pre-mixed Taq-DNA polymerase, MgSO$_4$, primer pairs for *iagA* and buffer); negative and positive controls (reagents D and E, respectively).

3. Methods

The real-time PCR system described below comprises the following steps: (i) artificial inoculation of foods with desired cell density and pre-enrichment of samples; (ii) DNA extraction from pure and mixed-cultures of various *Salmonella* serovar strains and from pre-enrichment broths; (iii) setting up the real-time PCR and data acquisition; and (iv) data analysis and confirmation of occurrence of *Salmonella* in food samples.

3.1. Artificial Inoculation of Foods and Pre-Enrichment

1. One day prior to the experiment, inoculate two to three colonies of *S. enterica* serovar Typhimurium (and other serovars, as needed) in 10 mL LB broth and grow in a shaker incubator at 37°C for 18–22 h at 220 rpm.
2. Begin with cleaning the work area with 5% bleach, and use powder-free gloves to keep PCR tube tops optically clean.
3. Dilute the overnight culture 1:10 and read A_{600} on a spectrophotometer. Take cells equivalent to 1.0 OD_{600} and make final volume to 1 mL with sterile saline, this represents approximately 10^9 cells mL^{-1}. Make appropriate 10-fold serial dilutions in saline to get 10^3 cells mL^{-1} or roughly 1 cell per 1 µL. Take a dry LB agar plate and spot several 5 and 20 µL spots, let the spots dry in a laminar flow. Incubate the plates overnight at 37°C and count colonies to determine actual inoculum level in food samples.
4. Simultaneously, take 25 g food materials and place in a stomacher bag 400 with incorporated filter. Add 5–10 µL diluted cell suspension to represent 5–10 *Salmonella* spp. cells. Perform several inoculations in triplicate to represent 10-fold inoculation on desired food matrices. Include multiple sample aliquots with and without addition of 50 µL saline per 25 g food sample to represent known negative-control samples.
5. Add 225 mL buffered peptone water and homogenize for 2 minutes at 120 strokes per min. Incubate the stomacher bags *without shaking* for 18–22 h at 37°C.

3.2. DNA Extraction

3.2.1. DNA Extraction from Enriched Food Samples

1. After 18–22 h of incubation, carefully remove 1 mL liquid from the top 1–3 cm portion of the stomacher bag *without disturbing* the food debris (for oily foods, *see* **Note 3**) and place it in a screw cap Eppendorf centrifuge tube.
2. Centrifuge at 10,000 *g* for 2 min at room temperature and discard the supernatant.
3. Resuspend the pellet by vortexing briefly and add 200 µL lysis reagent A (*see* **Note 2**). Heat the tightly capped tube in a boiling/dry bath for 15 min at 100°C, and briefly chill on ice. Collect the supernatant in a new tube by centrifugation (10,000 *g* for 10 min at room temperature) and store at −20°C till further use.

3.2.2. DNA Extraction from Pure Salmonella spp. Cultures

1. Perform 10-fold serial dilution in saline from 1.0 OD (A_{600}) of cell suspension and perform a viable cell count using LB agar plates.
2. Take 5 µL cell suspensions from each dilution tube and add to a screw capped Eppendorf tube containing 195 µL of either lysis reagent A or Instagen matrix. Heat in a boiling water or dry bath for 15 min and proceed with DNA isolation as outlined in Section 3.2.1.
3. To determine detection limits using mixed serovar strains two different mixtures were prepared by pooling 1.0 OD_{600} cells from individual cultures. Mixture-1 contained serotypes Agona, Anatum, Dublin, Haifa, and Choleraesuis and Mixture-2 was made from serovars that do not appear to be responsive to attachment-mediated

acid tolerance (serotypes Pullorum electrophoretic types Pu3 and Pu4, and Para-typhi A). A 10-fold serial dilution was carried out before taking 5 μL aliquot for DNA isolation.

4. The following day, calculate genome equivalents per microliter of the lysis reagent (i.e., colony forming units per microliter) using viable cell counts data.

3.3. Performing Real-Time PCR

3.3.1. Thermal Cycle Parameters

1. Set the thermal cycle for the use of iQ check *Salmonella* kit as follows: cycle 1: 50°C for 2 min; cycle 2: 95°C for 10 min; cycle 3 (repeat 50 times) [95°C for 20 s, 55°C for 30 s (set fluorescence detection at this stage), 72°C for 30 s]; cycle 4: 72°C for 5 min; cycle 5: 4°C hold.

2. For fluorescence measurements select appropriate filters for FAM (excitation wavelength of 490 nm and an emission wavelength of 530 nm) and Texas Red (excitation wavelength of 575 nm and an emission wavelength of 620 nm) for each sample.

3. Turn on the optical system and thermal cycler at least 1 h before use.

4. Most real-time PCR instruments allow '*plate set up*' and have '*sample identifiers*' which allow assignment of microtiter plate position with individual samples, such as sample replicates, positive control, negative control, etc. It is necessary to define a plate setup that corresponds exactly to where the samples are loaded in PCR tubes.

3.3.2. PCR Mix Preparation

1. Prepare a master mixture (5 μL reagent B and 40 μL PCR amplification mix per tube) according to the number of reactions to be carried out.

2. To prepare a standard curve of log-genome equivalent versus threshold cycle (C_t), take 5 μL DNA from each of the 10-fold dilution series samples (from Section 3.2.2., step 1) and place into tubes (in duplicate). Add 45 μL of the master mix and incubate in ice till other samples are ready.

3. For analyzing food samples, prepare the ratio of 1:10 dilution of sample DNA by adding 18 μL water to 2 μL DNA which was originally isolated from samples incubated in the enrichment broth (from step 3.2.1.). Transfer 5 μL diluted and 5 μL undiluted DNA in individual tubes (in duplicate). To each tube, add 45 μL PCR master mix. Gently mix by tapping on the bench-top of flicking with fingers.

4. At least one positive and one negative control must be included in each PCR run.

5. Close the flat-top PCR caps on each tube and wipe clean the lid surface with a tissue paper.

6. Centrifuge all the PCR tubes at room temperature for 2 min at 4000*g* to get rid of bubbles and the PCR mix at the bottom of the tube (*see* **Note 4**).

7. Place all the tubes in the thermocycler block and orient tubes as per the 'design set up plate' file (Section 3.3.1., step 4).

8. Start the PCR program and the data collection should begin automatically.

Table 1.
Interpretation of Sample Results by MB PCR iQCheck *Salmonella* Spp. Assay

Sample	*Salmonella* spp. detection (FAM)	Internal control dNA detection (Texas Red)	Interpretation
Negative control	$C_t = N/D$	$C_t > 20$	Experimental set-up free of *Salmonella* spp.
Positive control	$C_t > 10$	Not significant	PCR kit components in good condition
Negative test sample	$C_t = N/D$	$C_t > 20$	Absence of *Salmonella* spp., PCR successful
Positive test sample	$C_t > 10$	Not significant	Positive identification for *Salmonella* spp.
False negative test sample	$C_t = N/D$	$C_t = N/D$	Inhibition of PCR N/D, none detected.

3.3.3. Data Analysis

1. The data can be analyzed directly at the end of the PCR run. It is necessary to analyze the positive and negative controls before sample analyses. Select the fluorophore to be analyzed, i.e., FAM for *Salmonella* and Texas Red for the internal control.
2. One can select to analyze a subset of the samples at one time, however, it is recommended to select a negative and positive control with each analysis.
3. Choose 'PCR baseline subtracted' data option (instead of background subtracted option) to analyze selected wells. Note the cycle number located just before the positive-control curve rises significantly (increase in the fluorescence reading) above the background noise. The calculation of threshold cycles can be done by the software or by entering user-defined cycle numbers (*see* **Note 5**). Select 'calculate threshold cycle' option and the threshold line will appear and it should cross all curves above all the background traces.
4. It is necessary to analyze the data with the second fluorophore, Texas Red. Repeat steps 3.3.3., -2 and -3 and analyze all the samples by selecting the second fluorophore (Texas Red).
5. Data for each fluorophore can be viewed and saved in a plate format or as a list of C_t values with the associated graphs. For the experiment to be valid, the controls must have results as summarized in **Table 1**. Otherwise the PCR needs to be repeated.
6. A positive *Salmonella* sample must have a C_t value ≥ 10 for FAM fluorophore. If no C_t value is obtained for FAM, then the interpretation of the result depends on the internal control: (a) the food aliquot is consider free of *Salmonella* spp. if there is no C_t value for FAM, and the internal control in Texas Red has a $C_t \geq 10$; (b) if the internal control also has no C_t value, then it is not possible to interpret the data. In most instances, the data indicate an inhibition of the PCR. The 1:10 diluted DNA

Amplification plot of a 10-fold serial dilution series of *S. typhimurium*

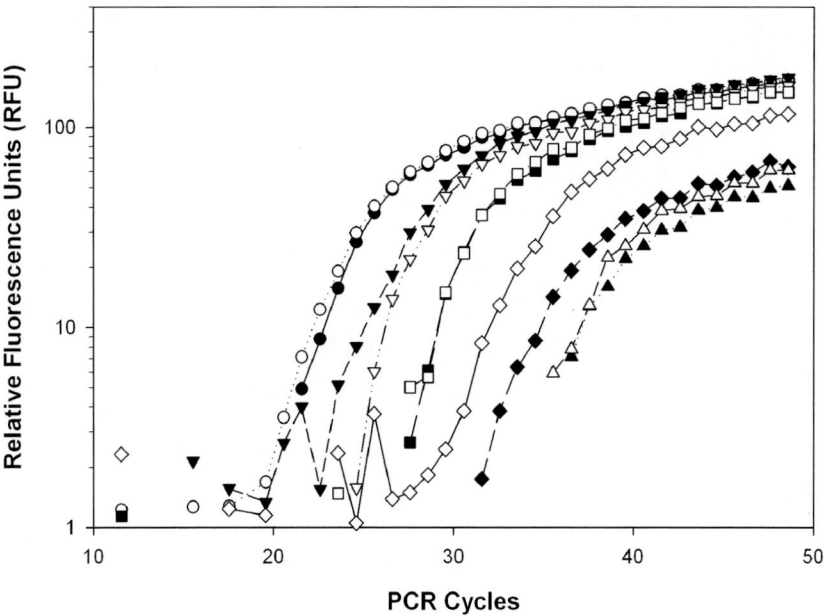

Fig. 1. Amplification plot of a 10-fold serial dilution series of *Salmonella enterica* serovar Typhimurium. Sample replicates (open and closed symbols) with cells per assay 10^0 (Δ, ▲); 10^1 (◊, ♦); 10^2 (□, ■); 10^3 (▽, ▼); and 10^4 (○, •) using a molecular beacon probe and real-time PCR assay. Real-time detection was done by measuring fluorescence of FAM during the annealing step of each PCR cycle (*X*-axis). Relative fluorescence units are plotted on the *Y*-axis. (Data obtained from Ref. *(10)*, With permission.)

   samples are useful in this respect, and occasionally one may have to dilute the DNA sample 1:25 to get successful amplification of the internal control.
7. We observed that the relative fluorescence from the MB probe increased with number of PCR cycles when cells per assay of serovar Typhimurium increased from 10^0 to 10^4 (**Fig. 1**). MB probe was also able to detect both mixtures of different serovars as effectively as serovar Typhimurium, and the C_t values decreased linearly with increasing target quantity per PCR assay (**Fig. 2**).
8. The variability for detection and quantitation between various serovar mixtures was minimal and correlation coefficient values were 0.98 and 0.94 were observed for mixtures 1 and 2, respectively. The correlation coefficient for serovar Typhimurium was 0.93 (**Fig. 2**).
9. The MB probe used in this study was able to detect *Salmonella* spp. from variety of fresh and fresh-cut produce at a very low level of contamination (i.e., at 1–3 CFU per 25 g of produce) (**Table 2**) *(10)*, as well as from poultry *(23)* and ready-to-eat meats (Patel and Bhagwat, unpublished work).

Standard curve for a 10-fold serial dilution series of *Salmonella* strains

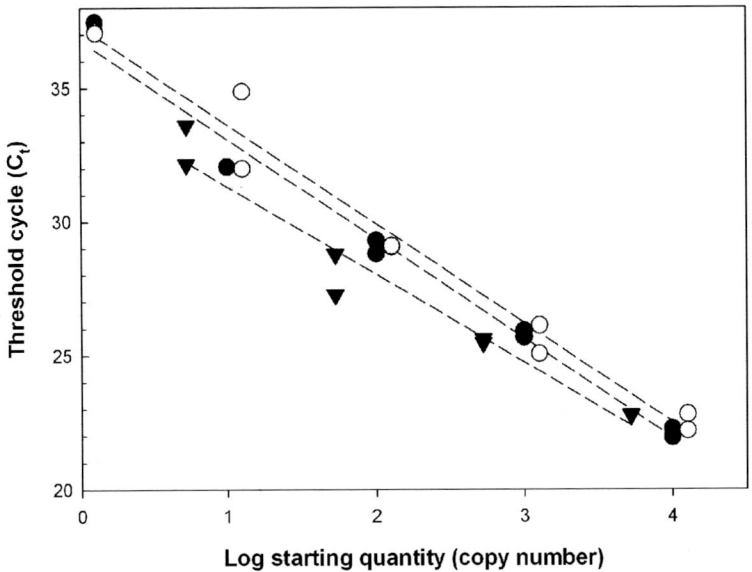

Fig. 2. Standard curve for a 10-fold serial dilution series of *Salmonella* strains (10^6 to 10^0 cells per assay, in duplicate) plotted as the threshold cycle (C_t) on the *Y*-axis using *S. enterica* serovar Typhimurium. Symbols: ○, serovar Typhimurium (data from **Fig. 1**); •, mixture of serovars Agona, Anatum, Dublin, Haifa, and Choleraesuis; or ▼, serovars Pullorum electrophoretic types Pu3, and Pu4, and Paratyphi A. The target copy number per assay is on the *X*-axis. (Data obtained from Ref. *(10)*, With permission.)

10. We also observed variation in enrichment of *Salmonella* spp. inoculated on different food matrices *(10)*. For example, the C_t values for the detection of *Salmonella* in alfalfa and cilantro were much higher (25.37 ± 4.6 and 32.66 ± 2.86, respectively) than the C_t values observed for cantaloupe and mixed-salad (17.74 ± 0.28 and 19.66 ± 2.02, respectively), even though all produce was inoculated at 3–4 CFU per 25 g (**Table 2**).

4. Notes

1. We prefer round-bottom 1.8 mL centrifuge tubes over conical tubes since round-bottom tubes tend to decant solutions cleaner, do not trap air bubbles and retain cell pellets reproducibly.

2. The lysis reagent A provided in the PCR-kit is not enough to perform DNA extractions from several reference strains to generate reference quantitative DNA templates. Among other lysis procedures, Instagel matrix (cat. no. 730-6030,

Table 2.
Comparative Analyses of Detection Frequencies of *Salmonella enterica* serovar Typhimurium from Artificially Inoculated Fresh Produce and Ready-to-Eat Meats by the MB Probe Real-Time PCR and Using Conventional Selective Media

Food matrices used for artificial inoculation	Contamination level (CFU/25 g)	Detection of *S. enterica* serovar typhimurium after one step enrichment (18 h)		
		Detection frequency (conventional) selective media[1]	Molecular beacon PCR	
			Detection frequency	Quantitation (C_t value[2])
Fresh produce				
Alfalfa sprouts	3.7 ± 1.2	9/9	9/9	25.37 ± 4.60
Cilantro	4.1 ± 2.0	8/8	8/8	32.66 ± 2.86
Parsley	8.5 ± 3.5	4/6	6/6	25.9 ± 2.20
Celery	4.9 ± 1.2	4/6	6/6	26.3 ± 3.50
Cauliflower	4.9 ± 1.2	2/4	4/4	25.4 ± 2.40
Fresh-cut produce				
Green onion	8.5 ± 3.5	6/6	6/6	29.5 ± 2.40
Mixed salad	3.3 ± 1.2	6/6	6/6	19.16 ± 2.02
Bell peppers	4.9 ± 1.2	6/6	6/6	32.6 ± 4.9
Cantaloupe	3.4 ± 1.5	6/6	6/6	17.74 ± 0.28
Ready-to-eat meat				
Turkey	3.4 ± 1.5	11/11	11/11	20.32 ± 2.55
Chicken	2.22 ± 0.11	8/8	8/8	18.9 ± 1.66
Ham	3.4 ± 1.5	12/12	12/12	18.16 ± 2.49
Bologna	3.4 ± 1.5	12/12	12/12	17.36 ± 1.57

[1]Standard microbiological procedure with selection on *Salmonella-Shigella* agar media was followed *(25)*.

[2]C_t value is defined as the cycle at which a significant increase in fluorescence is first recorded.

Bio-Rad Laboratories) closely matches lysis reagent A for DNA extraction and performance during the real-time PCR assay *(7)*.

3. In case of fatty foods, there may be a layer of oil/fat floating at the top. Immerse the pipette tip underneath the oil/fat layer that may have accumulated depending on food material being analyzed.

4. It is important to avoid bubbles at the bottom of each PCR tube (or PCR plate) by performing careful pipetting. To eliminate any bubbles from a PCR plate, hold the sealed plate in hand and force the solution to the bottom of the wells with a sharp single action movement.

5. In order to visualize correctly an amplification curve and set the threshold, it is recommended to change the graph properties from default linear scale to a semi-log scale (fluorescence units on the *Y*-axis on log-scale; PCR cycles on a linear scale on the *X*-axis).

References

1. Bhagwat, A. A. (2006) In *Microbiology of Fresh Produce* (Matthews, K. R., ed.), pp. 121–165, American Society for Microbiology, Washington, DC.
2. Bhagwat, A. A. and Lauer, W. (2004) Food borne outbreaks in raw produce can be prevented. *Food Quality* **11,** 62–63.
3. NACMCF (1999) National Advisory Committee on microbiological criteria for foods. Microbiological safety evaluations and recommendations on fresh produce. *Food Control* **10,** 117–143.
4. Feng, P. (1997) Impact of molecular biology on the detection of foodborne pathogens. *Molec. Biotechnol.* **7,** 267–278.
5. Bhagwat, A. A. (2003) Simultaneous detection of *Escherichia coli* O157:H7, *Listeria monocytogenes* and *Salmonella* strains by real-time PCR. *Int. J. Food Microbiol.* **84,** 217–224.
6. Hill, W. E. (1996) The polymerase chain reaction: applications for the detection of foodborne pathogens. *Crit. Rev. Food Sci. Nutr.* **36,** 123–173.
7. Bhagwat, A. A. (2004) Rapid detection of *Salmonella* from vegetable rinse-water using real-time PCR. *Food Microbiol.* **21,** 73–78.
8. Hines, E. (2000) PCR-based testing: unraveling the mystery. *Food Quality* **7,** 22–28.
9. Hoorfar, J. and Radstorm, P. (2000) Automated 5′ nuclease PCR assay for identification of *Salmonella enterica*. *J. Clin. Microbiol.* **38,** 3429–3435.
10. Liming, S. H. and Bhagwat, A. A. (2004) Application of a molecular beacon—real-time PCR technology to detect *Salmonella* species contaminating fruits and vegetables. *Int. J. Food Microbiol.* **95,** 177–187.
11. Liming, S. H., Zhang, Y., Meng, J., and Bhagwat, A. A. (2004) Detection of *Listeria monocytogenes* in fresh produce using molecular beacon- real-time PCR technology. *J. Food Sci.* **69,** 240–245.
12. Chen, W., Martinez, G., and Mulchandani, A. (2000) Molecular beacons: a real-time polymerase chain reaction assay for detecting *Salmonella*. *Anal. Biochem.* **280,** 166–172.
13. Tyagi, S. and Kramer, F. R. (1996) Molecular beacon: probes that fluoresce upon hybridization. *Nat. Biotechnol.* **14,** 303–308.
14. Nogva, H. K. and Lillehaug, D. (1999) Detection and quantification of *Salmonella* in pure cultures using 5′-nuclease polymerase chain reaction. *Int. J. Food Microbiol.* **51,** 191–196.
15. Kimura, B., Kawasaki, S., Fujii, T., Kusunoki, J., Itoh, T., and Flood, S. J. A. (1999) Evaluation of TaqMan PCR assay for detecting *Salmonella* in raw meat and shrimp. *J. Food Protect.* **62,** 329–35.
16. McKllip, J. L. and Drake, M. A. (2000) Molecular beacon polymerase chain reaction detection of *Escherichia coli* O157:H7 in milk. *J. Food Protect.* **63,** 855–859.

17. Higuchi, R., Fockler, C., Dollinger, G., and Watson, R. (1993) Kinetic PCR analysis: real-time monitoring of DNA amplification reactions. *Biotechnology* **11,** 1026–1030.

18. Chen, S., Yee, A., Griffiths, M., et al. (1997) The evaluation of a fluorogenic polymerase chain reaction assay for the detection of *Salmonella* species in food commodities. *Int. J. Food Microbiol.* **35,** 239–250.

19. Liao, C.-H. and Shollenberger, L. M. (2003) Detection of *Salmonella* by indicator agar media and PCR as affected by alfalfa seed homogenates and native bacteria. *Lett. Appl. Microbiol.* **36,** 152–156.

20. Miller, D. N. (2001) Evaluation of gel filtration resins for the removal of PCR-inhibitory substances from soils and sediments. *J. Microbiol. Methods* **44,** 49–58.

21. Heller, L. C., Davis, C. R., Peak, K. K., et al. (2003) Comparison of methods for DNA isolation from food samples for detection of Shiga toxin-producing *Escherichia coli* by real-time PCR. *Appl. Environ. Microbiol.* **69,** 1844–1846.

22. Shearer, A. E. H., Strapp, C. M., and Joerfer, R. D. (2001) Evaluationof a polymerase chain reaction- based system for detection of *Salmonella* Enteritidis, *Escherichia coli* O157:H7, *Listeria* spp., and *Listeria monocytogenes* on fresh fruits and vegetables. *J. Food Protect.* **64,** 788–795.

23. Patel, J. R., Bhagwat, A. A., Sanglay, G. C., and Solomon, M. B. (2006) Rapid detection of Salmonella from hydrodynamic pressure-treated poultry using molecular beacon real-time PCR. *Food Microbiol.* **23,** 39–46.

24. Gawande, P. V. and Bhagwat, A. A. (2002) Inoculation onto solid surfaces protect *Salmonella* spp. during acid challenge: a model study using polyethersulfone membranes. *Appl. Envron. Microbiol.* **68,** 86–92.

25. Andrews, W. H., June, G. A., Sherrod, P. S., Hammack, T. S., Amaguana, R. M., and Andrews, W. H. (1998) In *Bacteriological Analytical Manual* (Jackson, G. A., ed.), pp. 5.01–5.20, AOAC International, Gaithersburg, MD.

4

Identification of Homozygous Transgenic Mice by Genomic Real-Time PCR

Gunter Schmidtke and Marcus Groettrup

Abstract

The 26S proteasome is the executing protease of the ubiquitin-dependent degradation system. It consists of one or two 19S regulatory sub-complexes and one 20S proteolytic sub-complex *(1)*. The 20S proteasome is a barrel-shaped cylinder which consists or four stacked rings *(2)*. Each of the two outer rings consists of seven different α-subunits, whereas each of the two inner rings is formed by seven different β-subunits *(3)*. Only three of these β-subunits bear a catalytically active N-terminal threonine *(4,5)*. Under normal conditions, these are β1 (delta), β2 (Z), and β5 (mb1). However, by induction of some cytokines, e.g., interferon-γ, these subunits are exchanged against β1i(LMP2), β2i (Mecl1), and β5i (LMP7) and the so-called immunoproteasome is formed *(6,7)*. To investigate the role of LMP7 in MHC class I-restricted immunology, we decided to generate a transgenic mouse which constitutively expresses LMP7 in all tissues. To get the highest possible expression, we bread the mice to be homozygous for the transgene LMP7. These mice cannot be identified by conventional polymerase chain reaction (PCR). So far, Southern blotting was the only possible method to quantify the DNA content. Here, we describe the analysis of these mice by quantitative PCR using sequence specific fluorescence resonance energy transfer-primers to reliably detect a difference in DNA content as small as a factor of 2 or only one PCR cycle.

Key Words: MHC class I antigen presentation; proteasome; LMP7 (β5i); homozygous transgenic mouse; quantitative PCR; FRET primers.

1. Introduction

The ubiquitin—proteasome system is responsible for the targeted degradation of regulatory proteins as well as for the removal of misfolded, aged, and oxidized polypeptides *(8)*.

Proteasomes are relatively abundant and can account for about 1% of the soluble proteins of a cell. The so-called constitutive proteasome and the immuno-

From: *Methods in Molecular Biology, Vol. 429: Molecular Beacons: Signalling Nucleic Acid Probes, Methods and Protocols* Edited by: A. Marx and O. Seitz © Humana Press Inc., Totowa, NJ

proteasome, which is formed by induction with interferon-γ, differ not only in their subunit composition, but also in their specificity and activity *(9,10)*. Therefore, after incorporation of immunosubunits, a different set of peptides is generated out of a digested protein and displayed on MHC class I. Many reports presented evidence that LMP7 enhances the activity of the proteasome and improved antigen presentation *(11–14)*, but a viral epitope has been reported to be less well presented in the presence of immunoproteasome subunits *(15)*. We are interested to study the immune system under constitutive expression of LMP7. To do so, we generated a LMP7-transgenic mouse. Successful transfer of LMP7 into the genome was easily detected by conventional polymerase chain reaction (PCR). Because of the abundance of the proteasome, the amount of generated protein should be as high as possible to successfully compete with MB1 for the incorporation into the 20S complex. In a homozygous mouse, the expression should be twice as high as in a heterozygous mouse. The success of the generation of a homozygous was so far possible by Southern blotting of the transgene in comparison with a reference gene. We employed quantitative PCR with fluorescence resonance energy transfer (FRET) primers for our analysis. In normal quantitative PCR, the amount of DNA present after each amplification step is measured by the fluorescent signal, emitted by a dye, which is incorporated into the DNA, usually SYBR Green. This method does not allow for the simultaneous analysis of two genes, because the signal is not sequence-specific. To analyze the amplification of two genes simultaneously, each amplified DNA must emit a distinguishable signal. This is possible with FRET-PCR, which is performed with two conventional primers per DNA to amplify the target. In addition two hybridization probes, which can bind towards the 3′-end of the template sequence, are necessary for each DNA. They are designed to bind to the same strand with a maximum of 4 bp distance from each other (*see* **Fig. 1**). Probe 1 has a donor fluorophore (F1) at its 3′-site, probe 2 carries the acceptor fluorophore (F2 or F3) at its 5′-site. During the annealing step of the PCR, the primers and the hybridization probes bind to the single-stranded DNA. The F1 of probe 1 is now in close proximity of F2 (or F3) of probe 2. If F1 is now exited by the light source of the PCR machine it does not emit light, but passes the energy to the adjacent F2 (or F3) via dipole–dipole interaction. F2 and F3 emit measurable light at a different wavelength. Therefore, the signal of F2 is proportional to one amplified sequence, and F3 to another one. During the amplification, which follows annealing and measurement, the probes are displaced by the newly formed DNA strand. This strand is denatured by high temperature and forms the new template for primers and probes to anneal. Because the signal depends on the hybridization to a specific sequence, many problems of normal quantitative PCRs can be avoided. The signal is usually not compromised by primer "dimers" or unwanted amplification products due to miss priming.

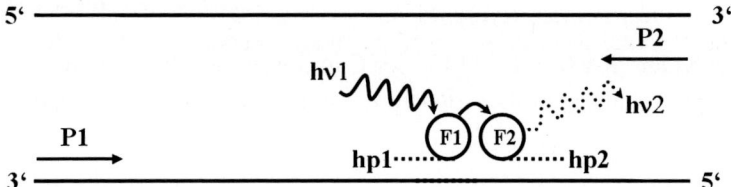

Fig. 1. The FRET principle. During the annealing step of the PCR, the primers P1 and P2 bind to the template at the 3′-site leaving their own 3′-site available for elongation by the polymerase. On one strand, the two hybridization probes hp1 and hp2 bind as well. Hp1 carries the donor fluorophore F1 hp2 the acceptor fluorophore F2 (or F3). If F1 is excited by light of wavelength 1 (hv1) it passes the energy to F2 (or F3) without emitting light by itself. F2 (or F3) will emit light of a different wavelength (hv2).

2. Materials

2.1. DNA Preparation

Isofluran, DNeasy tissue Kit (Qiagen), solution of 5 *M* sodium chloride (NaCl) (Merck), 100% and 70% ethanol (Riedel de Haen), Milly Q-water, optional RnaseA.

2.2. Agarose Gel Electrophoresis of DNA

Biorad mini subcell GT, Agarose (Serva),TAE-buffer 0.04 *M* Tris-acetate (Sigma), 0.001 *M* EDTA (Merck), glacial acetic acid (Roth), ethidiumbromide solution 10 mg/mL (Sigma). Sample buffer 6 × 0.25% bromophenol blue, 0.25% xylene cyanol, and 30% glycerol. Molecular weight standard marker (Eurogentec).

2.3. PCR Analysis

2.3.1. Primer Design

Any program can be used, the supplier of our PCR machine and PCR-kits recommends Roche molecular biochemicals primer probe design. Make sure that the primers bind to a "balanced" region (means nearly equal amount of all four bases), that they contain neither monotonous nor repetitive sequences, and that they are not self-complementary. The length of the product should be about 400 bp.

2.3.2. Hybridization Probe Design

These probes should bind at the 3′-end of the amplified template, avoid monotonous and repetitive sequences, self-complementary sequences, clusters of Cs and Gs, sequences rich in purines (G and A) and regions, which can hybridize to the 3′-termini of the PCR primers.

2.3.3. Instruments and Reagents

Light Cycler (Roche), Light Cycler Capillaries (Roche), Light Cycler DNA Master Hybridisation Probes (Roche), Light Cycler Color Compensation kit (Roche).

3. Methods

3.1. DNA Preparation (16)

1. Please read the instructions and important notes sections of the manual of your DNA-preparation kit before you begin. About 6- to 8-week-old mice were anesthetized with isofluran and a small 1 mm to a maximum of 2.5 mm piece was cut from the tip of the tail with sharp scissors. Ear mark the mouse or use another type of labeling to allow later identification. For controls, take non-transgene and a heterozygous mice as well.
2. Place the cut tip into a labeled 1.5 mL Eppendorf tube and add 180 µL Buffer ATL provided by the DNeasy tissue Kit (Quiagen). Add 20 µL Proteinase K and incubate overnight at 56°C. Vortex for 15 s.
3. Add 200 µL Buffer AL to the sample, and mix thoroughly by vortexing. Then add 200 µL ethanol (96–100%), and mix again thoroughly by vortexing. Pipette the mixture from step 3 (including any precipitate) into the DNeasy Mini spin column placed in a 2 mL collection tube (provided). Centrifuge at 6000g (8000 rpm) for 1 min. Discard flow-through and collection tube.
4. Place the DNeasy Mini spin column in a new 2 mL collection tube (provided), add 500 µL Buffer AW1, and centrifuge for 1 min at 6000g (8000 rpm). Discard flow-through and collection tube. Place the DNeasy Mini spin column in a new 2 mL collection tube (provided), add 500 µL Buffer AW2, and centrifuge for 3 min at 20,000g (14,000 rpm) to dry the DNeasy membrane. Discard flow-through and collection tube.
5. Place the DNeasy Mini spin column in a clean 1.5 mL microcentrifuge tube (not provided), and pipette 150 µL Buffer AE directly onto the DNeasy membrane. Incubate at room temperature for 1 min, and then centrifuge for 1 min at 6000g (8000 rpm) to elute.
6. Place the DNeasy Mini spin column in a new clean 1.5 mL microcentrifuge tube (not provided), and pipette 100 µL Buffer AE directly onto the DNeasy membrane. Incubate at room temperature for 1 min, and then centrifuge for 1 min at 6000g (8000 rpm) to elute.
7. Combine eluates from step 5 and 6 and precipitate the DNA. Add 35 µL of 5 M NaCl and 1 mL 100% ethanol. Incubate at –20°C for 15 min and centrifuge at 4°C for 15 min at maximum speed.
8. Remove the supernatant and wash with 500 µL 70% ethanol. Centrifuge as described above.
9. Remove supernatant and let the pellet air dry for 15 min at room temperature.
10. Redissolve the DNA of the non-transgenic and the heterozygous controls in 20 µL, all other samples in 50 µL water (taken from the PCR-Kit).

11. Take 8 µL of each sample, dilute to 80 µL with water and measure the absorption at 260 nm with water as blank. Typical values are between 0.1 and 0.8. Adjust the concentration of an aliquot of the initial sample by dilution with water that a repetition of the measurement would lead to an absorption between 0.1 and 0.25. Do not dilute the control samples.

3.2. Analyze 5 µL of the sample by agarose electrophoresis (17)

Boil 0.5 g agarose in 50 mL TAE-buffer until the agarose is dissolved. Let it cool down to about 60°C and add 2 µL of 10 mg/mL ethidiumbromide. Place the tray with a suitable comb into the pouring chamber. Pour the warm agarose solution into the tray that an approximate 5-mm thick gel is formed. After it is completely cooled down, put the gel in the electrophoresis tank and cover it with TAE-buffer to a depth of about 2 mm. Carefully remove the comb. Mix 5 µL sample with 1 µL sample buffer and load it in the slots. Add 5 µL of molecular weight standard markers to an empty slot. Run the gel at 100 V for about 30 min. The dark blue band should have moved about 5–6 cm. Analyze the gel with a UV light box. If DNA appears as a band of about 20–30 kbp, the preparation is suitable for PCR.

3.3. PCR Analysis

3.3.1. Primer Design (18)

Design of primers and hybridization probes for the reference gene. To decide whether or not LMP7 is in one or two copies in the genome we need a reference gene which occurs twice in the genome. We decided to use Psme 1 gene encoding the α-subunit of the PA28αβ proteasome regulator as reference. The computer program of the supplier of primers (TIB molbiol) suggested the following primers and probes with the given melting temperatures (T_m):

Psme 1 F	ctc agt cct gtc ttc ctc cct ac	T_m 57.1°C
Pmse 1 S	ggg acg aag acg aca aag gta c	T_m 58.1°C
Pmse 1 R	cgg aag agc tca cca gat tga	T_m 58.1°C
Pmse 1 A	act gcc tgg agc cga gac	T_m 58.6°C
Psme 1 F1	cag cct gga agc tct ggg tca cc F1	T_m 67.2°C
Psme 1 F2	tgt ctt ccc agc acc tcg ctt cc F2	T_m 67.3°C

Please see **Fig. 2A** for the binding regions.
For the gene of interest LMP7, the following suggestions were tested:

LMP7 Pe	tca tgg cgt tac tgg atctg	T_m 61.0°C
LMP7 S	ctc cgt gtc tgc agc atc ca	T_m 59.0°C
LMP7 A	tcc act ttc acc caa ccg tc	T_m 59.5°C
LMP7 mB	ata gtg gat cct cac aga gcg gcc tc	T_m 66.2°C
LMP7 F1	agc cac aga tca tgc tgc cca tg F1	T_m 66.1°C
LMP7 F3	gag gcc cat ccc ccg gta ctg F3	T_m 66.7°C

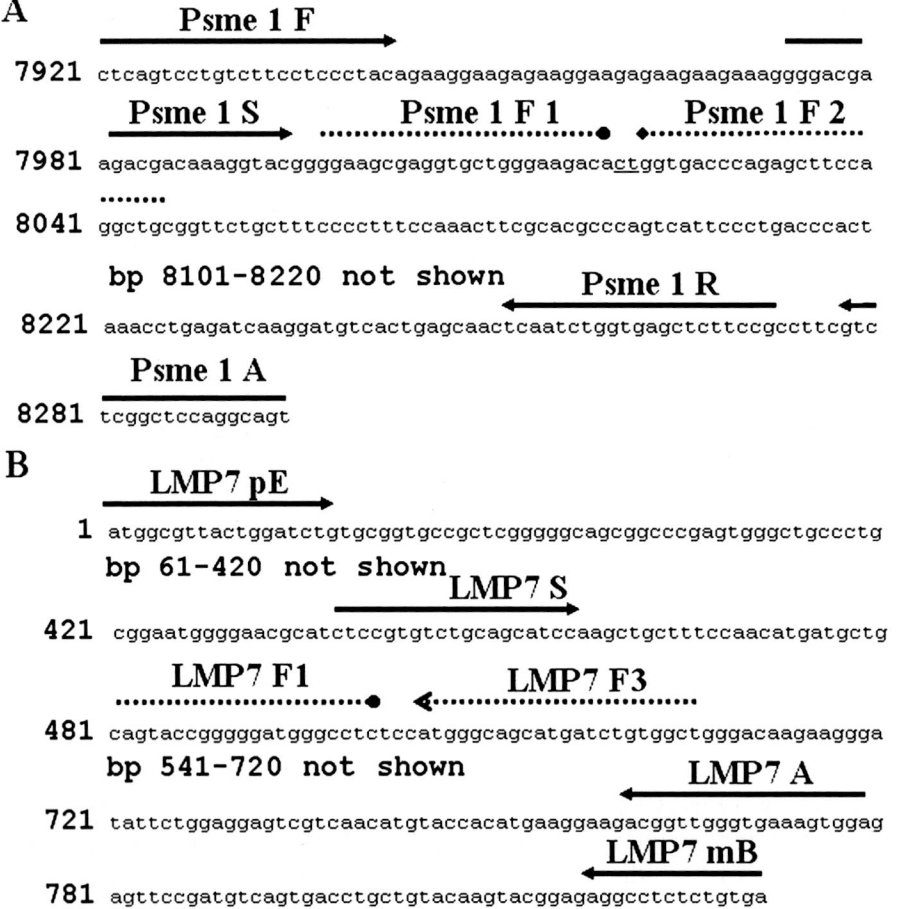

Fig. 2. The binding regions of the tested primers and hybridization probes are shown in 2A for Psme1 and 2B for LMP7.

Please see **Fig. 2B** for the binding regions.

3.3.2. Running the PCR and Adjusting the Parameters (19)

Generation of positive and negative controls. At first-time use, it is recommended to practice with the control reagents provided by the kit first. If you feel competent, prepare at least 100 µL of each dilution of the heterozygous and the nontransgene DNA 1:10, 1:100, 1:1000, 1:10,000. Both are used to optimize the PCR parameters. Create a color compensation file exactly as written in the manual of the color compensation kit.

Table 1
Parameters for Programming a Light Cycler Run

Analysis mode	Cycles	Segment	Target temperature[a]	Hold time	Acquisition mode
None	1	Denaturation	95°C	30s	None
		Amplification			
		Denaturation	95°C	5s	None
Quantification	45	Annealing	Primer dependent[b]	5–15 s[c]	Single
		Extension	72°C[d]	=Product (in bp)/25 = time in seconds[c]	None
None	1	Cooling	40°C	30s	None

[a]Temperature at end of segment.
[b]For initial experiments set 5°C below the calculated Tm.
[c]In same cases it can be advantageous to choose longer annealing and extension times to improve quantification.
[d]If annealin temperature is below 55°C, reduce transition rate/slope to 2–5°C/s.

Program the LightCycler Experimental Protocol before preparing the reaction mixes.

3.3.3. Programming a Run

Normally, a LightCycler protocol that uses the *LightCycler DNA Master HybProbe* contains the following parts:

Denaturation of the template DNA.
Amplification of the target DNA.
Cooling the rotor and thermal chamber.

1. **Table 1** shows the PCR parameters that must be programmed for a normal LightCycler PCR Run with LightCycler DNA Master HybProbe.
   Setup of the PCRs to find optimal temperature for annealing, optimal times for denaturation, annealing and elongation, and optimal concentrations of primers and $MgCl_2$.
2. Preparation of the PCR Mix and starting the run.
   Prepare each 20 µL standard reaction as described below.
   Important: Do not touch the surface of the capillaries. Always wear gloves when handling the capillaries.
   Depending on the total number of reactions, place the required number of LightCycler Capillaries in pre-cooled centrifuge adapters.
   Prepare a 10× concentrated solution of the PCR primers and a 10× concentrated solution of HybProbe pairs.

Thaw the LightCycler DNA Master HybProbe, 10× concentration (vial 1, green cap) and MgCl$_2$ stock solution, 25 mM (vial 2, blue cap), mix gently, and store on ice.

In a 1.5 mL reaction tube on ice, prepare the Master Mix for one reaction by adding the following components in the order mentioned below, then mix gently by pipetting up and down:

H$_2$O, PCR grade (vial 3, color less cap) × μL H$_2$O to reach 13 μL.

MgCl$_2$ stock solution (vial 2, blue cap) × μL MgCl$_2$ for required concentration of MgCl$_2$.

PCR Primer Mix, 10× concentration; 1.5 μL (0.2–1.0 μM each) (recommended concentration is 0.5 μM).

HybProbe probe mix, 10×; 1.5 μL (0.2 μM each).

LightCycler DNA Master HybProbe (vial 1, red cap) 1.5 μL.

Final volume (1×) 13 μL.

Pipet 13 μL Master Mix into the pre-cooled LightCycler Capillary.

Add 2 μL DNA template.

Use up to 500 ng complex genomic DNA or 101–1010 copies plasmid DNA.

Seal each capillary with a stopper and place the adapters, containing the capillaries, into a standard benchtop microcentrifuge.

Centrifuge at 700g for 5 s (3000 rpm in a standard benchtop microcentrifuge).

Place the LightCycler Sample Carousel with capillaries into the LightCycler Instrument and then start your protocol.

3.3.4. Optimization of Parameters

1. Because we have two forward and two reverse primers for each gene, we tried all four possible combinations of primers for each gene. For the initial experiments, we used only three different concentrations of DNA and three different annealing temperatures all with 2.5 mM MgCl$_2$. We used 1, 0.5, and 1 μL of a 1:10 dilution of DNA, and an annealing temperature 2.5°C below the theoretical T_m, the temperature at T_m, and at 2.5°C above T_m. After the run we took the capillaries, removed the caps, and put them upside down in an Eppendorf tube. Slow centrifugation at 200 × g for 1 min empties the capillaries into the tube. This reaction is analyzed by agarose gel electrophoresis for the presence of a DNA product of the correct size. The presence of other products does not harm the analysis as long as the hybridization probes do not bind to it.

   We found the combination Psme 1F/Psme 1A and LMP7S/LMP7A to give the best results.

   Because several parameters have to be optimized, e.g., annealing temperature, time for denaturation, annealing and elongation, and optimal concentrations of DNA, primers and MgCl$_2$, one should work according to the following system.

   First optimize the temperature for annealing of the primers. Use 2.5 mM MgCl$_2$ and 0.5 μM primers at the beginning. Add the following amount of both control DNAs (OD should be above 0.5 at 260 nm if measured as described earlier) to different capillaries containing the master mix: 2, 1, 0.5, and 1 μL of 1:10 dilution, 1 μL of 1:100 dilution, 1 μL of 1:1000 dilution, and 1 μL of 1.10,000 dilution. Adjust to 2 µL by adding water. Include a negative control for each primer pair/hybridization probe combination containing no DNA.

Prepare each sample at least six times. Use one set at a time and store the others on ice.

2. A good choice for optimizing the primer binding is 5°C below the calculated T_m according to the following formula: $T_m = 2°C (A+T) + 4°C (C+G)$. We started with 52°C and tested 54, 56, 58, 60, and 62°C as well. Label your samples as standards in the sample table (can be done during the run) and add the right relative concentration (2, 1, 0.5, 0.1, 0.01, 0.001, and 0.0001 in our case).

After the run, use the automatic analysis function of the software of the light cycler. Open the DATA ANALYSIS window. Select the file, open it by double clicking. Select F2/F1 or F3/F1 in the small window next to the fluorescence signal. The recorded signal (either F2/F1 or F3/F1) will be displayed. Select QUANTIFICATION. A new window will open with two different displays. Find your standards and samples with name, color code, concentration (as you entered before), and calculated concentration on the left site. The curves of the measured fluorescence can be found in the right panel. Make sure you selected Second Derivative Maximum as analysis method and Arithmetic as method for baseline calculation by clicking the appropriate buttons. Hit the analysis button on top of the right panel. The Second Derivative Maximum method does not require any adjustments to be done. A third panel appears in the lower left site showing the calculated concentration curve. With your included standards, the program should generate a linear calibration line. The calculated relative concentrations should match the one you entered for your standards. If not, the analysis can be done by the Fit Points method and with any of the four provided methods of baseline adjustment. If still no straight line is obtained, the PCR must be optimized further.

We found 60°C to be the optimal temperature for annealing. This reproducibly yielded the earliest detectable signal of amplification, e.g., detectable fluorescence. We found 5 s to be the best time for annealing rather than 10 or 15 s, probably also because of the relative position of the hybridization probes to the primers. The polymerase is also active at 60°C and will start to generate a new strand, which will displace the hybridization probes. This will lower the signal measured and may lead to nonlinear measurements either in the same PCR over time, or in different PCRs of the calibration curve. The data generated might not be useful for quantification anymore.

3. Next we tried different concentrations of $MgCl_2$. The master mix always contains 1 mM $MgCl_2$. We tested 1, 2, 3, 4, and 5 mM $MgCl_2$ final concentration. No amplification could be obtained with 1 mM $MgCl_2$. Increasing the concentration of $MgCl_2$ up to 4 mM yielded not only higher signals at the end of the exponential phase, the signal also started to rise earlier. With 5 mM $MgCl_2$ we detected again a decline in performance. Therefore, 4 mM $MgCl_2$ is the optimal concentration.

4. The time for elongation and the concentration of the primers are usually less critical. We use the suggested initial concentration and time and, because of satisfactory performance, further optimization was found not to be necessary.

The DNA should be used in the concentration given before, but make sure, that standards of both, higher and lower concentrations are included. If your sample is out of the calibration curve, it cannot be analyzed.

5. The PCRs for our two genes, LMP7 and psme1 both worked optimally under almost identical conditions, but only in separate tubes. This way we obtained a linear calibration curve for both over the whole range of dilutions.

   As soon as we tried to run both PCRs as a duplex reaction, no satisfactory results were obtained. Also changes in primer- and hybridization probe concentration did not solve the problem. At this point a decision has to be made. Either starting again with new primers and hybridization probes with no guarantee for success, or running the PCRs in separate capillaries. We decided in favor of the second option.

6. To test the performance one can use the DNA of the heterozygous mouse and a 1:1 dilution of this DNA with DNA from a non-transgene. The dilution should be done according to the amount of the reference gene as determined in the calibration curve. Now you run both samples for the reference gene and for the gene of interest. Divide the relative concentrations for both genes as determined by the calibration curve. The ratio of LMP7/psme 1 should be twice as high in the heterozygous sample as in the sample heterozygous plus non-transgene.

3.4. Analysis of the Transgenic Mice

In **Table 2** you find an overview of the first results obtained with our setup for the two genes LMP7 and Psme1. The quotient between the relative concentrations of LMP7 divided by Psme1 should either be close to 1 for heterozygous mice or close to 2 for homozygous mice. The value 1.5 would be the worst case, because no decision would be possible. The two values closest to 1.5 are 1.28 for a heterozygous and 1.95 for a homozygous mouse. We also observed values below 1 and above 2. The variations you get in the standards between different runs allow you to judge about the quality of your pipetting and your setup. Here, we got values between 0.89 and 1.11, so a variation of 0.22. So our value of 1.95 for a homozygous is within this range. The variations within the heterozygous are also acceptable, but not important, because they are excluded from further breeding. To proof our quantitative PCR results, we bred all putative homozygous male mice (one of this was 1.95) with C57BL/6 mice. If our results are right, the F1 generation should be heterozygous for LMP7. If the transgenic male is heterozygous, also LMP7-negative mice will be found in the F1 generation. Our test yielded only LMP7-positive mice in all cases, meaning all transgenic mice were indeed homozygous.

4. Notes

4.1. DNA Preparation

1. Start to optimize your PCR at least 3 to 4 weeks before you want to analyze your transgenic mice. Optimization takes time and sometimes the design of new primers

Table 2
Relative Concentrations of LMP7 and Psme1 mRNA

Number	Standard	Relative concentration LMP7	Relative concentration Psme1	Quotient	Homozygous
1		0.607	0.212	2.8	Y
2		0.186	0.144	1.28	
3		0.264	0.277	0.95	
4		0.243	0.203	1.2	
5		0.695	0.333	2.08	Y
6		0.355	0.296	1.19	
7		1.21	0.49	2.45	Y
8		1.21	0.492	2.46	Y
9		0.5	0.448	1.1	
10		1.6	0.628	2.55	Y
11		0.205	0.103	2	Y
12		0.17	0.151	1	
13		0.308	0.329	0.94	
14		0.96	0.398	2.4	Y
15		0.586	0.3	1.95	Y
16		0.135	0.064	2.12	Y
17		0.756	0.35	2.12	Y
a	1	0.936	1.042	0.89	
b	0.5	0.517	0.466	1.11	
c	0.2	0.211	0.205	1.03	
d	0.1	0.107	0.1	1.07	
e	0.05	0.046	0.051	0.9	

   and probes is necessary. Do not use too much material as this leads to decreased yields and worse PCR performance.

2. It is essential that the sample, Buffer AL, and ethanol are mixed immediately and thoroughly by vortexing or pipetting to yield a homogeneous solution. Buffer AL and ethanol can be premixed and added together in one step to save time when processing multiple samples.

4.2. PCR Analysis

1. *Primer design*: Make sure the primers of the reference gene bind to the genomic sequence and the amplified part is not too long because of an exon.

Make sure the primers of the gene of interest bind to the cDNA only, preferably on two exons which are separated by an intron to amplify the transgene only and avoid amplification of the genomic copies. Cloning primers are usually not suitable for quantitative PCR.

If you want to run a duplex PCR, meaning the amplification of two genes in the same reaction and analyze with two different fluorophores, the primers and hybridization probes of both PCRs must have about the same melting temperatures and must not anneal to each other, particularly not at the 3′-ends.

Here in our case, the hybridization probes do not bind in the 3′-region as suggested. However, the sequence requirements overruled the preferred binding position, because there are no suitable binding sites available in the more distant 3′-region. Because the hybridization probes bind close to the 3′-end of the primers a slightly elevated annealing temperature must be used to avoid displacement of the probes by the polymerase and the newly generated strand.

2. *Running the PCR and adjusting the parameters*: Use extreme care as you prepare the dilutions. Any mistake will be seen later, because the calibration curve will not be linear if the dilution is not properly done. So you will not be able to find the optimal parameters for your PCR.

3. *Programming a run*:
    1. Temperature transition rate/slope is 20°C/s, except where indicated.
    2. For initial experiments, set the target temperature (i.e., the primer annealing temperature) 5°C below the calculated primer T_m.
    3. If the primer annealing temperature is low (<55°C), reduce the transition rate/slope to 2–5°C/s.
    4. For greater precision in target quantification experiments, it can be advantageous (in some cases) to choose longer annealing and extension times for the amplification cycles.

    Set all protocol parameters not listed in the table to '0'.

4. The protocol is designed for a final reaction volume of 20 μL. For volumes <20 μL, the reaction and cycle conditions must be optimized. We use 15 μL reaction volume to save money.

    To prepare the Master Mix for more than one reaction, multiply the amount in the "Volume" column above by z, where z is the number of reactions to be run plus two additional reactions.

    Place the centrifuge adapters in a balanced arrangement within the centrifuge. Alternatively, centrifuge the capillaries within the sample carousel in the LightCycler Carousel Centrifuge.

5. This saves time and limits the variations from experiment to experiment. The PCR is finished in short time, usually shorter than it takes to pipette a new set of reactions. Therefore after testing the parameter annealing for example at 62°C, the next batch can run to test annealing at 60°C.

    It is possible to optimize the PCR for two genes (the transgene and the reference gene) at the same time in the same reaction capillary, because the aim of the experiment is to do two different PCRs in one tube, and the primers were designed accordingly. We did the optimization in separate reactions.

6. This relatively high temperature might be due to the unusual binding site of the hybridization probes.

7. Always run a calibration curve with your samples. Even if curves from former runs can be used, they do not tell you about the quality of your current run. Only standards run with the same master mix at the same time allow judgment about the performance.

8. A duplex PCR saves time and some money, but one of the most expensive things, the hybridization probe is required anyway. So we performed the PCR in different capillaries and compared the relative amounts for each gene as determined by the calibration curve.

9. Extreme care must be taken when pipetting small amounts. The unwanted carry over amount can be larger than the pipetted amount if for instance the pipette is dipped to deep into the DNA. Dip the pipette not more than 1 mm deep into all solutions. Always pipette up and down for three times to ensure equal mixing.

10. These variation can probably not be avoided, because in theory this means we pipetted once 1.1 μL and once 0.9 μL instead of 1 μL. This can be easily explained by to deep dipping of the pipette in the first case, and not complete pipetting up and down in the second case.

Acknowledgments

This work was supported by Deutsche Forschungsgemeinschaft grant GR1517/4-2.

References

1. Eytan, E., Ganoth, D., Armon, T., and Hershko, A. (1989) ATP-dependent incorporation of 20S protease into the 26S complex that degrades proteins conjugated to ubiquitin. *Proc. Natl Acad. Sci. U. S. A.* **20,** 7751–7755.
2. Zwickl, P., Grziwa, A., Puhler, G., Dahlmann, B., Lottspeich, F., and Baumeister, W. (1992) Primary structure of the Thermoplasma proteasome and its implications for the structure,function,and evolution of the multicatalytic proteinase. *Biochemistry* **31,** 964–972.
3. Lowe, J., Stock, D., Jap, B., Zwickl, P., Baumeister, W., and Huber, R. (1995) Crystal structure of the 20S proteasome from the archaeon T. acidophilum at 3.4 A resolution. *Science* **268,** 533–539.
4. Seemuller, E., Lupas, A., Stock, D., Lowe, J., Huber, R., and Baumeister, W. (1995) Proteasome from Thermoplasma acidophilum: a threonine protease. *Science* **268,** 579–582.
5. Fenteany, G., Standaert, R. F., Lane, W. S., Choi, S., Corey, E. J., and Schreiber, S. L. (1995) Inhibition of proteasome activities and subunit-specific amino-terminal threonine modification by lactacystin. *Science* **268,** 726–731.
6. Hisamatsu, H., Shimbara, N., Saito, Y., et al. (1996) Newly identified pair of proteasomal subunits regulated reciprocally by interferon gamma. *J. Exp. Med.* **183,** 1807–1816.
7. Coux, O., Tanaka, K., Goldberg, A. L. (1996) Structure and functions of the 20S and 26S proteasomes. *Annu. Rev. Biochem.* **65,** 801–847.

8. Hershko, A., Ciechanover, A. (1998) The ubiquitin system. *Annu. Rev. Biochem.* **67,** 425–479.

9. Cerundolo, V., Kelly, A., Elliott, T., Trowsdale, J., and Townsend, A. (1995) Genes encoded in the major histocompatibility complex affecting the generation of peptides for TAP transport. *Eur. J. Immunol.* **25,** 554–562.

10. Groettrup, M., Ruppert, T., Kuehn, L., et al. (1995) The interferon-g-inducible 11S regulator (PA28) and the LMP2/LMP7 subunits govern the peptide production by the 20S proteasome in vitro. *J. Biol. Chem.* **270,** 23,808–23,815.

11. Meidenbauer, N., Zippelius, A., Pittet, M. J., et al. (2004) High frequency of functionally active Melan-a-specific T cells in a patient with progressive immuno-proteasome-deficient melanoma. *Cancer Res.* **64,** 6319–6326.

12. Sijts, A. J. A. M., Standera, S., Toes, R. E. M., et al. (2000) MHC class I antigen processing of an Adenovirus CTL epitope is linked to the levels of immunoproteasomes in infected cells. *J. Immunol.* **164,** 4500–4506.

13. Chen, W., Norbury, C. C., Cho, Y., Yewdell, J. W., and Bennink, J. R. (2001) Immunoproteasomes shape immunodominance hierarchies of antiviral CD8(+) T cells at the levels of T cell repertoire and presentation of viral antigens. *J. Exp. Med.* **193,** 1319–1326.

14. Palmowski, M. J., Gileadi, U., Salio, M., et al. (2006) Role of immunoproteasomes in cross-presentation. *J. Immunol.* **177,** 983–990.

15. Basler, M., Youhnovski, N., Van Den Broek, M., Przybylski, M., and Groettrup, M. (2004) Immunoproteasomes down-regulate presentation of a subdominant T cell epitope from lymphocytic choriomeningitis virus. *J. Immunol.* **173,** 3925–3934.

16. Quiagen (2006) DNeasy Blood and Tissue Handbook.

17. Sambrook, J., Fritsch, E. F., and Maniatis, T. (1989) Molecular cloning, a loboratory manual.

18. Roche Molecular Biochemicals (2000) LightCycler Operator's Manual, Version3.5.

19. Roche Molecular Biochemicals (2000) LightCycler- DNA Master Hybridisation Probes, instruction manual, version 3.

5

Quantitative Analysis of Gene Expression Relative to 18S rRNA in Carcinoma Samples Using the LightCycler® Instrument and a SYBR GreenI-based Assay: Determining FAT10 mRNA Levels in Hepatocellular Carcinoma

Sebastian Lukasiak, Kai Breuhahn, Claudia Schiller, Gunter Schmidtke, and Marcus Groettrup

Abstract

Due to the fact that mutations and up- or downregulation of genes can lead to the development of cancer, quantitative comparison of relative gene expression in healthy and cancerous tissue can gain valuable insights into tumorigenesis. While the semi-quantitative DNA microarrays are being used to identify differentially expressed genes on a genomic scale, real-time RT-PCR provides a powerful tool for quantitative measurement of gene expression. Presently, it is the most sensitive method available. Here we describe in detail a SYBR GreenI-based assay using the LightCycler® instrument to measure the levels of mRNA for the ubiquitin-like protein FAT10 relative to 18S rRNA in human hepatocellular carcinoma tissue. This method can be easily adapted to any tissue (human or mouse, rat, etc.) and any gene.

Key Words: Relative gene expression; quantitative RT-PCR; SYBR GreenI; co-application reverse transcription (Co-RT); normalization to 18S rRNA; highly pure RNA from tissue.

1. Introduction

FAT10 is a ubiquitin-like protein that is encoded in the MHC class I locus *(1)* and is synergistically inducible with interferon-γ and tumor necrosis factor-α *(2,3)*. It was previously found to be expressed in dendritic cells and mature B cells *(4)*. Ectopic expression of FAT10 leads to apoptosis in a caspase-dependent manner *(5)*. Recently, it was reported that FAT10 is highly upregulated on mRNA

From: *Methods in Molecular Biology, Vol. 429: Molecular Beacons: Signalling Nucleic Acid Probes, Methods and Protocols* Edited by: A. Marx and O. Seitz © Humana Press Inc., Totowa, NJ

level (in 90% of samples) in hepatocellular carcinoma (HCC) and other gastro-intestinal cancers *(6)*. The HCC is one of the most frequent visceral cancers worldwide, with nearly 600,000 new cases each year and almost as many deaths *(7)*.

To better understand which genes could play a central functional role in disease, particularly in human carcinogenesis, cDNA microarrays are being used to screen for differences in gene expression in healthy and cancerous tissue. This technique provides information on whether genes are up- or downregulated on a genomic scale. However, microarray data are only semi-quantitative and need to be verified using a more quantitative method. Today, real-time RT-PCR is the most sensitive method to measure gene expression.

Real-time RT-PCR *(8)* is a very sensitive and reproducible method with a low error rate *(9–11)*. It can produce quantitative data with an accurate dynamic range of seven to eight log orders of magnitude *(12)* and can even detect single copy transcripts *(13)*. Highly pure and intact RNA and the reverse transcription (RT) step are critical for sensitive and accurate quantification, since the amounts of cDNAs produced must correctly reflect the input amounts of the mRNAs *(14)*. Real-time PCR allows fast and specific detection of deletions or duplications of exons or whole genes, point mutations (e.g., single nucleotide polymorphisms *(15,16)*) which qualifies the technique for diagnostic purposes *(17–20)*. Real-time RT-PCR assays have lower coefficients of variation (SYBR GreenI at 12.4%; TaqMan® at 24%) than end point assays such as band densitometry (44.9%) and probe hybridization (45.1%) *(11)*.

In our study, we investigated FAT10 expression in different HCC tissue samples with quantitative RT-PCR, using the LightCycler® instrument and a SYBR GreenI-based assay.

2. Materials

2.1. Tissue Homogenization

Mixer mill (MM200, Retsch, Haan, Germany), grinding jars (stainless steel; Retsch), and grinding balls (stainless steel, Ø 1.2 mm, Retsch). Chill grinding jars and grinding balls in liquid nitrogen before use.

2.2. Preparation of Pure RNA from Tissue Samples

1. For the preparation of a 4 *M* guanidine isothiocyanate (GITC) solution 500 g GITC (Applichem, Darmstadt, Germany), 26.5 mL of 1 *M* Na-citrate (pH 7.0), and 8.5 mL of 2-mercaptoethanol (Merck, Hohenbrunn, Germany) were dissolved in 1058 mL autoclaved Milli-Q water, stirred overnight at room temperature and pH was adjusted to 7.0 using sodium hydroxide or citric acid. Aliquots (50 mL) were stored at −20°C.
2. For preparation of the cesium chloride (CsCl) solution 479.85 g CsCl (Applichem) and 4.15 mL sodium acetate (3 *M*, pH 5.0) were dissolved in 500 mL Milli-Q water. After sterile filtration using a syringe (50 mL) and syringe filters (0.2 µm) under a safety workbench the CsCl solution can be stored at room temperature.

3. For preparation of the RNasin-mix 60 μL RNasin (Promega, Mannheim, Germany), 6.9465 mL diethylpyrocarbonate (DEPC)-treated Milli-Q water, and 193.5 μL of 0.1 *M* dithiothreitol (dissolved in 0.01 *M* sodium acetate, pH 5.2) were mixed carefully. Aliquots (1.4 mL) were stored at −20°C.
4. 3 *M* sodium acetate (pH 5.2) in DEPC-Milli-Q water.
5. SW41Ti swing bucket rotor and appropriate polyallomer centrifugation tubes (PA thin wall, 13.2 mL; Beranek Service and Supplies, Weinheim, Germany).
6. Sterile cotton pads and forceps.

2.3. cDNA Synthesis

For the synthesis of single-stranded cDNA from pure total RNA, the "Reverse Transcription System"-Kit (Promega) was used. Additionally to an oligo(dT) primer from the kit, the specific 18S rRNA-RT primer gagctggaattaccgcggct *(21)* was used for reverse transcription.

2.4. Primers for Real-Time PCR

The following primers were used for real-time PCR:

18S rRNA forward: gaggtagtgacgaaaaataacaat *(21)*
18S rRNA reverse: ttgccctccaatggatcct *(21)*
Product length: 99 bp
FAT10 forward: ttgtttcttgtggagtcaggtg
FAT10 reverse: agtaagttgccctttctgatgc
Product length: 200 bp

All primers were purchased desalted (Microsynth AG, Balgach, Switzerland) and dissolved in sterile nuclease-free water to a final concentration of 50 μ*M*. Aliquots (200 μL) were stored at −20°C.

2.5. Real-Time PCR

For determination of relative gene expression, the "LightCycler® FastStart DNA Master SYBR GreenI"-Kit (Roche, Mannheim, Germany) was used in conjunction with the LightCycler® Instrument (Roche) and the LightCycler® Software Version 3.5 (Roche). Store kit at −20°C.

2.6. Determining Relative Gene Expression

Relative gene expression was evaluated using the Pfaffl-method *(22)* and the Excel-based software tool REST® (relative expression software tool) *(23)*.

3. Methods
3.1. Tissue Homogenization and RNA Preparation

1. Insert liquid nitrogen frozen tissue sample (~pea-size; *see* **Notes 1–3**) and chilled grinding ball into the chilled grinding jar. Close the grinding jar immediately, fix it in the mixer mill (use cold protection gloves!), and shake for 1 min at 30 cycles/s.

Detach and open grinding jar very (!) carefully (protection gloves!), since residual nitrogen in the jar may cause overpressure which may catapult the jar lid. Keep the grinding jar strongly fixed in your hand until you opened the grinding jar with the other hand and the pressure vanished.

2. Transfer the entire frozen tissue powder immediately into 6 mL GITC solution (in 50 mL cap) and store on ice until further processing. Shake carefully (*see* **Note 4**).
3. Centrifugation: 832,5g, 15 min, room temperature (e.g., in Heraeus Varifuge).
4. In the meantime transfer exactly 4 mL CsCl-solution in polyallomer centrifugation tubes for each sample.
5. Transfer (overspread) the lysate supernatant (Step 3) carefully on (!) the CsCl-solution (Step 4). Use forceps for centrifugation tube handling. Adjust all samples (normally six samples/tubes for a SW41Ti rotor) to the highest centrifugation tube weight (accuracy < 0.005 g) and insert them in the rotor. Take care, that no fluid sticks outside the centrifugation tube!
6. Centrifugation: 174020g, 16 h, 15°C in SW41Ti.
7. Carefully remove lipid layer on each sample using cotton pads.
8. Carefully (!) decant the entire supernatant (Attention: hazardous waste). Invert the tube and let the transparent RNA pellet dry for 30 min at 4°C.
9. Resuspend the pellet with 100 µL RNasin-mix and transfer it into a new cap. Samples are stored on ice. Dissolve the residual pellet again with 100 µL RNasin-mix. Pool both fractions (sum: 200 µL).
10. Add exactly 20 µL of 3 M sodium acetate (pH 5.2). Mix samples thoroughly and store them on ice.
11. Add 500 µL ice-cold ethanol (100%) and mix thoroughly. Store samples at −80°C for 30 min to precipitate nucleic acid.
12. Centrifugation: 18492,6g, 10 min, 4°C; discard supernatant.
13. Wash pellet with ice-cold ethanol (70%).
14. Centrifugation: 18492,6g, 10 min, 4°C; discard supernatant.
15. Dry RNA pellet for 10 min at room temperature.
16. Resuspend pellet in 150 µL RNasin-mix; shake solution for 10 min at room temperature.
17. Quantify RNA yield by spectrophotometric analysis (OD260/OD280 usually > 1.9) and check integrity on an agarose/formaldehyde gel (500 ng total RNA). Store RNA at −70°C.

3.2. cDNA Synthesis (see Note 6)

There are three possibilities to prime a cDNA synthesis: random hexamers, oligo(dT), and a gene (sequence)-specific primer. For measuring gene expression (steady-state levels of different mRNAs), random hexamers and oligo(dT) primers can be used. However, random hexamers can overestimate mRNA copy numbers *(24)*. Therefore, it has been described as the least reliable method for priming cDNA *(25)*. For that reason an oligo(dT) primer was used for cDNA priming.

To normalize expression data endogenous genes are used as internal standards (e.g., GAPDH, HPRT, β-actin, etc.). Ideally, these genes should be expressed at a constant level in different tissues and should be unaffected by experimental treatment *(26)*. Constant expression of a reference gene should be validated in the system that is investigated. Some of the most reliable reference genes are the rRNAs. Several recent studies report that the use of rRNA, e.g., 18S rRNA as an internal control was consistently the best choice *(27–29)*. 18S rRNA is ribosomal and does not contain a 3′-poly-A tail and can therefore not be reverse transcribed with an oligo(dT) primer. To combine the advantages of using oligo (dT) primer for mRNA reverse transcription and 18S rRNA for normalization of gene expression, a method called "coapplication reverse transcription" (Co-RT) was applied *(21)*.

1. Thaw $MgCl_2$ solution (25 m*M*), reverse transcription 10 × buffer, dNTP mixture (10 m*M*), oligo(dT), and 18S rRNA-RT primer. Mix by vortexing and spin down all components briefly in a benchtop centrifuge. Keep solutions on ice.
2. Prepare autoclaved 0.5 mL reaction tubes for every RNA sample, label them, and place them on ice.
3. Thaw RNA samples at room temperature and incubate them at 70°C for 10 min (*see* **Note 7**). Spin briefly and place them on ice.
4. During RNA incubation (Step 3), prepare a master mix for the RT reaction, mix it by pipetting up and down (*do not vortex enzymes!*) and distribute it in every 0.5 mL reaction tube with a separate tip (*see* **Note 8**). Volumes for one sample:

$MgCl_2$	4 μL
Reverse transcription 10 × buffer	2 μL
dNTP mixture (10 m*M* each)	2 μL
oligo(dT)15 primer (0.5 μg/mL)	1 μL
18S rRNA-RT primer (50 μM)	1 μL
RNasin® Ribonuclease Inhibitor	0.5 μL
AMV reverse transcriptase (24 U/μL)	15 U (0.625 μL)
	11.13 μL (11.125 μL)

5. Pipette 1 μg total RNA (from Step 3) according to its measured concentration into separate 0.5 mL reaction tubes containing the master mix. All RNA samples are normalized for reverse transcription at this step (*see* **Note 9**). After pipetting RNA, place it immediately back at −70°C.
6. Add nuclease-free water to a final volume of 20 μL and mix by pipetting up and down several times.
7. Incubate the reaction tubes for 10 min at room temperature and place them into a PCR thermocycler. Run following program: 1 h at 42°C, 5 min at 95°C to inactivate reverse transcriptase, and 5 min at 4°C.
8. Spin the RT-reactions briefly in a benchtop centrifuge and place them on ice for a real-time PCR run or store them at −20°C (*see* **Note 10**).

3.3. Primer Design

Real-time PCR primers should be designed in such a way that the amplicon length is between 100 and 300 bp. Short amplicons have much higher amplification efficiencies (PCR efficiency) than longer ones. This is very important for evaluating gene expression data. Additionally, more SYBR GreenI molecules bind to a longer amplicon (*see* **Note 11**), which would result in earlier C_t-values. This could change the relative expression data significantly. When comparing gene expression from several different genes, amplicons should have approximately the same length.

There are many programs and websites for primer design such as:

Primer3: http://biotools.umassmed.edu/bioapps/primer3_www.cgi (Whitehead Institute of MIT, USA)

PCR Now (especially for real-time PCR): http://pathogene.swmed.edu/rt_primer/ (UT Southwestern Medical Center, Texas, USA)

1. To design primers for the gene of interest, first get the sequence from Entrez Gene database.
2. Get the REFSEQ of your gene (e.g., NM_006398 for human FAT10 mRNA, 971 bp).
3. Copy only the open reading frame of the sequence (bp 225–722; *see* **Note 12**) and paste the sequence into the program.
4. Set the product size as followed, min: 100, optimal 200: and max: 300. The melting temperature (T_m) optimum should be 60°C (*see* **Notes 13** and **14**). Then let the program design the primers (*see* **Notes 15** and **16**).
5. It is also very important to run a BLAST search with all primer sequences to make sure they do not bind to other genes in the genome.
6. To test new primers run simply a real-time PCR following a melting curve analysis (*see* Section 3.4.) with a cDNA sample from commonly used cell lines (e.g., HeLa or Hek293), or perform a conventional PCR.
7. Verify correct product length and existence of possible primer–dimers also by agarose gel-electrophoresis.

3.4. Real-Time PCR Assay (see Note 6)

The measurements of relative gene expression in this study were conducted with the fluorescence dye SYBR GreenI which binds only to double-stranded DNA (dsDNA) *(12)*. The dye has virtually no fluorescence in free solution. Upon binding to dsDNA, the specific fluorescence of SYBR GreenI increases 100-fold. The more transcripts were amplified in the sample during PCR-cycling, the earlier the signal reaches a detectable threshold where it is significantly higher than background levels. The cycle at which this occurs is called the C_t or C_p (crossing point) value, which is always in the exponential phase of the PCR *(9,30)*. The advantage of dsDNA-binding dyes is that expansive sequence-specific fluorescence labeled probes (e.g., TaqMan®, Molecular Beacons, Scorpions) are not needed. A disadvantage of SYBR GreenI is that this dye binds to all kind of dsDNA such as primer–dimers, which can occur during amplification. This means that the specificity of the real-time PCR is mostly dependent on primer design. Therefore, it is necessary to perform a melting curve analysis after the PCR to discriminate between specific PCR products and possible primer–dimers, which have a lower melting temperature and a wider peak *(31)*.

We basically use the instructions of the "LightCycler FastStart DNA Master SYBR GreenI"- kit for preparing the master mix:

1. Thaw cDNA samples, magnesium chloride solution (vial 2) and primers, mix them by vortexing, and spin them briefly down in a 4°C cold benchtop centrifuge. Keep all substances on ice.
2. Prepare a serial dilution (undiluted, 1:10, 1:100, and 1:1000; *see* **Note 17**) of the cDNA samples with nuclease-free or PCR grade water. First pipette 1 μL cDNA into a new 0.5 mL reaction tube (autoclaved) and mix it with 9 μL PCR grade water by pipetting up and down several times. Then pipette 1 μL from the diluted cDNA into a new reaction tube and mix it with 9 μL H_2O and so forth for further dilutions.
3. Pipette 10 μL from vial 1a (LightCycler Fast-Start Enzyme) into vial 1b (LightCycler Fast-Start Reaction Mix SYBR GreenI). Now this mix comprise a ready-to-use hot start PCR mix ($10 \times$ concentrated), which contains FastStart Taq DNA polymerase, reaction buffer, dNTP mix, SYBR GreenI dye, and magnesium chloride (10 mM). Keep the vial with SYBR GreenI away from light because it is light sensitive. The ready-to-use PCR mix can be stored protected from light at 4°C for a month without loosing efficacy.
4. Prepare a master mix for all cDNA samples, including the negative control (H_2O instead of cDNA) to reduce pipette inaccuracy. Measure each cDNA sample at least in duplicate or triplicate. Pipette scheme for one sample:

$MgCl_2$ (25 mM)	2.4 μL (final concentration 4 mM)
Primer 1	0.5 μL (final concentration 1.25 μM)
Primer 2	0.5 μL (final concentration 1.25 μM)
SYBR GreenI PCR mix	2.0 μL (final concentration 1×)
PCR grade H_2O (vial 3)	12.6 μL
	18 μL

Mix all components by pipetting up and down several times. *Do not vortex*!
5. Precool LightCycler capillaries for each cDNA sample, including one for the negative control (H_2O instead of cDNA) in the LightCycler centrifuge adapters.
6. Pipette 18 μL from the master mix into each precooled LightCycler capillary. The same pipette tip can be used for the same primer pair.
7. Add 2 μL of each cDNA dilution into a separate capillary. Use a new tip each time. Seal capillaries with a stopper.
8. Place the adaptors with the capillaries into a benchtop centrifuge and spin the samples at 700*g* for 5 s or briefly to ~5000 rpm.
9. Transfer the capillaries carefully into the sample carousel of the LightCycler instrument.
10. Run PCR with the corresponding program (*see* **Notes 18 – 20**):

	18S rRNA	FAT10	Fluorescence acquisition
Polymerase activation	95°C for 10 min	95°C for 10 min	None
Amplification 40 cycles	95°C for 10 s	95°C for 10 s	None
	57°C for 5 s	60°C for 5 s	None
	72°C for 5 s	72°C for 9 s	Single
Melting curve	95°C for 0 s		None
	65°C for 15 s		None
	95°C for 0 s; 0.1°C/s		Continuous
Cooling	40°C		None

11. After the run (*see* **Fig. 1**), take a look at the melting curves (*see* **Fig. 2**) of the generated real-time PCR products. The primer pairs are specific, if there is only one sharp peak (melting temperature of specific PCR products is the maximum of the negative first derivative of the melting curve, *see* **Note 21**).

12. T_m of the 18S rRNA real-time PCR product: 80.2–80.4°C (*see* **Note 22**).
    T_m of the FAT10 real-time PCR product: 85.9–86.3°C (*see* **Note 22**).

13. Perform an agarose gel electrophoresis with the PCR products for quality check (*see* **Note 23**).

14. Take the capillaries *carefully* out of the sample carousel, uncap them, and put the capillaries inversely into a 1.5 mL reaction tube; spin briefly down (~700*g*) in a benchtop centrifuge.

15. Mix with DNA loading dye and apply the samples (5 µL are enough) onto a 1.5% agarose gel.

3.5. Evaluation of Relative Gene Expression

Obtained data were analyzed with a new mathematical model for relative gene quantification in real-time RT-PCR, called the Pfaffl-method *(22)*, which relies on the C_t-values and the PCR efficiencies of the respective primer pair. The corresponding PCR efficiency (*E*; *see* **Note 24**) for a transcript is determined using the slope given by C_t-values of a cDNA serial dilution and the equation $E = 10^{[-1/\text{slope}]}$. Efficiency for 18S rRNA in HCC tissue from five different patients in **Fig. 1** is $10^{[-1/-3.507]} = 1.928$ ($r = 0.99$) (*see* **Note 25**). The Excel-based program REST® *(23)* calculates the relative expression ratio (*R*) of a target gene, in comparison to a reference gene, based on *E* and the C_t deviation (ΔC_t) of an unknown sample versus a control.

Simply enter raw data (C_t-values) of the reference gene and the target gene(s) to REST®. PCR efficiencies are automatically calculated and the relative gene expression is expressed as *x*-fold up- or downregulation of the target gene, in comparison to untreated/normal tissue. Normalization was performed by calculating the expression of the reference gene in normal/untreated and cancer/treated (e.g., cytokine-treated cells) tissue.

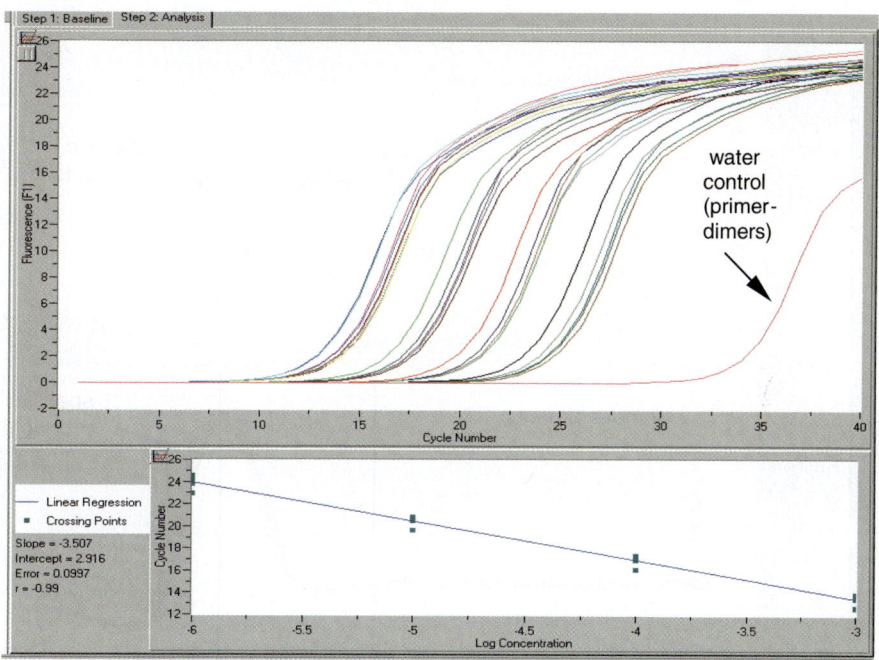

Fig. 1. 18S rRNA serial dilution ($1:10^3$ to $1:10^6$) of five HCC samples from five different patients. The C_t-values should yield a straight line when plotted on a logarithmic scale. The LightCycler software calculates the slope of this line. When the PCR efficiency is ideal (100%; doubling DNA amount every cycle), the slope should be -3.32 ($2^{3.321} = 10$). Therefore, the slope from this run is needed to calculate the actual PCR efficiency.

4. Notes

1. To obtain high quality total RNA always use fresh-frozen tissue samples for preparation. Repeated freeze/thaw cycles and even brief storage at room temperature significantly decreases the nucleic acid quality.

2. For every sample use a new grinding jar and new grinding balls. To avoid contamination incubate grinding jars and grinding balls in H_2O_2 (30%) overnight after every application. Wash carefully in DEPC-treated water and bake for 6 h at 200°C.

3. The size of the sample used for homogenization strongly depends on the type of tissue that has to be analyzed. Liver, e.g., contains high amounts of nucleic acid, hence only small tissue samples are necessary (edge length of 0.5 cm).

4. Since defrosted tissue powder sticks at the cap wall, pay attention that the entire sample is covered by GITC at the bottom of the cap when transferring the pulverized tissue sample into the GITC solution.

5. After supernatant has been discarded, carefully remove residual alcohol with a pipette-tip. Avoid extensive drying of the pellet.

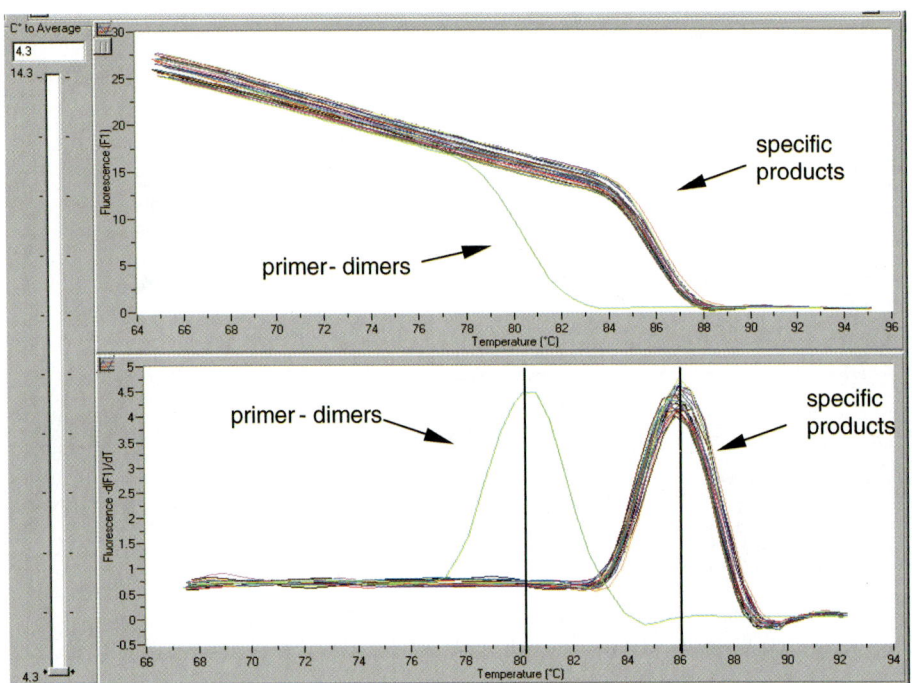

Fig. 2. Melting curve analysis of FAT10. Specific FAT10 products from serial dilutions can be easily distinguished from primer–dimers that were formed in the water control sample (no cDNA).

6. Wear always gloves and use only autoclaved reaction tubes and pipette tips to avoid contamination of RNA and cDNA when pipetting RT or real-time PCR.

7. Incubation at 70°C resolves secondary structures of RNA, which could hamper the reverse transcriptase. Without this step the efficacy of transcription is only very poor and the cDNA is not suitable for gene expression analysis or for other applications like cloning of full-length cDNA.

8. Enzymes for the reverse transcription (RNasin Ribonuclease Inhibitor and the AMV-RT) should be pipetted at last and should be taken out of the freezer in a precooled (–20°C) benchtop cooler, or at least on ice. Pipette quickly and put them immediately back to the freezer.

9. To avoid pipette inaccuracy at the first normalization step do not pipette volumes under 1 μL (samples with high RNA concentration). To avoid contaminations of stock solutions with RNases, take an aliquot of RNA (at least 3–5 μL for accurate pipetting depending on concentration) and mix it with sterile nuclease-free water prior to incubation at 70°C (Step 3).

10. For long time storage (≥1 year) of cDNA take Safe-Lock reaction tubes, otherwise H_2O can sublimate (at –20°C) and the concentration of cDNA is not correct any more. Additionally, wrap the reaction tube with parafilm to be sure.

11. For low-copy transcripts, it can be advantageous to design primer pairs that produce longer amplicons (400–700 bp) so more SYBR GreenI can bind to it and amplify the weak signal.
12. Primers should not lie in the 5′-untranslated region (UTR). It is possible that the reverse transcription does not extend mRNA to its 5′-UTR (e.g., very long mRNA, secondary structures may hamper reverse transcriptase activity). Another possibility is that a gene may have different transcription initiations sites in different tissues; hence, identical primer pairs might not work in all tissue-derived samples.
13. The T_m of the two primers should be within 2°C.
14. The T_m can also be determined using the rule of thumb: 4°C for C/G and 2°C for A/T. The optimal annealing temperature in the PCR is the sum in °C of the primer minus 5°C.
    Example for a 21 bp primer: 4°C × 11 C/G + 2°C × 10 A/T = 64°C–5°C = 59°C. But this is just an approximate value. Nevertheless, the PCR can work very well with an annealing temperature of 60°C, but this should be simply tried out.
15. One primer should reside on an exon–exon boundary to prevent amplification from possible residual genomic DNA, which could produce an overestimation of transcript levels. Generally, a conventional DnaseI digest efficiently removes DNA during RNA purification. This problem can be equally prevented with intron-spanning primers.
16. It is not necessary to order HPLC purified oligonucleotides for real-time PCR. These are much more expensive (approximately three times the price). The purity (>95%) of desalted primers is absolutely sufficient for real-time PCR but this depends also on the quality of the respective commercial supplier.
17. To measure 18S rRNA, a serial dilution up to the ratio of $1:10^6$ is needed, because 80–85% of total RNA in cells is rRNA. High template amount can inhibit the PCR. C_t-values (approximate cycle 3–4) for serial dilutions from undiluted to 1:1000 diluted cDNA would be too early for reliable results. The recommended starting dilution for real-time PCR of 18S rRNA is 1:1000. Generally, the C_t values should be after the 10th PCR cycle to get reliable data.
18. Fluorescence of SYBR GreenI is measured in fluorescence channel 1. The gain value is set automatically by the LightCycler Software Version 3.5 and higher.
19. During amplification, fluorescence is measured once for every sample at the end of the elongation step (at 72°C). During the melting curve step, fluorescence is measured continuously and the temperature transition rate/slope is set to 0.1°C/s otherwise it is set to 20°C/s. The complete run takes 45–50 min.
20. Elongation time depends on amplicon size (amplicon in bp/25) sec (+1 s for safeness).
21. The specific fluorescence signal of samples with very low copy numbers (C_t-value ≥ 30) could be superimposed from unspecific primer–dimers signals. If the melting temperature of primer–dimers differs by more than 2°C, a forth step in the amplification cycle can be added to overcome this problem *(32)*. In case of FAT10, fluorescence acquisition can be simply changed to this additional step (*see* **Fig. 3**):

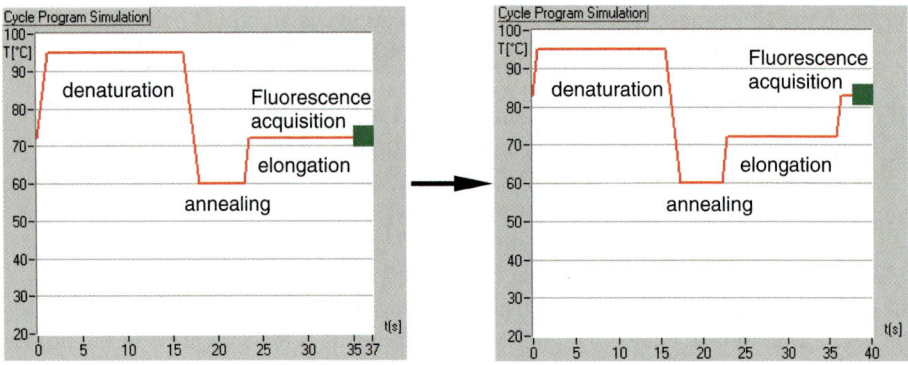

Fig. 3. Scheme of a modified real-time PCR program. At 83°C all primer–dimers ($T_m \sim 80°C$) are melted but not the specific FAT10 products ($T_m \sim 86°C$), therefore the measured fluorescence results only from the specific PCR product.

	FAT10	Fluorescence acquisition
Amplification 40 cycles	95°C for 10 min	None
	95°C for 10 s	None
	60°C for 5 s	None
	72°C for 9 s	None
	83°C for 3 s	Single

22. There is a slight difference in the melting temperature of cDNA dilutions, which can be explained by the different yield of resulting PCR product and by the inaccuracy of the LightCycler instrument itself (±0.4°C).

23. Analysis of PCR products is only necessary when establishing new primer pairs and new cycling parameters. After this was done, agarose gel-electrophoresis of PCR products is not required, except if there are some additional/unexpected peaks in the melting curve. Obtained PCR products can also be verified by digestion with restriction enzymes.

24. Determination of the PCR efficiency for a primer pair is important, because amplification during the PCR is exponential. Already small differences in efficiency change the result dramatically. A PCR with 90% efficiency, compared to ideal 100%, results in a 4.66-fold difference in the final ratio of PCR products after 30 cycles. Therefore, corrections for PCR efficiencies should be used when determining gene expression.

25. The calculated correlation for a serial dilution from one patient is in most cases $r = 1.00$.

Acknowledgments

This work was supported by grant Az.10.05.1.145 from the Fritz Thyssen Foundation.

References

1. Fan, W., Cai, W., Parimoo, S., Schwarz, D. C., Lennon, G. G., and Weissman, S. M. (1996) Identification of seven new human MHC class I region genes around the HLAF locus. *Immunogenetics* **44,** 97–103.
2. Liu, Y.-C., Pan, J., Zhang, C., et al. (1999) A MHC-encoded ubiquitin-like protein (FAT10) binds noncovalently to the spindle assembly checkpoint protein MAD2. *PNAS* **96,** 4313–4318.
3. Raasi, S., Schmidtke, G., de Giuli, R., and Groettrup, M. (1999) A ubiquitin-like protein which is synergistically inducible by interferon-gamma and tumor necrosis factor-alpha. *Eur J Immunol* **29,** 4030–4036.
4. Bates, E. E., Ravel, O., Dieu, M. C., et al. (1997) Identification and analysis of a novel member of the ubiquitin family expressed in dendritic cells and mature B cells. *Eur J Immunol* **27,** 2471–2477.
5. Raasi, S., Schmidtke, G., and Groettrup, M. (2001) The Ubiquitin-like Protein FAT10 Forms Covalent Conjugates and Induces Apoptosis. *J. Biol. Chem.* **276,** 35,334–35,343.
6. Lee, C. G., Ren, J., Cheong, I. S., et al. (2003) Expression of the FAT10 gene is highly upregulated in hepatocellular carcinoma and other gastrointestinal and gynecological cancers. *Oncogene* **22,** 2592–2603.
7. Kern, M. A., Breuhahn, K., and Schirmacher, P. (2002) Molecular pathogenesis of human hepatocellular carcinoma. *Adv. Cancer Res.* **86,** 67–112.
8. Higuchi, R., Fockler, C., Dollinger, G., and Watson, R. (1993) Kinetic PCR analysis: real-time monitoring of DNA amplification reactions. *Biotechnology (NY)* **11,** 1026–1030.
9. Heid, C. A., Stevens, J., Livak, K. J., and Williams, P. M. (1996) Real time quantitative PCR. *Genome Res.* **6,** 986–994.
10. Wittwer, C. T., Herrmann, M. G., Moss, A. A., and Rasmussen, R. P. (1997) Continuous fluorescence monitoring of rapid cycle DNA amplification. *Biotechniques* **22,** 130–131, 34–38.
11. Schmittgen, T. D., Zakrajsek, B. A., Mills, A. G., Gorn, V., Singer, M. J., and Reed, M. W. (2000) Quantitative reverse transcription-polymerase chain reaction to study mRNA decay: comparison of endpoint and real-time methods. *Anal. Biochem.* **285,** 194–204.
12. Morrison, T. B., Weis, J. J., and Wittwer, C. T. (1998) Quantification of low-copy transcripts by continuous SYBR Green I monitoring during amplification. *Biotechniques* **24,** 954–958, 60, 62.
13. Palmer, S., Wiegand, A. P., Maldarelli, F., et al. (2003) New real-time reverse transcriptase-initiated PCR assay with single-copy sensitivity for human immunodeficiency virus type 1 RNA in plasma. *J Clin Microbiol* **41,** 4531–4536.
14. Kubista, M., Andrade, J. M., Bengtsson, M., et al. (2006) The real-time polymerase chain reaction. *Mol. Aspects Med.* **27,** 95–125.
15. Carles, J., Monzo, M., Amat, M., et al. (2006) Single-nucleotide polymorphisms in base excision repair, nucleotide excision repair, and double strand break genes as markers for response to radiotherapy in patients with Stage I to II head-and-neck cancer.*Int. J. Radiat. Oncol. Biol. Phys* **66,** 1022–1030.

16. Sasaki, H., Endo, K., Konishi, A., et al. (2005) EGFR Mutation status in Japanese lung cancer patients: genotyping analysis using LightCycler. *Clin. Cancer Res.* **11,** 2924–2929.

17. Boensch, M., Oberthuer, A., Fischer, M., et al. (2005) Quantitative Real-Time PCR for Quick Simultaneous Determination of Therapy-Stratifying Markers MYCN Amplification, Deletion 1p and 11q. *Diagn. Mol. Pathol.* **14,** 177–182.

18. Hoebeeck, J., van der Luijt, R., Poppe, B., et al. (2005) Rapid detection of VHL exon deletions using real-time quantitative PCR. *Lab. Invest.* **85,** 24–33.

19. Ruiz-Ponte, C., Carracedo, A., and Barros, F. (2005) Duplication and deletion analysis by fluorescent real-time PCR-based genotyping. *Clin. Chim. Acta* **363,** 138–146.

20. Tan, E. K., Shen, H., Tan, J. M., et al. (2005) Differential expression of splice variant and wild-type parkin in sporadic Parkinson's disease. *Neurogenetics* **6,** 179–184.

21. Zhu, L. J. and Altmann, S. W. (2005) mRNA and 18S-RNA coapplication-reverse transcription for quantitative gene expression analysis. *Anal. Biochem.* **345,** 102–109.

22. Pfaffl, M. W. (2001) A new mathematical model for relative quantification in real-time RT-PCR. *Nucleic Acids Res.* **29,** 2002–2007.

23. Pfaffl, M. W., Horgan, G. W., and Dempfle, L. (2002) Relative expression software tool (REST) for group-wise comparison and statistical analysis of relative expression results in real-time PCR. *Nucleic Acids Res.* **30,** e36.

24. Zhang, J. and Byrne, C. D. (1999) Differential priming of RNA templates during cDNA synthesis markedly affects both accuracy and reproducibility of quantitative competitive reverse-transcriptase PCR. *Biochem. J.* **337 (Pt 2),** 231–241.

25. Lekanne Deprez, R. H., Fijnvandraat, A. C., Ruijter, J. M., and Moorman, A. F. (2002) Sensitivity and accuracy of quantitative real-time polymerase chain reaction using SYBR green I depends on cDNA synthesis conditions. *Anal. Biochem.* **307,** 63–69.

26. Bustin, S. A. (2000) Absolute quantification of mRNA using real-time reverse transcription polymerase chain reaction assays. *J. Mol. Endocrinol.* **25,** 169–193.

27. Goidin, D., Mamessier, A., Staquet, M. J., Schmitt, D., and Berthier-Vergnes, O. (2001) Ribosomal 18S RNA prevails over glyceraldehyde-3-phosphate dehydrogenase and beta-actin genes as internal standard for quantitative comparison of mRNA levels in invasive and noninvasive human melanoma cell subpopulations. *Anal, Biochem.* **295,** 17–21.

28. Aerts, J. L., Gonzales, M. I., and Topalian, S. L. (2004) Selection of appropriate control genes to assess expression of tumor antigens using real-time RT-PCR. *Biotechniques* **36,** 84–86, 88, 90–91.

29. Schmid, H., Cohen, C. D., Henger, A., Irrgang, S., Schlondorff, D., and Kretzler, M. (2003) Validation of endogenous controls for gene expression analysis in microdissected human renal biopsies. *Kidney Intz.* **64,** 356–360.

30. Gibson, U. E., Heid, C. A., and Williams, P. M. (1996) A novel method for real time quantitative RT-PCR. *Genome Res.* **6,** 995–1001.

31. Ririe, K. M., Rasmussen, R. P., and Wittwer, C. T. (1997) Product differentiation by analysis of DNA melting curves during the polymerase chain reaction. *Anal. Biochem.* **245,** 154–160.

32. Ball, T. B., Plummer, F. A., and HayGlass, K. T. (2003) Improved mRNA quantitation in LightCycler RT-PCR. *Int. Arch. Allergy Immunol.* **130,** 82–86.

6

Absolute Quantitation of Feline Leukemia Virus Proviral DNA and Viral RNA Loads by TaqMan® Real-time PCR and RT-PCR

Valentino Cattori and Regina Hofmann-Lehmann

Abstract

Sensitive TaqMan® real-time polymerase chain reaction (PCR)-based methods have been developed recently for the detection and quantitation of feline leukemia virus (FeLV) proviral DNA in infected cats. In this chapter, we outline the design and implementation of a TaqMan® real-time PCR assay to quantify total FeLV proviral and viral RNA loads in infected cats. The assay is designed to amplify all three FeLV subtypes (A–C), but not FeLV-related endogenous retroviral sequences. The system is tested and optimized using proviral DNA or viral RNA from cells infected with reference strains. The sequence used to produce the standard DNA and RNA is amplified, subcloned into a vector, and sequenced. cRNA is synthesized from the linearized plasmid DNA. Standard DNA and RNA are quantified, diluted and used to determine efficiency, sensitivity, linear amplification range, and precision of the quantitative TaqMan® real-time PCR assays.

Key Words: TaqMan®; quantitative real-time PCR; quantitative real-time RT-PCR; feline leukemia virus; proviral loads; viral RNA loads; gammaretrovirus.

1. Introduction

The feline leukemia virus (FeLV) is a gammaretrovirus of felids *(1)*, which induces fatal diseases in domestic cats and some close relatives. The pathogenesis and outcome of the FeLV infection might be associated or even causally linked with plasma viral RNA loads, as well as proviral loads. Recently, sensitive TaqMan® real-time polymerase chain reaction (PCR) assays based on the 5′–3′ nuclease activity of DNA polymerases *(2,3)* and the Förster-type energy transfer *(4)* have been developed for the detection and quantitation of FeLV proviral DNA and viral RNA in infected cats *(5–7)*. Real-time PCR assays offer linear quantitation of input template over a broad linear range, low sample consumption,

From: *Methods in Molecular Biology, Vol. 429: Molecular Beacons: Signalling Nucleic Acid Probes,
Methods and Protocols* Edited by: A. Marx and O. Seitz © Humana Press Inc., Totowa, NJ

rapid throughput of large sample numbers, and low risk of contamination *(8–10)*. Exposure of cats to FeLV can be detected with greater sensitivity using FeLV TaqMan® real-time PCR than previously described detection methods, such as FeLV p27 antigen ELISA *(11)*, immunofluorescence, and virus isolation *(12,13)*: vaccinated cats which are constantly p27 ELISA negative are positive for FeLV TaqMan® real-time PCR *(1)*, and a significant proportion of Swiss field cats are FeLV TaqMan® real-time PCR positive but p27 ELISA negative *(5)*.

In this chapter, we outline the design and implementation of a TaqMan® real-time PCR assay (**Fig. 1**) to quantify total FeLV proviral and viral RNA loads in infected cats. The assay is designed to amplify all three FeLV subtypes (A, B and C), but not endogenous, FeLV-related sequences. The system is tested and optimized using 10-fold serial dilutions of proviral DNA or viral RNA from cells infected with the reference strains of FeLV-A, -B, or -C. The sequence to be used to produce the standard DNA and RNA is amplified from the reference strain FeLV-A/Glasgow-1, subcloned into a vector and sequenced. After linearization, the plasmid DNA is quantified and diluted to determine efficiency, sensitivity, linearity and range of the amplification, and reproducibility of the quantitative TaqMan® real-time PCR system. For the quantitative TaqMan® real-time RT-PCR system, cRNA is synthesized from the linearized vector, and the same parameters are assessed as for the real-time PCR. Using the strategy described, which can also be applied to design quantitative real-time PCR assays for other retroviruses, we could reach a lower detection limit of one copy per reaction (PCR) and 180 copies per reaction (RT-PCR), with mean efficiencies of 0.94 (PCR) and 0.98 (RT-PCR) *(6)*.

2. Materials (*see* Note 1)

2.1. Reagents

1. Primers are delivered in solution at a concentration of 100 μM (Microsynth GmBH, Balgach, Switzerland). A portion of the solutions is diluted to 20 μM upon arrival ("working solution"). All primer solutions are kept at −20°C.
2. The TaqMan® probe is labeled at the 5′-end with the fluorescent reporter dye FAM (6-carboxyfluorescein) and at the 3′-end with the fluorescent quencher dye TAMRA (5-(6)carboxytetramethylrhodamine) (Eurogentec, Seraing, Belgium). The probe is delivered lyophilized and diluted to 100 μM upon arrival. A portion of the solution is diluted to 10 μM ("working solution"). All the probe solutions are kept at −20°C (*see* **Note 2**).
3. Conventional PCR: *Pfu* DNA proofreading polymerase (Promega, Madison, WI, USA), delivered with 10× *Pfu* reaction buffer (200 mM Tris–HCl (pH 8.8 at 25°C), 100 mM KCl, 100 mM (NH$_4$)$_2$SO$_4$, 20 mM MgSO$_4$, 1.0% Triton® X-100, and 1 mg/ml nuclease-free BSA); 10 mM dNTP's (Sigma-Aldrich, St Louis, MO, USA); molecu-

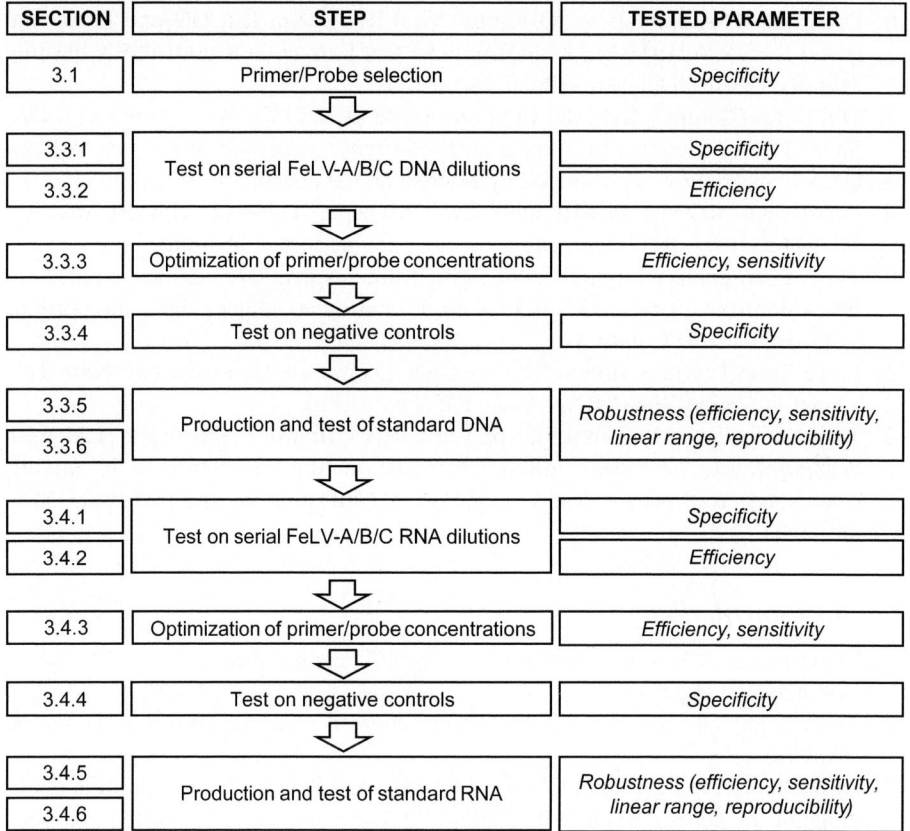

Fig. 1. Workflow for the design and test of the real-time PCR and RT-PCR assays. The major steps in the implementation are listed (central column), as well as the reaction parameters that are assessed by these steps (right column). The sections in the text that cover the steps are referenced in the left column.

lar biology–grade agarose (Sigma-Aldrich), TAE buffer (0.04 *M* TRIS acetate, pH 8.2 ± 0.1 (25°C); 0.001 *M* EDTA-Na$_2$), 10 mg/mL ethidium bromide solution (Sigma-Aldrich).

4. Real-time PCR: qPCR™ 2× Mastermix (Eurogentec) (*see* **Note 3**); SuperScript™ III Platinum® One-Step Quantitative RT-PCR System (Invitrogen, Carlsbad, CA, USA); RNase inhibitor, 40 U/μL (RNasin®, Promega).

5. Cell culture: medium for feline embryonic fibroblast (FEA) cells: Dulbecco's-modified Eagle's medium containing HEPES buffer (Invitrogen) with 10% heat-inactivated fetal calf serum (Bioconcept, Allschwil, Switzerland), 1 m*M* sodium pyruvate, 2 m*M* L-glutamine, and 400 IU/mL penicillin–streptomycin; trypsin-EDTA; phosphate-buffered saline (PBS; all Invitrogen).

6. DNA and RNA extractions: QIAamp® Viral RNA Mini Kit; DNeasy Tissue and Blood Kits; RNeasy® Mini Kit; QIAquick® Gel Extraction Kit; QIAGEN Plasmid Midi Kit (all from Qiagen, Hilden, Germany).
7. TOPO TA Cloning® Kit Dual Promoter (with pCR®II-TOPO® vector) with One Shot® TOP10 Chemically Competent *Escherichia coli* (Invitrogen) (*see* **Note 4**).
8. DNA sequencing was performed by Microsynth (Balgach, Switzerland).
9. Restriction enzymes *Bam*HI and *Not*I (Promega, delivered with 10× buffers, MULTI-CORE™ 10× buffer, and acetylated bovine serum albumin).
10. Thirty micrograms per milliliters of carrier salmon sperm DNA diluted in nuclease-free water from a 10 mg/mL stock solution (Invitrogen), aliquoted in 2 mL portions and frozen at −20°C until use.
11. Large Scale T7 Transcription Kit (Novagen, Darmstadt, Germany) (*see* **Note 4**).
12. Quant-iT™ RiboGreen® RNA Assay Kit (Invitrogen).
13. Thirty micrograms per milliliter of yeast tRNA (transfer ribonucleic acid from *Saccharomyces cerevisiae*) diluted in nuclease-free water from a 10 mg/mL stock solution (Sigma-Aldrich), aliquoted in 2 mL portions and frozen at −20°C until use.

2.2. Instrumentation

1. Conventional PCR is performed on a Biometra T Personal thermal cycler (Biometra, Göttingen, Germany).
2. Agarose gels are run in a Bio-Rad electrophoresis chamber (Bio-Rad, Hercules, CA, USA).
3. All the real-time PCR and real-time RT-PCRs, and the cRNA concentration determination are performed on an ABI Prism 7700 sequence detection system (Applied Biosystems, Foster City, CA, USA).
4. Spectrophotometrical DNA concentration determination is performed on a GeneQuant™ (Amersham-Pharmacia Biotech—GE Healthcare Ltd, Buckinghamshire, UK).
5. Agarose gel electrophoresis evaluation and DNA quantitation are performed using a ChemiGenius2 Gel Documentation System with GeneTool imaging software (Syngene, Cambridge, UK).

3. Methods
3.1. PCR Primer Design and Selection

One of the most suitable regions for the selection of primers that are identical in FeLV-A, -B, and -C, but not cross-hybridizing to feline endogenous FeLV-related sequences is the unique region (U3) of the long-terminal repeat (LTR) *(14)*.

1. Retrieve GenBank FeLV long-terminal repeat (LTR) database sequences in FASTA format using the NCBI BLASTN web interface (http://www.ncbi.nih.gov/blast).

The accession numbers of the sequences containing the LTR of FeLV-A/Glasgow-1 strain, FeLV-B and FeLV-C are M12500; X00188 and M14331, respectively.
2. Align the sequences using the GCG Wisconsin Package version 10 (Accelrys, San Diego, CA, USA) programs from fasta, pileup, and prettybox.
3. Design TaqMan® Probe and primer sequences for real-time PCR and for production of the standard templates within U3 of the LTR of FeLV using Primer Express™ 2.0 (Applied Biosystems).
4. Select the primers and probe on the basis of following criteria:
   1. Predicted fit with FeLV-A, -B, and -C and lack of fit with feline endogenous FeLV-related sequences.
   2. Melting temperature of the probe of 69°C.
   3. Lack of predicted dimer formation with corresponding primers.
   4. Lack of self-annealing.
   5. Probe melting temperature 10°C higher than primer melting temperature.
   6. Compare the sense and antisense sequence of the probe and choose the one with more C_s.
   7. Maximal size of the system 150 bp, the ideal length is approximately 80 bp.
   8. No G at 5′-end of the probe (*see* **Note 5**).
   9. No more than two C_s/G_s in the last five 3′ bases (*see* **Note 5**).
   10. For the primers used to amplify the template region for the standard DNA and RNA, the length of the insert should be at least 300 bases and encompass the region covered by the PCR system (*see* **Note 6**).

5. The sequences of the selected primers are: for real-time PCR: forward primer (U3-exo-F) 5′-AACAGCAGAAGTTTCAAGGCC-3′; reverse primer (U3-exo-R) 5′-TTATAGCAGAAAGCGCGCG-3′; probe (U3-P) 5′-CCAGCAGTCTCCAG-GCTCCCCA-3′; and for the amplification of the template region for the standard DNA and RNA: forward primer standard F 5′-CTACCCCAAAATTTAGCCAGC-TACT-3′; reverse primer standard R 5′-AAGACCCCCGAACTAGGTCTTC-3′. The size of the real-time PCR system is 131 bp, the size of the amplified standard template 468 bp (*see* **Note 7**).

3.2. Cell Culture and Extraction of Template DNA and RNA

1. Culture FEA cells containing molecularly cloned FeLV-A (Glasgow-1 strain), FeLV-B or FeLV-C at 37°C, 4.5% CO_2 until confluence in a 75 cm² cell culture flask using FEA culture medium.
2. Collect the cell culture supernatant and extract viral RNA using the QIAamp® Viral RNA Mini Kit. Store RNA at −80°C.
3. Wash the cells two times with PBS, incubate them 5 min at room temperature with trypsin-EDTA, collect the cells in FEA culture medium and centrifuge at 1000*g* for 10 min. Discard the supernatants and tap the tube several times to loosen the cell pellet (*see* **Note 8**).
4. Extract FeLV genomic DNA from the cell pellet using the DNeasy® Tissue Kit. Store the DNA at −20°C.

3.3. Development of the TaqMan® Fluorogenic Real-Time PCR Assay

3.3.1. Initial Setup

1. Prepare real-time PCRs with 12.5 µL qPCR™ Mastermix, a final concentration of 900 nM of both U3-exo forward and reverse primers, 250 nM of fluorogenic U3 probe and 5 µL of template (FEA-FeLV-A) in a 25 µL total reaction volume.
2. The thermal cycling conditions are: an initial denaturation of 10 min at 95°C followed by 45 cycles of 95°C for 15 s and 60°C for 1 min.
3. Run the assay in triplicates of serial 10-fold dilutions of the template.

3.3.2. Amplification Efficiency

Determine the amplification efficiency of the real-time PCR assay in triplicate using serial 10-fold dilutions of genomic DNA from FEA/FeLV-A, -B, and -C. Amplification efficiencies are calculated as $10^{1/-s}-1$ *(15)*, where s is the slope of the curve (threshold cycle (C_t) vs. dilution). The efficiencies of the assays for different targets are considered equal if the difference of the slopes (Δs) is smaller than 0.1 (**Fig. 3A**).

3.3.3. System Optimization

1. Optimize the system for primer concentration by using a 3×3 matrix with 50, 300, and 900 nM end concentration; each sample is measured in quadruplicate (**Fig. 2**). Use a FEA/FeLV-A genomic DNA dilution that gives a C_t-value of approximately 30–32 in the initial set-up experiment.
2. Optimize the system for probe concentration using three different probe end concentrations (50, 100, and 250 nM).
3. Determine again the amplification efficiency (**Figs. 3A** and **4**, and **Table 1**).
4. The optimal concentrations for the quantitative real-time PCR assays are 480 nM of each primer and 160 nM of the probe (*see* **Note 9**).

3.3.4. Specificity

Confirm the specificity of the system by lack of amplification of endogenous FeLV-related sequences: analyze the DNA extracted with the DNeasy Blood Kit from blood samples collected from specific pathogen-free cats. The test of 80 cats yielded negative results.

3.3.5. Construction and Production of DNA Standards for Absolute Quantitation

1. Generate the DNA standard template from FEA-FeLV-A genomic DNA *(6)* using a proofreading polymerase to minimize the amount of erroneously integrated bases.
2. The PCR conditions are: 2.5 µL of 10× *Pfu* polymerase buffer, 0.5 µL of 10 mM dNTP's, 1.0 µL of each 20 µM standard primer (final concentration 800 nM), 12.7

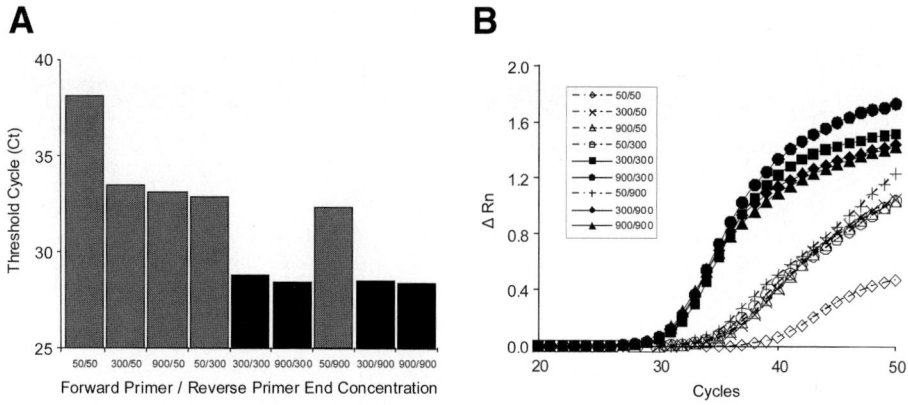

Fig. 2. Effect of primer optimization on the quality of quantitative real-time PCR assays. Representative results from the RT-PCR optimization are shown. (**A**) Effect of different primer concentrations on the C_t-value. The final concentration of the forward/reverse primers are given on the *x*-axis. The best results are given by 300/300, 900/300, 300/900, and 900/900 n*M*. (**B**) Effect of different primer concentrations on the reporter dye fluorescence (ΔR_n): ΔR_n versus amplification cycle is depicted. The best result is given by the combination 900/300 (continuous line with closed circles).

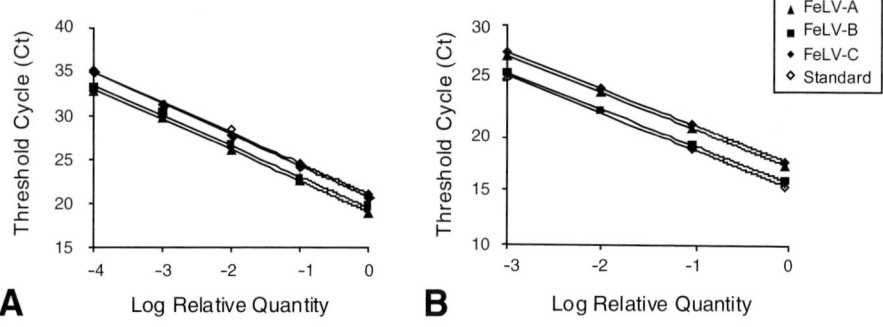

Fig. 3. Specificity and amplification efficiency of the FeLV real-time PCR (**A**) and RT-PCR assays (**B**). The designed assays readily amplified FeLV-A, -B, and -C grown in FEA cells. To assess the amplification efficiency of the assays, 10-fold serial dilutions of the concerned DNA or RNA were analyzed. The means of the assays run in triplicates are depicted. The slopes of the curves (logarithmic dilution vs. threshold cycle) for FeLV-A, -B, and -C and the corresponding standard templates did not differ by more than 0.1 (modified from Ref. *[6]*).

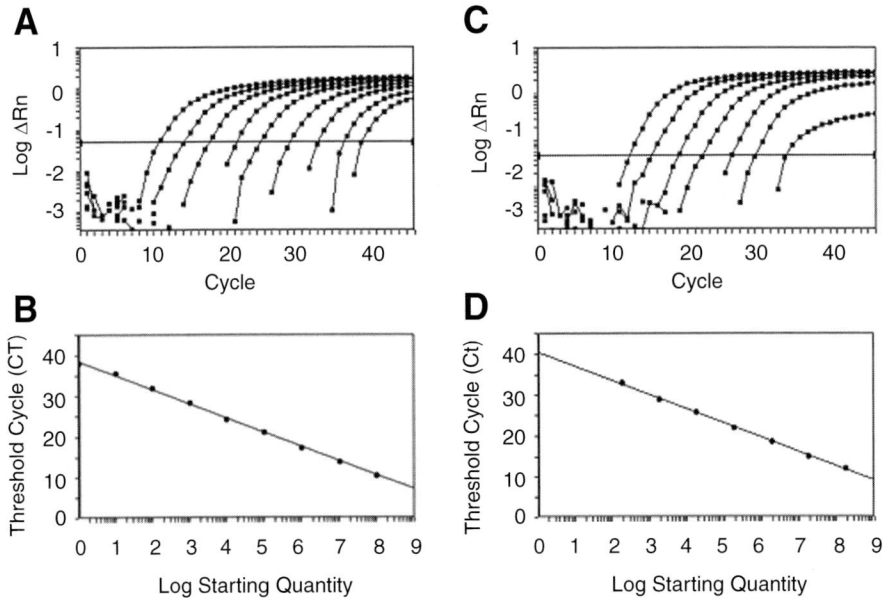

Fig. 4. Linearity of the real-time PCR assay (**A** and **B**) and RT-PCR assay (**C** and **D**) and absolute quantitation of amplification of FeLV provirus and virus, respectively. Amplification plots of representative real-time PCR (**A**) and RT-PCR assays (**C**) depict the cycle number versus the measured logarithmic reporter dye fluorescence (ΔR). Standard curves of representative real-time PCR (**B**) and RT-PCR assays (**D**) depict the logarithmic starting input quantity (copies per reaction) of a 10-fold serial dilution of standard template versus the measured threshold cycle. The earlier the fluorescence ΔR_n increases, the lower the threshold cycle, and the higher the starting quantity of template in the sample. For the DNA assay, linearity of the assay was found over nine logs, while for the RT-PCR, linearity was observed over a seven-log magnitude. Correlation coefficient of the curves (**B** and **D**): 0.999 (modified from Ref. *[6]*).

    µL nuclease-free water, 0.31 µL (1 U) of *Pfu* polymerase and 2 µL of genomic DNA in a 25 µL total reaction volume.

3. The thermal cycling conditions are: an initial denaturation of 10 min at 95°C, followed by 30 cycles of 95°C for 30 s, 65°C for 30 s, and 72°C for 45 s, with a final elongation step of 5 min at 72°C.

4. Separate the product by electrophoresis on a 0.7% ethidium bromide-stained agarose gel, excise the resulting amplicon, purify it with the QIAquick® Gel Extraction Kit and clone it into the pCR®II-TOPO® vector.

5. Select positive colonies by kanamycin resistance and verify the sequence of the cloned FeLV-A U3 DNA.

6. Expand the colonies, and isolate plasmid DNA with the QIAGEN Plasmid Midi Kit.

7. Quantify the isolated plasmid using the GeneQuant™.

Table 1
Efficiency of the PCR and RT-PCR Assays (Modified from Ref. [6])

Target	Efficiency PCR	Efficiency RT-PCR
FeLV-A	0.94	0.97
FeLV-B	0.94	0.99
FeLV-C	0.91	0.97
Standard DNA	0.95	0.96

8. Linearize 20 µg plasmid DNA by overnight restriction digestion using the restriction enzyme *Bam*HI.
9. The linearized plasmid is separated by agarose gel electrophoresis on a 0.7% agarose gel prepared in TAE buffer, stained with 0.5 µg/mL ethidium bromide in TAE buffer, visualized with ChemiGenius2, excised from the gel with a sterile scalpel, purified with the QIAquick® Gel Extraction Kit and quantitated with GeneQuant®.
10. Calculate the copy numbers by both spectrophotometer (GeneQuant™) and agarose gel electrophoresis analysis (Gene Tools) (*see* **Note 10**).
11. Tenfold serial dilutions of the FeLV U3 standard DNA are made in 30 µg/mL carrier salmon sperm DNA and frozen at −20°C in 50 µL aliquots.
12. Determine efficiency, linear range, and sensitivity of the amplification. The efficiency is determined as described above (*see* Section 3.3.3.) using 10-fold serial dilutions of the standards. Use dilutions ranging from 10^9 to 10^{-2} copies/5 µL (*see* **Note 11**). The linear range of quantitation of the real-time PCR assay is determined by the points of the curve of the threshold cycle versus dilution (the same used for the efficiency), which fit a linear regression. The quality of the fit, measured by the coefficient of correlation (r^2), is usually automatically given by the evaluation of the standard curve in the ABI Prism 7700 (Fit R). The sensitivity of the system is determined by an endpoint dilution experiment: test 10 replicates of the last positive dilution and the next, negative one. The result is given by the dilution in which at least 64% of the replicates are still positive *(16)* (**Fig. 4A** and **B**, **Tables 1** and **2**).

3.3.6. Reproducibility

Assess within- and between-run precision of the quantitative real-time PCR assay by multiple measurements of several dilutions of the DNA standard plasmid (**Table 2**).

1. Within-run precision: evaluate 10 replicates of three dilutions of the DNA standard (10^2, 10^4, and 10^6 molecules per reaction).
2. Between-run precision: the three dilutions are evaluated in five separate experiments.
3. Calculate, for all measurements: mean value, standard deviation, and coefficients of variation (CV) of the C_t values (CV_{CT}) and of the absolute copy numbers (CV_{abs}) (**Table 2**). CV_{CT} values up to 4.5% are considered acceptable *(5,17)*.

Table 2
Sensitivity and Reproducibility of the PCR
and RT-PCR Assays (Modified from Ref. [6])

Parameter	PCR	RT-PCR
Sensitivity (copies/5 µL)	1	180
CV_{CT} (%)		
Within run	0.5–0.6	0.5–0.9
Inbetween run	1.9–3.1	1.6–1.9
CV_{abs} (%)		
Within run	6.8–13.6	7.1–18.7
Inbetween run	8.9–29.2	14.3–19.6
Linearity (logs)	9	7

3.4. Development of the TaqMan® Fluorogenic Real-Time RT-PCR Assay

3.4.1. Initial Setup

1. Real-time RT-PCRs: amplify 5 µL of viral RNA from FEA-FeLV-A cell supernatant in a total reaction volume of 25 µL using the SuperScript™ III Platinum® One-Step Quantitative RT-PCR System, with a final concentration of 900 nM of both forward and reverse U3-exo primer, 250 nM of fluorogenic U3 probe and 25 U RNase inhibitor.
2. Cycling conditions: reverse transcription for 30 min at 50°C, followed by a denaturation step of 95°C for 2 min and 45 cycles of 95°C for 15 s and 60°C for 30 s (*see* **Note 12**).
3. Run the assay in triplicates of serial 10-fold dilutions of the template.

3.4.2. Amplification Efficiency

The amplification efficiency of the real-time RT-PCR assay is determined as described above (*see* Section 3.3.2.) (**Figs. 3B** and **4**, and **Table 1**).

3.4.3. System Optimization

The RT-PCR system is optimized by analogy with the real-time PCR system using FeLV-A/Glasgow-1 viral RNA extracted from infected FEA cell culture supernatant as a target:

1. Optimize primer concentration by using a 3×3 matrix with 50, 300, and 900, nM end concentration, each sample measured in quadruplicate (**Fig. 2**). Use a dilution that gives a C_t-value of approximately 30–32 in the initial set-up experiment.
2. Optimize the system for probe concentration using three different probe end concentrations (50, 100, and 250 nM).
3. Determine again the amplification efficiency of the system (**Fig. 3B** and **Table 1**).

4. The optimal conditions for the quantitative real-time RT-PCR were of 900 nM U3-exo forward primer, 300 nM U3-exo reverse primer, and 200 nM for the U3 probe (*see* **Note 9**).

3.4.4. Specificity

To confirm the absence of cross-reactivity of the system with endogenous retroviruses, RNA extracted with the QIAamp® Viral RNA Mini Kit from plasma samples collected from specific pathogen-free cats has to be tested. In this case, 80 SPF cats were tested and yielded negative results.

3.4.5. Construction and Production of RNA Standards for Absolute Quantitation

1. Linearize 20 µg FeLV U3 plasmid overnight using the restriction enzyme *Not*I (*see* **Note 13**).
2. The linearized plasmid is separated by agarose gel electrophoresis on a 0.7% agarose gel prepared in TAE buffer, stained with 0.5 µg/mL ethidium bromide in TAE buffer, visualized with ChemiGenius2, excised from the gel with a sterile scalpel, purified with the QIAquick® Gel Extraction Kit and quantitated with GeneQuant®.
3. Transcribe 5 µg of the plasmid in vitro using the Large Scale T7 Transcription Kit and purify the synthesized cRNA with the RNeasy® Mini Kit.
4. Quantitate three dilutions (1:16, 1:32, and 1:64 in nuclease-free water) of the cRNA with the Quant-iT™ RiboGreen® RNA Assay Kit, using the following protocol *(18)*:
   1. Dilute the rRNA stock solution (100 ng/µL) 50 times in 1× TE buffer to get a concentration of 2 ng/µL.
   2. Make a twofold serial dilution of the 2 ng/µL rRNA in 1× TE buffer to get an input concentration series of 1.0, 0.5, 0.25, 0.125, 0.0625, 0.0313, 0.0156, and 0.0078 ng/mL.
   3. Dilute the standard cRNA 1:100 in 1× TE buffer Dilute the Quant-iT™ RiboGreen® RNA reagent 1:200 in 1× TE buffer.
   4. Mix 15 µL diluted samples/standards and 15 µL 1:200 Quant-iT™ RiboGreen® RNA reagent. The assay volume is 30 µL.
   5. Run the samples and the standards in triplicate in an ABI PRISM 7700 sequence detection system using the following settings:

In the "Setup" view:
1. Run = Plate read
2. Dye Layer = Ribo (or Sybr)
3. Sample Type Reporter = Ribo (or Sybr)

In the "Analysis" view:
1. Unmark ROX in Instrument/Diagnostics/Advanced Options
2. Run "Post-PCR Read"
3. After analyzing: Display = $R_n/\Delta R_n$

4. Export the experiment results as an Excel worksheet.
5. Plot the mean fluorescence value (column "mean" of the Excel sheet) against the concentration (input RNA). The sample concentration can be read from the linear regression estimation of the curve.
6. Calculate the RNA copy numbers and make serial 10-fold dilutions ranging from 10^9 to 10^{-2} copies/5 µL of the FeLV U3 standard RNA in 30 mg/µL carrier yeast tRNA.
7. Store the standard RNA transcript dilutions in 50 µL aliquots at -80°C.
8. Determine efficiency, linear range, and sensitivity of the amplification reaction as described above (*see* Section 3.3.2.) (**Fig. 4C** and **D**). The sensitivity of the system is 180 copies/5 µL (**Tables 1** and **2**) (*see* **Note 14**).

3.4.6. Reproducibility

As for the PCR (*see* Section 3.3.6.) assess within- and between-run precision of the RT-PCR assay by multiple measurements of several dilutions of the standard RNA (**Table 2**).

1. Within-run precision: evaluate 10 replicates of three dilutions of the DNA standard (1.8×10^2, 1.8×10^4, and 1.8×10^6 molecules per reaction, *see* **Note 15**).
2. Between-run precision: the three dilutions are evaluated in seven separate experiments.
3. Calculate, for all measurements: mean value, standard deviation, CV_{CT}, and CV_{abs} (**Table 2**). As for the PCR assay, CV_{CT} values up to 4.5% are well within the values reported for other tests and considered acceptable (*17,18*).

4. Notes

1. Where not stated differently, manufacturer's instructions were followed.
2. "Working solutions" concentrations of 20 µ*M* for the primers and of 10 µ*M* for the probe are suitable to further dilute the primers/probes, i.e., when pipetting PCRs. Always use the "working solutions" and not the original solutions: if the original solution gets contaminated, it has to be synthesized again, *see* also (*19*); and Section 6 in (*20*). Please note that probes are light sensitive and exposure should be reduced to a minimum.
3. Contains all the reagents, includes a heat-activated enzyme; allows pipetting of the samples at room temperature.
4. The synthesis of cRNA standards for RNA quantitation can only occur from one side of the amplicon (the cRNA has to be a copy of the positive strand to be identical to the retroviral RNA). To be on the safe (and faster) side, we recommend to clone the amplicons in a vector containing a RNA polymerase-binding site on both sides of the Multiple Cloning Site (the PCRII-TOPO vector contains a T3-binding site on one side and a T7-binding site on the other) and chose the promoter to use after sequencing of the insert. Use of vectors lacking RNA polymerase-binding sites is possible as well, but the insert has to be subcloned into a vector containing a binding site or using primers containing the binding site, thus slowing down the process. The choice of the in vitro transcription kit depends on the RNA polymerase promoter.

5. This is to prevent strong binding at the 3'-end of the primer or probe, which can cause unspecific amplification.

6. The cutoff size for RNeasy® Mini Kit used to clean-up the cRNA synthesized for the RNA standards is 200 bp. The advantage of this approach is that a cut-off near to the size of the insert allows the purification of predominantly full-length products, thus allowing for a more accurate quantitation of the purified RNA.

7. The major requirement of our systems is the specificity for FeLV-A, -B and -C; and the lack of cross-reactivity to the endogenous FeLV-related sequences. This restricts the variety of regions that can be included in choice of the system so that the sequences of our primers and the probe do not respect all the rules listed above.

8. DNA extraction may be less efficient from cell pellets that are not "solubilized" in the few remaining microliters of medium after discarding the supernatant.

9. The current standard optimization of our laboratory include 50, 300, and 900 nM end concentration for both primers in a 3×3 matrix, and 50, 100, and 250 nM for the probes after the primer optimization. Usually, higher concentrations of the probe do not significantly improve the efficiency or the sensitivity of the system. The parameters we use to assess the system are: (i) the best (lowest) C_t, (ii) the best (highest) reporter dye fluorescence (ΔR_n). Some of the optimal concentrations given for our systems depart from the values given above because the systems were developed before standardized optimization was introduced in our laboratory. For quality control of the established assays during routine analyses, data are always evaluated using the same baseline and threshold settings and the performance of the systems (C_t values of the standards) is constantly registered and followed over time in order to recognize problems caused by erroneous manipulations, product degradations, or defective reagents.

10. There are a lot of calculators to help you find the molecular weight (mw) of DNA and RNA molecules, which is needed to calculate the copy numbers from concentrations. Access the internet and search for "DNA molecular weight calculator" in a search engine, e.g., Google (www.google.com) or Yahoo (www.yahoo.com). As a role of thumb, the mw of a double-stranded DNA standard molecule is equal to the number of nucleotides × 660 (average DNA base pair mw). Do not forget to take in account the length of the vector. The mw of a RNA standard molecule is equal to the number of ribonucleotides × 340 (average ribonucleotide mw). The length of the transcript is calculated from the promoter to the cutting site of the restriction enzyme.

11. To avoid contaminations clone and pipette the reaction involving standard DNA/RNA in a separate room/building. Also, test the dilutions starting from maximum 10^9 copies (C_t-values of approximately 10) and down to 10^{-2} copies to validate the concentration determination. If 10^{-2} copies are still positive and the plot of the log dilution versus C_t-value is linear, there might be something wrong in your calculations. Running the standard templates with 10^{-2} and 10^{-1} copies per reaction is done only for evaluation purposes and these dilutions are not used when running a standard curve in routine analysis.

12. Reminder: a PCR control without reverse transcription step is mandatory here to ensure that the samples contain no or little DNA contamination.

13. The choice of the enzyme to use depends on the orientation of the insert (*see* **Note 4**). The enzyme should cut 3′ of the coding strand (a copy of which will be produced by in vitro transcription), and not produce 3′ overhangs that could lead to the generation of aberrant transcripts caused by transcription of the non-template DNA strand *(21)*. If a restriction enzyme has to be used that leaves 3′ overhangs, you have to prepare blunt ends prior to transcription. This is usually achieved by treatment with Klenow DNA polymerase: if the restriction digest is performed in a compatible buffer containing 5–10 mM MgCl$_2$, the procedure involves adding 1 U Klenow/µg DNA and 25 µM of all four dNTPs directly to the reaction upon completion, and incubating at room temperature for 5 min.

14. In this case, the sensitivity of the system is not maximal, even after optimization. Testing of alternative enzymes or other brands, which had been shown to considerably enhance the sensitivity of an other retroviral TaqMan assays *(17)*, as well as optimization of magnesium concentrations, reverse transcription temperature, and primer/probe annealing temperature did not significantly improve the sensitivity of the system.

15. The reason for the use of these quantities is that the synthesized cRNA had a concentration of 1.8×10^{11} copies/5 µL, and 10-fold dilutions were made directly from this solution without calculating the exact amount needed to have a "round" number.

References

1. Hofmann-Lehmann, R., Tandon, R., Boretti, F. S., et al. (2006) *Vaccine* **24,** 1087–1094.
2. Holland, P. M., Abramson, R. D., Watson, R., and Gelfand, D. H. (1991) *Proc. Natl Acad. Sci.* **88,** 7276–7280.
3. Lyamichev, V., Brow, M. A., and Dahlberg, J. E. (1993) *Science* **260,** 778–783.
4. Förster, V. T. (1948) *Ann. Phys.* **2,** 55–75.
5. Hofmann-Lehmann, R., Huder, J. B., Gruber, S., Boretti, F., Sigrist, B., and Lutz, H. (2001) *J. Gen. Virol.* **82,** 1589–1596.
6. Tandon, R., Cattori, V., Gomes-Keller, M. A., et al. (2005) *J. Virol. Methods* **130,** 124–132.
7. Torres, A. N., Mathiason, C. K., and Hoover, E. A. (2005) *Virology* **332,** 272–283.
8. Gibson, U. E., Heid, C. A., and Williams, P. M. (1996) *Genome Res.* **6,** 995–1001.
9. Heid, C. A., Stevens, J., Livak, K. J., and Williams, P. M. (1996) *Genome Res.* **6,** 986–994.
10. Livak, K. J., Flood, S. J., Marmaro, J., Giusti, W., and Deetz, K. (1995) *PCR Meth. Appl.* **4,** 357–362.
11. Lutz, H., Pedersen, N. C., Durbin, R., and Theilen, G. H. (1983) *J. Immunol. Methods* **56,** 209–220.
12. Hardy, W. D. Jr., Geering, G., Old, L. J., de Harven, E., Brodey, R. S., and McDonough, S. K. (1970) *Bibl. Haematol.* **36,** 343–354.
13. Jarrett, O., Golder, M. C., and Weijer, K. (1982) *Vet. Rec.* **110,** 325–328.
14. Rohn, J. L. and Overbaugh, J. (1995) *Virology* **206,** 661–665.
15. Klein, D., Janda, P., Steinborn, R., Muller, M., Salmons, B., and Gunzburg, W. H. (1999) *Electrophoresis* **20,** 291–299.

16. Lockey, C., Otto, E., and Long, Z. (1998) *Biotechniques* **24,** 744–746.
17. Hofmann-Lehmann, R., Swenerton, R. K., Liska, V., et al. (2000) *AIDS Res. Hum. Retroviruses* **16,** 1247–1257.
18. Hofmann-Lehmann, R., Williams, A. L., Swenerton, R. K., et al. (2002) *AIDS Res. Hum. Retroviruses* **18,** 627–639.
19. McDonagh, S. (2003) Equipping and Establishing a PCR Laboratory. In *PCR Protocols* (Bartlett, J. M. S. and Stirling, D., eds.), Vol. 226, Humana Press, Totowa, NJ.
20. O'Connell, J. (2002) The basics of RT-PCR—some practical considerations. In *RT-PCR Protocols* (O'Connell, J., ed.), Vol. 193, Humana Press, Totowa, NJ.
21. Schenborn, E. T. and Mierendorf, R. C. (1985) *Nucleic Acids Res.* **13,** 6223–6236.

7

High-Throughput Real-Time PCR

Thomas D. Schmittgen, Eun Joo Lee, and Jinmai Jiang

Abstract

Real-time PCR is presently the gold standard of gene expression quantification. Configuration of real-time PCR instruments with 384-well reaction blocks, enables the instrument to be used essentially as a low-density array. While PCR will never rival the throughput of microchip arrays, in situations where one is interested in assaying several hundreds of genes, high throughput, real-time PCR is an excellent alternative to microchip arrays. By combining SYBR green detection and 5 μL reaction volume, the associated costs of high-throughput real-time PCR are comparable to microarrays. Described here is a complete protocol to perform real-time PCR in a 384-well configuration. Examples are provided to access numerous PCR primer sequences that may be used for high-throughput real-time PCR. Methods of analysis are described to present real-time PCR data as heat maps and clustered similar to the presentation of cDNA microarray data. An example is provided to profile the expression of over 200 microRNA precursors using high-throughput real-time PCR.

Key Words: Real-time PCR; gene expression, microRNA; RT-PCR.

1. Introduction

It is generally accepted that real-time PCR produces the most superior quantitative data due to the exquisite sensitivity and specificity of the PCR. This is exemplified by the fact that cDNA microarray data are validated by real-time PCR and not the other way around. Configuration of real-time PCR instruments with a 384-well reaction block, enables the instrument to be used as a low-density array. While the advantage of the conventional microchip array will always be the shear number of genes that may be profiled, microchip arrays suffer from high costs, batch-to-batch variability, poor sensitivity, and specificity. High-throughput profiling of a relatively small number of genes (e.g., 100–400) using real-time PCR generates high-quality data that do not require additional validation. Real-time

From: *Methods in Molecular Biology, Vol. 429: Molecular Beacons: Signalling Nucleic Acid Probes, Methods and Protocols* Edited by: A. Marx and O. Seitz © Humana Press Inc., Totowa, NJ

PCR data may be presented as heat maps and clustered similar to the presentation of microarray data.

We are aware of at least one laboratory, in addition to ours, that has published these types of analysis. Dittmer et al. has profiled the expression of the Rhesus monkey rhadinovirus *(1)* and Kaposi's Sarcoma-associated Herpes virus *(2)* genomes using real-time RT-PCR. These are appropriate applications of high-throughput real-time PCR profiling because these viral genomes contain <100 genes. We have applied real-time PCR to profile the expression of several hundred miRNA precursors *(3)*. This is another good application of high-throughput, real-time PCR because the number of known human miRNAs (presently at 326) is still relatively manageable to quantify using low-density arrays. Since micro RNAs are non-protein coding RNAs and are structurally different from traditional RNAs, they are not included on commercially available microarrays.

The term "high throughput" is reserved here to describe 384-well, rather than 96-well, real-time PCR. A real-time PCR protocol in a 96-well format has been previously described *(4)*. Advantages of 384- versus 96-well formats include increased throughput by fourfold and reducing the reaction volume by fourfold or more. Thus, the reduction in reagents and time with the 384-well when compared with the 96-well application is substantial. The only disadvantage of performing real-time PCR in 384-well plates is the pipetting. While pipetting into 384-well plates was jokingly described by a student rotating through my lab as the equivalent of pipetting through the eye of a needle, we have learnt that with proper training, proper tools, and a good protocol, any conscientious pipetter can generate high quality and consistent data without the need for robotics or other automated liquid dispensing systems. This chapter describes such a protocol for setting up 384-well real-time PCR assays. An assay to profile the expression of human microRNA precursors is used as an example. Information on retrieving sequences to numerous PCR primers and data analysis tools to present the data as heat-maps are provided.

2. Materials

2.1. RNA Isolation

1. Trizol® Reagent (Invitrogen, Carlsbad, CA, USA; Cat. No. 15596-018).
2. Molecular biology grade water (Sigma Chemical Co., St Louis, MO; Cat. No. W4502).

2.2. Reverse Transcription

1. RNase-free DNase I (10 U/µL) (Roche, Indianapolis, IN; Cat. No. 776 785).
2. RNA guard (GE Healthcare, Piscataway, NJ, USA; Cat. No. 27-0816-01).
3. ThermoScript Reverse Transcriptase (15 U/µL) (Invitrogen; Cat. No. 12236-014).
4. 100 m*M* dNTP set (Invitrogen; Cat. No. 10297-018).

2.3. Real-Time PCR

SYBR green PCR master mix (Applied Biosystems, Foster City, CA, USA; Cat. No. 4309155).

2.4. Consumables

1. Optical Adhesive Cover (Applied Biosystems; Cat. No. 4311971).
2. 384-well real-time PCR plate (Applied Biosystems; Cat. No. 4309849).
3. Aerosol resistant pipette tips: Rainin (Woburn, MA, USA; Cat. Nos RT-L10F, RT-L200F, and RT-L1000F).
4. Phase lock gel (2 mL) Eppendorf (Westbury, NY; Cat. No. 0032 005.152).
5. 12-well, 200 µL strip tubes and caps USA Scientific, Inc. (Ocala, FL; Cat. Nos 1402-3500 and 1402-1200).

2.5. Equipment

1. Applied Biosystems 7900 HT real-time PCR instrument equipped with a 384-well reaction block (Applied Biosystems).
2. ND-1000 micro spectrophotometer (Nanodrop Technologies, Wilmingon, DE, USA).
3. Pipettes: Rainin (Cat. Nos L-2, L-10, L-20, L-100, L-200, and L-1000).
4. Repeating 12-well multichannel pipettes: Rainin (Cat. Nos E12-20 and E12-50).
5. CapEase Tool: BioRad (Waltham, MA, USA; Cat. No. F-341).
6. Stainless steel mortar and pestle sets: Fisher Scientific (Pittsburg, PA, USA; Cat. No. 12-990).

3. Methods

3.1. RNA Isolation (Cultured Cells or Blood)

1. Remove frozen cell pellets from the −80°C freezer and place the tubes on dry ice to prevent thawing.
2. Remove one sample from the dry ice and add 1 mL Trizol®. It is not necessary to thaw the cells prior to adding Trizol®.
3. Lyse the cells using an L-1000 pipette by repeated pipetting until the solution is homogeneous. Incubate at room temperature for 5 min.
4. Add 200 µL of chloroform, shake vigorously by hand for 15 s and incubate at room temperature for 3 min.
5. Transfer the mixture to a 2 mL Eppendorf phase lock tube. The phase lock tube prevents mixing of the organic and aqueous layers.
6. Centrifuge the phase lock tube in the cold (e.g., cold room, refrigerated microcentrifuge or microcentrifuge placed in a refrigerated unit) for 15 min at 12,000g. Remove the supernatant and place into a 1.5 mL colored, microcentrifuge tube. Colored tubes will enhance visualizing the RNA pellet.
7. Add 500 µL of isopropanol and precipitate the RNA for 10 min at room temperature. This is a good stopping point. If necessary, the samples may be placed in the −20°C freezer overnight.
8. Centrifuge in the cold for 10 min at 12,000g.

9. Decant the supernatant into a 2 mL microcentrifuge tube. It is not necessary to remove all of the supernatant. Add 1000 µL of 75% ethanol. Centrifuge for 5 min at 7500*g* in the cold.

10. Decant the supernatant into the same 2 mL microcentrifuge tube used for the isopropanol. Briefly spin the tube containing the RNA for several seconds to bring the residual ethanol to the bottom of the tube.

11. Using a pipette, remove most of the residual ethanol, being careful not to disturb the pellet. Air dry to remove the remaining ethanol.

12. Dissolve the RNA pellet in 30–50 µL of molecular biology grade water and place the tubes on wet ice. RNA may be stored at −80°C, however, it is recommended that RNA be converted to cDNA on the same day of the isolation.

3.2. RNA Isolation (Tissue)

1. Remove the tissues from the −80°C freezer and place on dry ice.

2. Pre-chill the stainless steel mortar and pestle sets on dry ice. Place the frozen tissue into the cold mortar. Place the pestle onto the mortar and pulverize the tissue into a frozen powder by striking with a 2-pound hammer.

3 Transfer the frozen, pulverized tissue as quickly as possible into a 2 mL microcentrifuge tube that has been pre-chilled on dry ice. It is critical that the pulverized tissue does not thaw. A sterile disposable surgical blade is ideal for transferring the powdered tissue. Use one new blade per tissue sample.

4. Place the tubes containing the pulverized tissue on dry ice until all of the samples are ready for lysis.

5. Proceed with steps 2–12 above to isolate the RNA.

3.3. Reverse Transcription

1. Quantify the RNA by placing 1 µL of undiluted RNA onto the lower measurement pedestal of the ND-1000 micro-spectrophotometer. Lower the sample arm and take the reading.

2. Briefly treat the RNA with RNase-free DNase to remove any residual genomic DNA that may be present in the RNA. Add 0.9 µL of RNase-free DNase I, 0.15 µL of RNA guard, 1.2 µL of 25 m*M* $MgCl_2$, 1.2 µg of RNA, and molecular biology grade water to 12.5 µL (*see* **Note 1**).

3. Using a PCR thermal cycler, incubate at 37°C for 10 min and then 90°C for 5 min to inactivate the DNase.

4. Place reactions on wet ice immediately following the incubation.

5. Add 10.5 µL of the DNase-treated RNA into 200 µL tubes.

6. Add 1.5 µL of a cocktail containing 10 µ*M* of each of the antisense PCR primers for the miRNA precursors (*see* **Note 2**).

7. Heat the reaction for 5 min at 80°C to denature the RNA, and then incubate for 5 min at 60°C to anneal the primers. Finally, cool the reaction at room temperature.

8. Add 4 µL of 5× cDNA synthesis buffer, 2 µL of 10 m*M* dNTPs, 1 µL of 0.1 *M* DTT, 1 µL of RNA guard, and 1 µL of ThermoScript RT to the RNA.

9. Incubate for 45 min at 60°C and 5 min at 85°C.

3.4. Real-Time PCR

1. This protocol is for assaying 192 genes in duplicate per 384-well plate. One sample of cDNA is profiled per 384-well plate (*see* **Note 3**).
2. Prepare a master mix containing everything for the PCR except the primers. Prepare enough master mix for 5 µL reactions (384-well plus 10% extra). The master mix contains 2.5 µL of the 2× SYBR green master mix and 0.5 µL of diluted cDNA (dilute cDNA 1:50 in molecular biology grade water).
3. Add 105 µL of the master mix to each well of a 12-well strip tube.
4. Using a 12-well repeating multichannel pipette (Model E 12-20) transfer 3 µL of the master mix from the strip tube to each well of the 384-well plate.
5. In separate 12-well strip tubes have the primer pairs diluted to 2 µ*M* (*see* **Notes 4–8**).
6. Using a 12-well repeating multichannel pipette (Model E 12-20) add 2 µL of the 2 µ*M* primers to each well of the 384-well plate.
7. Add the optical adhesive cover and seal.
8. Perform a quick spin (up to 1500 rpm) on a centrifuge equipped with a 96-well plate adapter.
9. Perform PCR using the real-time instrument per the manufacturer's protocol. Incubation for 2 min at 50°C, 10 min at 95°C, and 40 cycles of 15 s at 95°C and 60 s at 60°C followed by the thermal denaturation protocol (*see* **Note 9**).

3.5. Data Analysis

1. Relative gene expression data are calculated using the formula *(5)* (*see* **Notes 10 and 11**).

$$\text{Relative fold change in gene expression} = 2^{-\Delta C_T}$$

where: $\Delta C_T = (\Delta C_{T\text{ target gene}} - C_{T\text{ reference gene}})$
2. The reference gene is typically a housekeeping gene such as 18S rRNA or U6 RNA.
3. Upload the raw data into MicroSoft Excel (*see* **Note 12**).
4. A sample spreadsheet is provided (**Fig. 1**) to demonstrate how the data are analyzed. We typically calculate the mean C_T for the duplicate PCRs and then calculate the ΔC_T from the means.
5. The data may be presented using heatmaps and clustered using clustering algorithms *(6)* (*see* **Note 13**).
6. A sample figure from the real-time PCR analysis of 201 miRNA precursors in 32 human cancer cell lines is provided (**Fig. 2**).

4. Notes

1. Prepare a master mix containing all of the reagents for the desired number of reactions, allowing for one to two additional reactions.
2. For microRNA precursor expression, it is necessary to prime the reverse transcription using gene-specific primers (antisense PCR primer) *(7)*. However for most situations, priming the reverse transcription with random primers is sufficient and preferred. Random hexamers (Invitrogen; Cat. No. 48190-011) and SuperScript II (Invitrogen; Cat. No. 18064-014) and the accompanying protocol is sufficient for generating

A

B

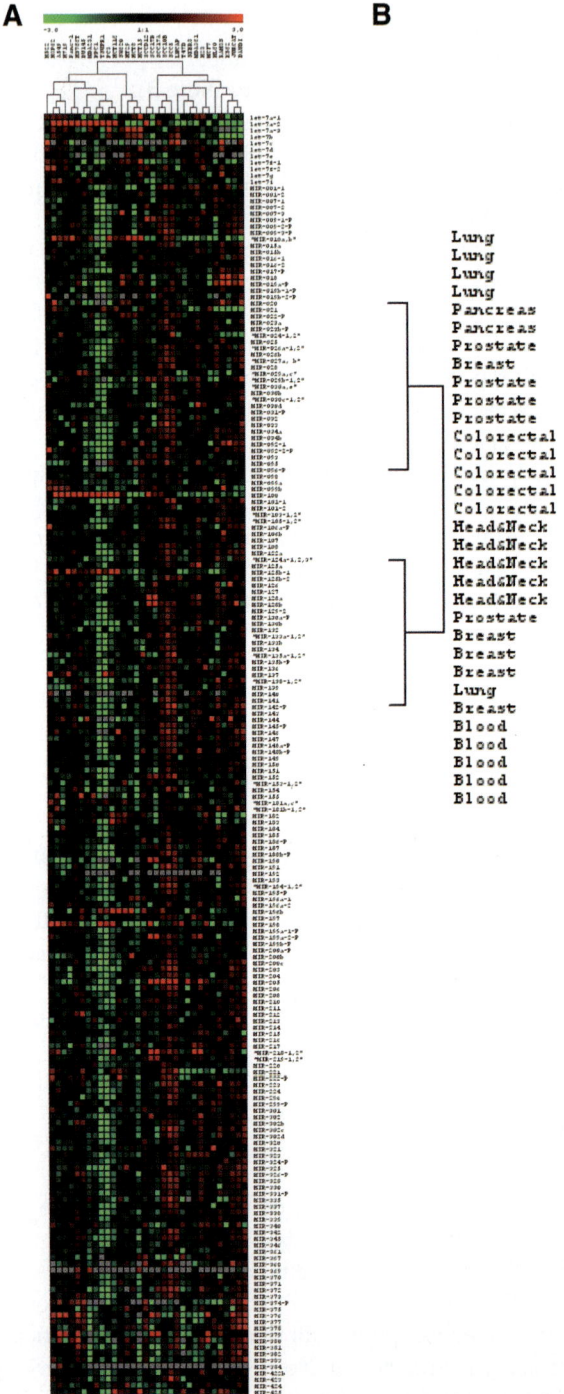

Lung
Lung
Lung
Lung
Pancreas
Pancreas
Prostate
Breast
Prostate
Prostate
Prostate
Colorectal
Colorectal
Colorectal
Colorectal
Colorectal
Head&Neck
Head&Neck
Head&Neck
Head&Neck
Head&Neck
Prostate
Breast
Breast
Breast
Lung
Breast
Blood
Blood
Blood
Blood
Blood

Well	Sample Name	Gene Name	Ct	Mean	ΔC_T	$2^{-\Delta C_T}$
1	A1	mir-010a,b	29.357738			
2	A2	mir-010a,b	29.453217	29.4054775	15.38873625	2.33093E-05
3	A3	mir-021	27.058971			
4	A4	mir-021	27.171118	27.1150445	13.09830325	0.00011403
5	A5	mir-023a	29.777506			
6	A6	mir-023a	29.696415	29.7369605	15.72021925	1.85243E-05
7	A7	mir-026b	28.779152			
8	A8	mir-026b	28.649353	28.7142525	14.69751125	3.76364E-05
9	A9	mir-027a, b	28.647537			
10	A10	mir-027a, b	28.732227	28.689882	14.67314075	3.82776E-05
11	A11	mir-093	30.283978			
12	A12	mir-093	30.288877	30.2864275	16.26968625	1.26572E-05
13	A13	mir-095	33.500263			
14	A14	mir-095	33.65004	33.5751515	19.55841025	1.29519E-06
15	A15	mir-183	31.729328			
16	A16	mir-183	31.884745	31.8070365	17.79029525	4.41151E-06
17	A17	U6	14.0185175			
18	A18	U6	14.014965	14.01674125		

Fig. 1. Sample spreadsheet of real-time PCR data. Shown is a partial listing of genes analyzed by real-time PCR using a 384-well plate. The individual and mean C_{Ts} of each gene are shown as are the C_T and $2^{-\Delta C_T}$ values relative to the U6 internal control gene.

    random primed cDNA. An advantage of random-primed cDNA is that it allows one to conceivably amplify any gene in the genome.

3. We have experienced that once the transition was made from performing 20–25 µL reactions in 200 µL tubes or plates to 5 µL reactions in 384-well plates, there was no interest in returning to the larger volume reactions. Since we are unaware of any suppliers who sell 25 µL PCR strip tubes, we routinely use only a portion of the 384-well plate (e.g., for 36 reactions). The used section of the plate is marked with a marking pen and a clean portion of the plate is reused. 384-well plates may be reused in this manner up to three times.

Fig. 2. Heatmap of microRNA precursor expression in 32 human cancer cell lines. The data were determined using high-throughput, real-time PCR, and then presented as a heatmap using the method of Eisen et al. (*6*). The names of the 32 cancer cell lines are listed on the top of the figure. The names of the microRNAs that were profiled in the cancer cell lines are listed to the right of the figure. The relative expression of each gene was determined by real-time PCR; data are presented as C_T. Unsupervised hierarchical clustering was performed using PCR primers to 201 miRNA precursors. Data were unfiltered prior to clustering. A median expression value equal to one was designated black; red increased expression; green, reduced expression; grey, undetectable expression. B, Dendrogram of clustering analysis. (Reproduced with permission of Oxford University Press from Ref. *[3]*.)

Table 1
Design Criteria for Real-Time PCR Primers, SYBR Green Detection

Both sense and antisense primers should have a $T_m \leq 2°C$ of each other
Primer T_m 50–60°C (ideal range 55–59°C)
% GC content of primers ~50%
Amplicon length <150 bp
Primer length 18–24 nt
No 3′ GC clamp on primers (i.e., GG, CC, CG, or GC)
≤ 2 GC in the last five nucleotides of the 3′-end of the primer

4. It is possible to design and validate hundreds of primers as evidenced by our experience with miRNA precursors *(3)* and Dittmer et al. experience with viral genomes *(1,2)*. A program such as Primers Express® (Applied Biosytems, Foster City, CA, USA) should be used to assist in the design of primers. One should have an idea where the exon borders are in the cDNA. One easy way to identify these borders is to BLAST the cDNA sequence against the genomic sequence of the organism you are working with. Exon-spanning primers (i.e., sense primer located in a different exon than antisense primer) are preferred, so that if any contaminating genomic DNA is present, it can be distinguished by different melt curves on the thermal denaturation protocol. We have found that using the defaults on Primer Express® (**Table 1**) yields useful real-time primers >95% of the time.

5. Thousands of published PCR primer sequences may be accessed and applied to a particular study. Several websites have organized real-time PCR primers sequences. A highly recommended database is Quantitative PCR Primer Database (http://web.ncifcrf.gov/rtp/GEL/primerdb/). This database provides several thousands of real-time PCR primers and lists only those primer sequences that have been published. Advantages of this site include: (i) the primers have been designed specifically for real-time PCR, (ii) primers are listed for both SYBR green and TaqMan assays, and (iii) the primers are validated since they have been published.

6. A clever algorithm has been developed to design over 35,000 primers to mouse and human genes *(8)*. The primer sequences are readily retrievable on the following website (http://pga.mgh.harvard.edu/primerbank/) by simply entering the RefSeq identifier of the gene of interest. Not all of these primers have been tested and validated, however, of 1000 primers synthesized and tested 99% were successful *(8)*. We have independently confirmed these results on several hundred primers designed by Primer Bank. Primers designed by Primer Bank are recommended for applications where one wants to do a rough screen of several hundreds of genes without going through the trouble of designing or searching though the literature or websites for primers. One disadvantage of the Primer Bank algorithm is that it does not specifically design primers that span an intron.

7. To assist in pipetting, primers should be dispensed in 96-well plates by the manufacturer following synthesis. Request that the primers are synthesized at a fixed concentration and volume (e.g., 100 µL of 500 µ*M*) and that they are organized

on the plate so that each pair of primers is directly next to one another. Thaw the plate of primers and perform a brief spin of the plate using a centrifuge that is equipped with a 96-well plate adapter. Remove the cover of the plate, throw it away, and replace with a new cover to avoid contamination. Using an 8-well multichannel pipette remove 50 µL of one primer and add to a strip tube or plate. Next add 50 µL of the second primer of the pair and mix. Repeat until all pairs of primers are used.

8. Dilute primers to 2 µM and place in a clean strip tubes. Store primers at −20°C. It is essential that the following protocol is adhered to each time these primers are used. Thaw, perform a brief on the centrifuge, remove the caps using the CapEase tool, and discard caps. Pipette in a class II hood. If you will be assaying several plates (we routinely assay up to eight plates per day), cover the plates to prevent dust from entering and evaporation. A clean, plastic cover from a pack of Rainin pipette tips works extremely well. This way you do not need to cap and recap throughout the day. When finished for the day, add new caps using the CapEase tool and store primers at −20°C.

9. An excellent alternative to pipetting into 384-well plates is the TaqMan® Low Density Array 384-well microfluidic cards (Applied Biosystems). These low-density arrays contain up to 384 individual primers and TaqMan probes spotted into individual wells of the plate. Microfluidics channels connect the wells. Since the primers are spotted into each individual wells of the plate, issues associated with pipetting multiple primers are eliminated. The user simply makes a master mix of cDNA and PCR reagents and pipettes the entire mixture into a single channel of the microfluidics card. The card is then centrifuged and the master mix will drop into each well of the cards. The card is then ready for real-time PCR analysis.

10. When presented alone, real-time PCR data are commonly presented as the expression relative to some internal control gene. Selection and validation of the internal control gene is not discussed here and has been reported (9). The relative gene expression is presented as $2^{-\Delta C_T}$, where $\Delta C_T = C_T\ _{\text{gene of interest}} - C_T\ _{\text{internal control gene}}$ *(5)*.

11. When comparing gene expression data to some other reference sample (e.g., untreated control or nondiseased tissue), the fold change in expression may be presented as $2^{-\Delta\Delta C_T}$, $\Delta\Delta C_T = (C_T\ _{\text{gene of interest}} - C_T\ _{\text{internal control gene}})_{\text{Treated}} - (C_T\ _{\text{gene of interest}} - C_T\ _{\text{internal control gene}})_{\text{Untreated}}$ *(5)*.

12. When profiling multiple plates using the same set of primers, a template of the plate's configuration should be set-up using the PCR instrument's software. This will conveniently link the data to the gene's name.

13. When presenting real-time PCR data as heatmaps *(6)*, we *(3)* and Dittmer et al. *(1)* have found it more appropriate to analyze the C_T data rather than the $2^{-\Delta C_T}$ data using the algorithm.

References

1. Dittmer, D. P., Gonzalez, C. M., Vahrson, W., DeWire, S. M., Hines-Boykin, R., and Damania, B. (2005) Whole-genome transcription profiling of rhesus monkey rhadinovirus. *J. Virol.* **79,** 8637–8650.

2. Dittmer, D. P. (2003) Transcription profile of kaposi's sarcoma-associated herpesvirus in primary kaposi's sarcoma lesions as determined by real-time PCR arrays. *Cancer Res.* **63,** 2010–2015.

3. Jiang, J., Lee, E. J., Gusev, Y., and Schmittgen, T. D. (2005) Real-time expression profiling of microRNA precursors in human cancer cell lines. *Nucleic Acids Res.* **33,** 5394–5403.

4. Schmittgen, T. D. (2006) Quantitative gene expression by real-time PCR: a complete protocol. In Real-time PCR (Dorak, M. T. ed.), Bios Scientific Publishers, Oxfordshire.

5. Livak, K. J. and Schmittgen, T. D. (2001) Analysis of relative gene expression data using real-time quantitative PCR and the 2(-Delta Delta C(T)) *Method Methods* **25,** 402–408.

6. Eisen, M. B., Spellman, P. T., Brown, P. O., and Botstein, D. (1998) Cluster analysis and display of genome-wide expression patterns. *Proc. Natl Acad. Sci. USA.* **95,** 14,863–14,868.

7. Schmittgen, T. D., Jiang, J., Liu, Q., and Yang, L. (2004) A high-throughput method to monitor the expression of microRNA precursors. *Nucleic Acids Res.* **32,** e43.

8. Wang, X. and Seed, B. (2003) A PCR primer bank for quantitative gene expression analysis. *Nucleic Acids Res.* **31,** e154.

9. Schmittgen, T. D. and Zakrajsek, B. A. (2000) Effect of experimental treatment on housekeeping gene expression: validation by real-time, quantitative RT-PCR. *J. Biochem. Biophys. Methods* **46,** 69–81.

8

Design and Use of Scorpions Fluorescent Signaling Molecules

Rachael Carters, Jennifer Ferguson, Rupert Gaut, Paul Ravetto, Nicola Thelwell, and David Whitcombe

Abstract

A number of probe systems exist for the real-time detection of PCR products. Scorpions are a unique method wherein primer and probe are combined in a single oligonucleotide. During the PCR, the probe element becomes linked directly to its complementary target site with beneficial consequences. In particular, the unimolecular mechanism of probe/target hybridization ensures rapid, reliable, and robust probing of a chosen amplicon. We discuss the design and use of Scorpions and compare their use with similar systems.

Key Words: Scorpions; fluorescent probe; PCR; real-time PCR; unimolecular probing.

1. Introduction

The ability to monitor PCRs in closed tubes throughout the process has massively improved the power of the PCR in a number of ways:

- Closed tube detection of PCR products is quicker and simpler than any methods that require post-PCR manipulation (such as agarose gels).
- Fluorescence monitoring is at least as sensitive as most post-PCR analysis methods, allowing detection of small amounts of product.
- Continuous monitoring of PCR through the process allows accurate quantitation of the initial number of templates introduced.
- Closed tube fluorescence reporting systems allow analysis to be performed without opening the reaction vessel, thus minimizing the potential for product contamination.
- Probe-based reporter systems, such as TaqMan *(1)*, Molecular Beacons *(2)*, and the one described in this Chapter (Scorpions) *(3)* offer enhanced levels of specificity because the product of the PCR is identified by an amplicon-specific probe; this is comparable to screening of product by dot blot or Southern blotting.

From: *Methods in Molecular Biology, Vol. 429: Molecular Beacons: Signalling Nucleic Acid Probes, Methods and Protocols* Edited by: A. Marx and O. Seitz © Humana Press Inc., Totowa, NJ

A   Stem Loop Format

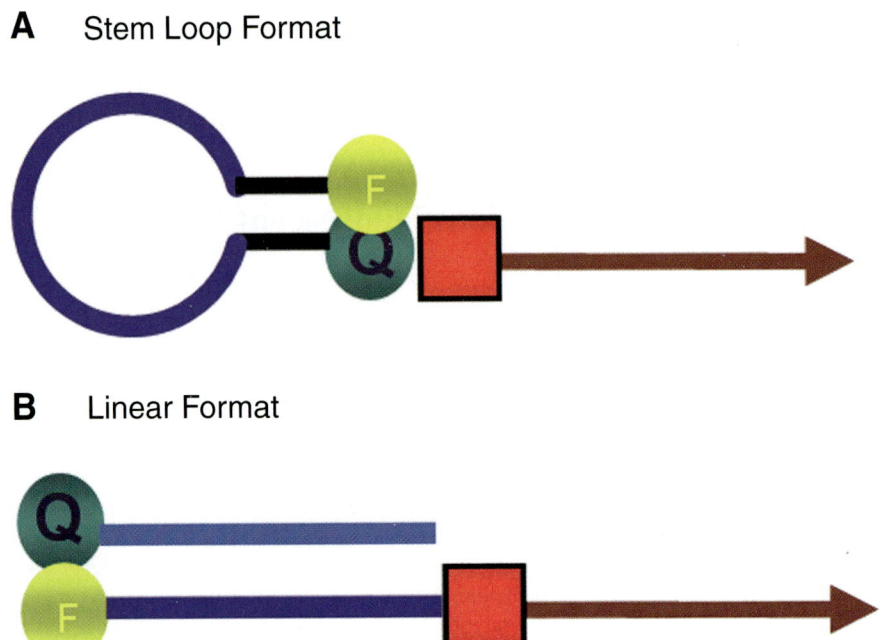

B   Linear Format

Fig. 1. (A) The stem-loop Scorpion comprises (3′–5′): Primer element, blocking group, quencher, stem 1, probe element, stem 2 and fluorophore. **(B)** The linear Scorpion comprises (3′ to 5′): Primer element, blocking group, probe element, fluorophore. A separate quencher molecule comprises a quencher dye attached to an oligonucleotide complementary to the probe element.

1.1. Scorpions—Structure and Mechanism (see Fig. 1)

Scorpions are bi-functional molecules that carry both probe and priming functions on the same oligonucleotide construct. The probe element is attached as a tail to the primer element through a PCR blocker that ensures the probe element does not get incorporated into the double-stranded product.

The Scorpions construct is arranged such that the probe element is complementary to the anticipated extension product of the primer element, i.e., it binds to the opposite strand than the primer element. The probe element also carries a reporter fluorophore that (in the initial conformation) is quenched with a "dark quencher" (usually DABCYL, Methyl Red or Black Hole Quenchers [BHQs]) through a collisional mechanism. The quencher is held in close proximity to the fluorophore by base pairing and this can be achieved by the introduction of a stem-loop in the probe (**Fig. 1A**), or through a second molecule, substantially complementary to the probe element (**Fig. 1B**).

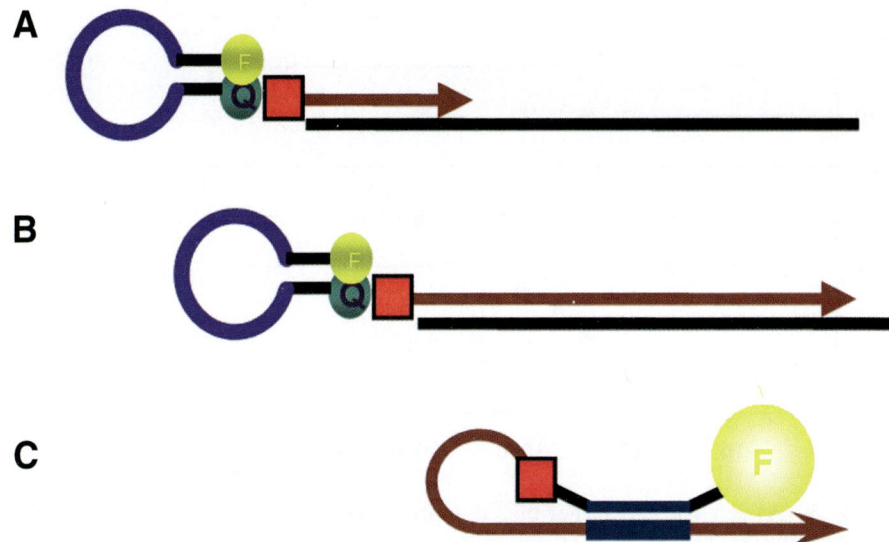

Fig. 2. (A) The primer element binds to the DNA target; the probe element is in the closed, non-fluorescent configuration. **(B)** The primer extends, incorporating a probe binding site into the new strand; the probe element remains in the closed, non-fluorescent configuration. **(C)** After a cycle of denaturation and re-annealing, the probe flips forward to bind its target site on the same molecule. The fluorophore and quencher are now separated and fluorescence increases.

After the primer element has been extended, the probe element is now on the same strand as its complementary target. Upon subsequent rounds of heating and cooling, the extension product becomes single stranded and the probe can bind to its complement in a rapid intramolecular rearrangement and the fluorophore becomes unquenched leading to an increase in specific fluorescence (**Fig. 2**). When complementary strand is synthesized, the probe becomes displaced and resumes its quenched conformation (either stem loop or double stranded). The blocker group prevents DNA synthesis from "copying through" the probe rendering the probe element double stranded and producing a nonspecific "always on" configuration. If that were to happen, nonspecific PCR products and primer dimers would all lead to apparently positive signals.

1.2. Benefits of the Scorpions Mechanism

1. *Speed*: because of the unimolecular rearrangement mechanism, the kinetics of probe binding is extremely rapid. This ensures that signal generation is not the rate-limiting step in a given PCR application.

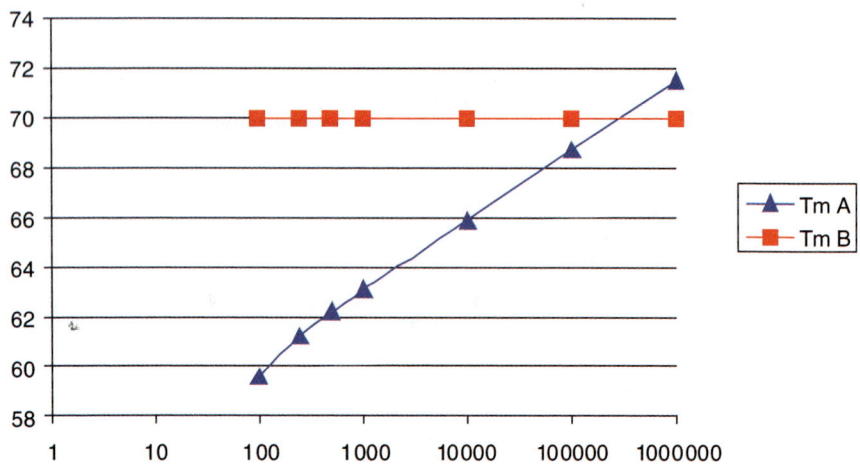

Fig. 3. (**A**) T_m: 100 nM target plus complementary probe at concentrations from 100 nM to 1 mM. The T_m was calculated for each probe concentration. The T_m is highly dependent upon an excess of probe. (**B**) T_m: The same sequences were modeled in the same way at the same concentrations. In the unimolecular configuration, the T_m remains identical throughout. This ensures equally good probe binding at all template and Scorpion concentrations.

2. *Reliability of probing*: with most probing systems, the thermodynamic design of the probe can be easily calculated, but kinetically the probe/target hybridization reaction is in competition with a number of other reactions such as:
   - The amplicon folding to assume a conformation in which the probe site is hidden.
   - Synthesis of the complementary strand by PCR.
   - Re-annealing of the two complementary strands in the mixture.
   - Folding up of the probe itself, making it unavailable for hybridization to its target.

3. *High signal-to-noise*: as a quenched probe system, background signals are low and the efficiency of probing ensures that signals are high. This allows relatively small amounts of labeled material to be used and is important for multiplexing where optical crosstalk is a limiting factor.

4. *Stoichiometry*: one amplicon gives one fluorescent signal. This is important in quantitative reactions.

5. *Concentration independent*: probing occurs irrespective of the concentration of the reagents since the reaction is zero order. This means that all probings are equally efficient and enhances the overall sensitivity of the detection (*see* **Fig. 3**).

6. T_m *bonus*: as a consequence of the unimolecular mode of action, the local concentration of the probe relative to the target binding site is essentially infinite, leading to an elevated T_m (**Fig. 4**).

A Constructs with various linkages

		Sequence	Tm
Bi-molecular			
	Probe	FAM-AGTACAGTC	
	Quencher	GACTGTACT-DAB	36°C
Construct 1		FAM-AGTACAGTC(A)$_{45}$GACTGTACT-DAB	55°C
Construct 2		FAM-AGTACAGTC(A)$_{15}$GACTGTACT-DAB	61°C
Construct 3		FAM-AGTACAGTC(A)$_{5}$GACTGTACT-DAB	73°C

B Quenching of FAM Fluorescence by DABCYL by hybridisation

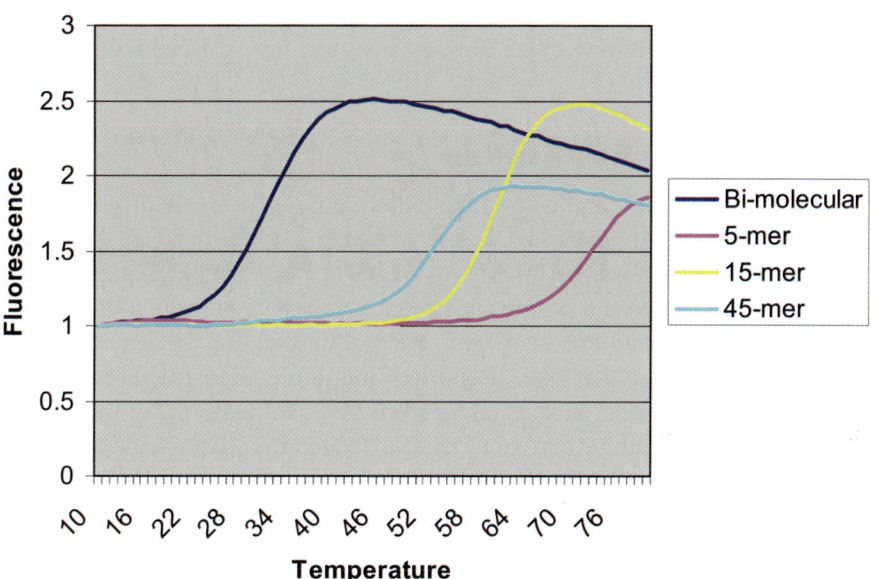

Fig. 4. (A) A nine base sequence with a FAM label was hybridized to a complementary nine base sequence with a DABCYL quencher. In the base case, the two sequences were on separate molecules; in the other cases, the two sequences were linked by different lengths of spacers (5, 15, or 45 bases). The 45 base spacer is a mimic of the distance between the probe and its complement in the amplicon. **(B)** Melting curves of the various constructs.

1.3. Aims

This chapter will discuss and illustrate the design and use of Scorpions and compare it to some alternative methodologies, using a model system based upon an exon in the BRCA1 gene.

2. Materials
2.1. PCR Reagents

Hot Gold Star Taq Polymerase (Eurogentec, Liege, Belgium); hot start enzyme.
Hot Gold Star Reaction Buffer (10×): 150 mM Tris–HCl (pH 8.0 at 25°C), 500 mM KCl, 0.1% Tween 20 plus $MgCl_2$ at 2.5 mM.
Scorpions and other probes were purchased from ATDBIO (Southampton, UK).
Primers were purchased from Invitrogen (Paisley, UK) or ATDBIO.
Human genomic DNA from cell lines was purchased from ECACC (Porton Down, UK).

2.2. Software

VisualOMP (version 5.05) was obtained from DNASoftware. Inc., Ann Arbor, USA.
OLIGO5 was from Molecular Biology Insights, Inc., Cascade, CO, USA.

3. Methods
3.1. Design of Scorpions Reactions

Scorpions reactions can be designed either *de novo*, or by modifying existing assays (probe based or otherwise). Equally, one can use readily available software such as the web-based PRIMER 3, OLIGO *(4)*, Primer Express, Primer Premier; or one can design primers and probes "by eye" and test out the resulting molecules using the above software. We prefer the use of somewhat more sophisticated software that takes account of intramolecular structures as well as bimolecular interactions, such as *mfold* from Michael Zuker *(5)* or the commercially available VisualOMP (from DNASoftware, Inc., Ann Arbor), which is based upon the latest thermodynamic parameters from John SantaLucia's Lab *(6)*.
The baseline conditions to which we design Scorpions are:
- 1× reaction buffer with $MgCl_2$ at 2.5 mM final.
- PCR annealing temperature of 60°C.

3.1.1. Primer Design Using Software to Check The Designs

1. Obtain the target sequence from Genbank databases or other in sources.
2. Identify the key region or design constraints:
   a. For cDNA assays, primer pairs should span at least one intron/exon boundary.
   b. For other assays, it will be important to avoid certain regions.

Suitable Primers Identified

```
TAATGATAGGCGGACTCCCAGCACAGAAAAAAAGGTAGATCTGAATGCTGATCCCCTG

TGTGAGAGAAAAGAATGGAATAAGCAGAAACTGCCATGCTCAGAGAATCCTAGAGATA

CTGAAGATGTTCCTTGGATAACACTAAATAGCAGCATTCAGAAAGTTAATGAGTGGTT

TTCCAGAAGTGATGAACTGTTAGGTTCTGATGACTCACATGATGGGGAGTCTGAATCA

AATGCCAAAGTAGCTGATGTATTGGACGTTCTAAATGAGGTAGATGAATATTCTGGTT

CTTCAGAGAAAATAGACTTACTGGCCAGTGATCCTCATGAGGCTTTAATATGTAAAAG

TGAAAGAG
```

Name	Sequence	Tm (250 nM primer)
Forward	ATGCTCAGAGAATCCTAGAGATACTGAA	64.3°C
Reverse	CCATCATGTGAGTCATCAGAACCTAA	64.3°C

Amplicon Size: 126 base pairs

Fig. 5. Primers that terminate with AA, separated by approximately 100 bp. The T_ms of the primers were modified by lengthening them at the 5′-end.

c. For yet others, it may be important to target a conserved region.

d. Amplicons should be small (80–150 bp) to maximize PCR efficiency.

3. Select primers according to the parameters below (*see* **Fig. 5**):

a. Matching T_ms (or better still by ΔG_0); aim for 5°C above annealing step—usually 63–65°C for a 60°C assay.

b. Relatively GC rich at 5′-end, relatively AT rich at 3′-end; improves specificity and minimizes extensible dimers; where possible, aim to include an -A or -AA bases at the 3′-end of each primer. This tends to minimize the opportunities for mispriming and/or primer–dimer formation.

c. No 3′-dimer prospects (self- or hetero-dimers).

d. Avoid long runs of a single base (especially 4Gs).

3.1.2. Probe Design Using Software to Check The Designs

3.1.2.1. GENERAL POINTS (*SEE* **FIG. 6**)

1. Place the probe close to one end of the amplicon; 1–40 bases from the end of the primer to which it will be attached (up to 100 bases is acceptable but diminishes the benefits of the Scorpions technology).

2. Assume the probe will be attached to the "closer" primer.

3. Ensure that the probe is designed against the opposite strand to the primer; i.e., forward primer carries a reverse probe and *vice versa*.

```
TAATGATAGGCGGACTCCCAGCACAGAAAAAAAGGTAGATCTGAATGCTGATCCCCTG

TGTGAGAGAAAAGAATGGAATAAGCAGAAACTGCCATGCTCAGAGAATCCTAGAGATA

CTGAAGATGTTCCTTGGATAACACTAAATAGCAGCATTCAGAAAGTTAATGAGTGGTT

TTCCAGAAGTGATGAACTGTTAGGTTCTGATGACTCACATGATGGGGAGTCTGAATCA

AATGCCAAAGTAGCTGATGTATTGGACGTTCTAAATGAGGTAGATGAATATTCTGGTT

CTTCAGAGAAAATAGACTTACTGGCCAGTGATCCTCATGAGGCTTTAATATGTAAAAG

TGAAAGAG
```

Name	Sequence	Predicted target Tm (250 nM)	Tm Stem (predicted)
Probe 1	**CAGCATTCAGAAAGTTAATGAGTG**	60°C	NA
Probe 2 (equivalent to a Molecular Beacon)	FAM-CCGCG**CAGCATTCAGAAAGTTAATGAGTG**CGCGG-DAB	63.8°C	73°C
Quencher 1Q	CACTCATTAACTTTCTGAATGCTG-DAB (2x excess)	61.8°C	NA
	CACTCATTAACTTTCTGAATGCTG-DAB (4x excess)	62.8°C	NA
Scorpion_P1 (linear)	FAM- **CAGCATTCAGAAAGTTAATGAGTG**-HEG-CCATCATGTGAGTCATCAGAACCTAA	70.0°C	NA
Scorpion_P2 (stem/loop)	FAM-CCGCG**CAGCATTCAGAAAGTTAATGAGTG**CGCGG-DAB-HEG-CCATCATGTGAGTCATCAGAACCTAA	71.7°C	73.0°C
TMP (longer than P1)	**CAGCATTCAGAAAGTTAATGAGTG**GTTTTCC	67.4°C	NA

Fig. 6. A probe element (Probe 1) with a calculated T_m of 62.2°C was designed according to the criteria in Section 3.1. Of the several possible designs, this one was selected because at either end, there is the beginning of a stem (CA….TG). Additional GC bases were introduced to increase the T_m of the stem, ensuring a 5′-terminal C base, adjacent to the fluorophore (Probe 2, which is equivalent to a Molecular Beacon). An oligonucleotide complementary to Probe 1 was also generated (Quencher 1Q). Probes 1 and 2 were coupled to the reverse primer to create Scorpion molecules (Scorpion_P1 and Scorpion_P2) and a somewhat longer design was also produced to create a TaqMan probe (TMP). Finally, predicted T_ms of the probe on target and any stem/loop structures were calculated. *Abbreviations*: FAM, 6-FAM; DAB, dUTP-DABCYL; HEG, hexethylene glycol (C18) spacer.

4. Calculate the T_m by standard nearest neighbor methods (Primer Express, Oligo, etc.) may be lower than primers, typically 55–65°C, ΔG_0 of −4 kcal/mol or less (more negative).
5. Model the probe hybridization using Visual OMP, *mfold*, or other software.

3.1.2.2. STEM-LOOP FORMAT

1. The probe may contain some bases (usually some *G*s and *C*s) to form the beginnings of a stem.
2. Divide the primer element from the stem by a C18 (also called HEG, hexethylene glycol) group.
3. Use fluorophores that can be added by phosphoramidite chemistry for optimum yields, although other labels can be attached through post-synthesis labeling chemistries.
4. Include a dark quencher adjacent to the spacer, at the end of the 5′-stem.
5. Avoid placing a *G* next to the fluorophore as this leads to lower fluorescence.
6. Design the stem to have a ΔG_0 of about −1.5 to −2 kcal/mol, stem strength will also depend upon the length of the intervening loop; short loops give higher T_ms.
7. Stems should be as short as possible, 5 or 6 bases preferred (seven or more may lead to baseline drift; *see* **Note 4.3**).
8. Stems will be mostly GC; a standard stem: CCGCGC-loop-GCGCGG may be used but the exact T_m of this stem will vary depending upon the length of the probe element (*see* **Fig. 4**).
9. Model the structures of:
   a. The unincorporated Scorpion to confirm the T_m of the stem (**Fig. 7A**).
   b. The extended Scorpion (essentially the amplicon, plus the probe element); ensure the correct strand of the amplicon is used: if the probe selected goes on the reverse primer, use the reverse-complement of the sequence, plus the forward strand probe. The ΔG_0 of the second construct MUST be more negative than that of the stem although the T_m may be higher for the unincorporated version than the extended product (**Fig. 7B**).

3.1.2.3. BI-PROBE FORMAT

1. Design the probe T_m to be ~60°C as above.
2. Incorporate a spacer group (HEG/C18) but no internal Quencher. There is no built-in secondary structure to the native probe.
3. Include the fluorophore on the 5′-end of the linear Scorpion; ensure the label is not adjacent to a *G*.
4. Design a Quencher molecule complementary to the probe element and carrying a dark quencher on the 3′-end (usually introduced via the solid support controlled pore glass column).
5. Quencher oligonucleotide/target hybridization is driven by [Quencher]; we usually use two- to fourfold excess.
6. Model the structures of:
   a. The unincorporated Scorpion; plus quencher (**Fig. 8A**).
   b. The extended Scorpion (essentially the amplicon plus the probe element; **Fig. 8B**) ensure the correct strand of the amplicon is used: if the probe selected goes on the reverse primer, use the reverse-complement of the sequence, plus the forward strand probe.

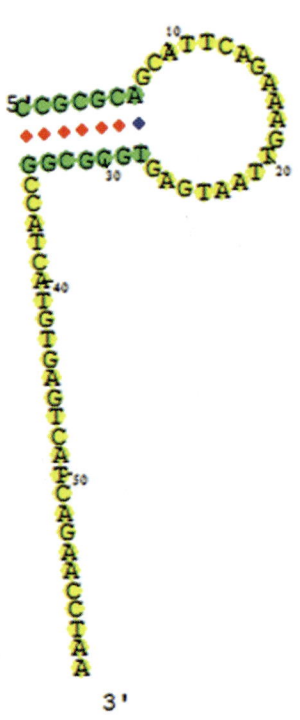

BRCA1 Scorpion P2

dG (60.00'C) = -2.42 kcal/mole
dH = -60.41 kcal/mole
dS = -174.07 cal/mole K
Tm = 73.90'C
[Na+] = 0.05 mol/L
[Mg+2] = 0.00 mol/L

BRCA1 Amplicon P2

dG (60.00'C) = -6.91 kcal/mole
dH = -203.60 kcal/mole
dS = -590.40 cal/mole K
Tm = 71.70'C
[Na+] = 0.05 mol/L
[Mg+2] = 0.00 mol/L

Fig. 7. Stem loop Scorpion-folding analysis. (**A**) Native Scorpon: using visual OMP, the stem loop Scorpion was folded (2.5 m*M* Mg, 250 n*M* oligonucleotide, 60°C). The most favored structure is shown and a numerical analysis indicated that >95% of the molecules would assume this configuration under the stated conditions. (**B**) Extended Scorpion: analysis was conducted as above on the extended Scorpion product. The T_m and ΔG_0 show this to be a favored structure and again >95% of the molecules would adopt this or related configurations.

A   With Quencher

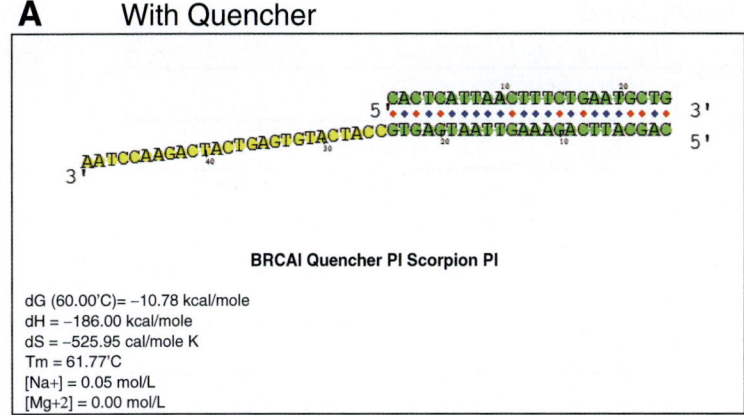

BRCAI Quencher PI Scorpion PI

dG (60.00'C)= –10.78 kcal/mole
dH = –186.00 kcal/mole
dS = –525.95 cal/mole K
Tm = 61.77'C
[Na+] = 0.05 mol/L
[Mg+2] = 0.00 mol/L

B   Incorporated into Amplicon

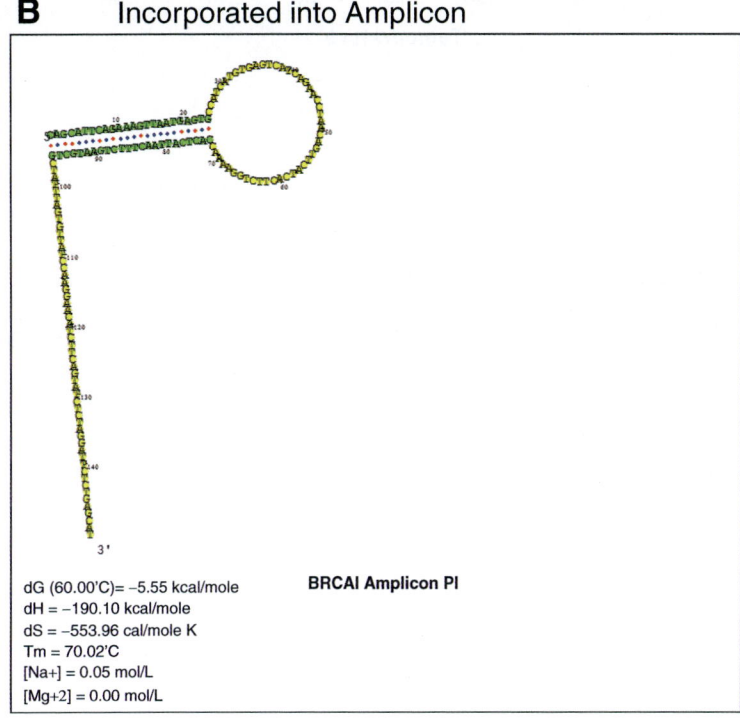

BRCAI Amplicon PI

dG (60.00'C)= –5.55 kcal/mole
dH = –190.10 kcal/mole
dS = –553.96 cal/mole K
Tm = 70.02'C
[Na+] = 0.05 mol/L
[Mg+2] = 0.00 mol/L

Fig. 8. Linear Scorpion-folding analysis. (**A**) Unincorporated Scorpion with 2×excess of quencher molecule. Analysis using Visual OMP (2.5 mM Mg, 250 nM Scorpion oligonucleotide, 500 nM quencher oligonucleotide, 60°C). The T_m and ΔG_0 show this to be a favored structure. (**B**) Extended Scorpion with 2× excess of quencher molecule. Analysis using visual OMP (2.5 mM Mg, 100 nM Scorpion extension product, 500 nM quencher oligonucleotide, 60°C). The T_m and ΔG_0 show this to be a favored structure.

A Melting Curve

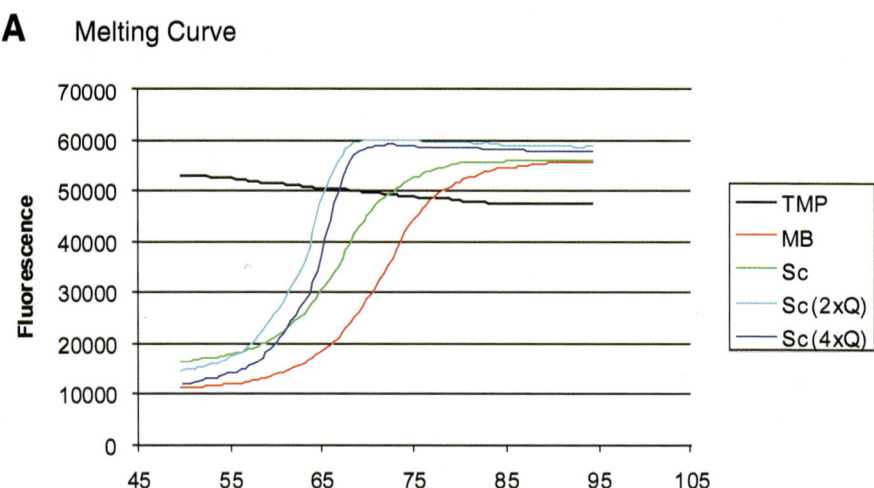

B First Derivative

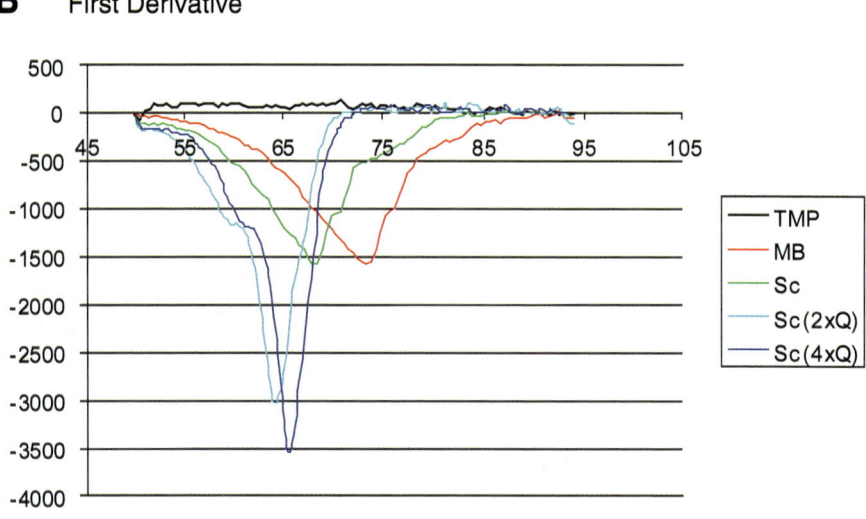

Fig. 9. Melting curve analysis of Scorpions and other probes (**A**) *Raw fluorescence curves*. The fluorescence was monitored between 55°C and 95°C. The TaqMan probe (TMP; black line) shows little or no change in fluorescence. The Molecular Beacon (MB; red line) shows high fluorescence at 95°C which quickly reduces to a low baseline; the equivalent stem-loop Scorpion (green line) shows similar initial high temperature fluorescence and equivalent quenching but at a lower temperature, which we believe reflects the effect of the HEG on the stem T_m. The linear Scorpion with two different concentrations of quencher molecules [Sc(2× Q), Sc(4× Q)] and showed similar high

3.2. Melt Curves

1. Create standard reaction mixes containing 1.1× reaction buffer with 2.5 mM MgCl$_2$, 200 μM dNTPs, and 250 nM primers/probes/Scorpions as appropriate and allowing 10% of the reaction volume for the addition of template or water control. In addition, it is advisable to prepare 10% excess of mix to allow for pipetting losses. The table below gives typical reaction set up volumes:

Reagent	Scorpion Based		Probe Based	
	1	12	1	12
10× reaction buffer	2.5	33	2.5	33
25 mM MgCl$_2$	2.5	33	2.5	33
2 mM dNTPs	2.5	33	2.5	33
2.5 μM forward primer			2.5	33
2.5 μM forward Scorpion	2.5	33		
2.5 μM reverse primer	2.5	33	2.5	33
2.5 μM probe				
(molecular beacon or TaqMan)			2.5	33
Taq polymerase	0.2	2.64	0.2	2.64
Water	9.8	129.36	7.3	96.36
Total	**22.5**	**297**	**22.5**	**297**

2. Dispense 22.5 μL of reaction mix to each tube/well.
3. Add 2.5 μL water (no template control) or template to each reaction and seal the tubes.
4. Centrifuge the tubes briefly to collect the liquid.
5. Perform a melt analysis on the real-time PCR instrument. For a Stratagene Mx3000, we use the following parameters in the default programme designed for SYBR Green Melt Analysis:
   a. Heat to 95°C for 1 min to denature the probe or Scorpion.
   b. Cool to 10°C below the anticipated PCR annealing temperature for 30 s.
   c. Heat to 95°C collecting data throughout the whole of the ramp between the two temperatures.
   d. The fluorescence data (**Fig. 9A**) are useful to directly visualize the melting characteristics.
   e. The first derivative of the fluorescence/temperature plot enables the determination of the T_m (**Fig. 9B**).

---

fluorescence and also quenching. The T_m of this hybridization was increased by the higher concentration of quencher oligonucleotide. (**B**) *First derivative curves.* The raw data were processed as a first derivative; i.e., the change in fluorescence against temperature. This shows the point of inflection of the melt curve and is a good indicator of the T_m of the quenching reaction. The T_ms were therefore: TMP- not applicable, MB- 74°C, Stem-loop Scorpion 68°C, Linear Scorpion with twofold excess quencher- 64°C, Linear Scorpion with fourfold excess quencher- 65.5°C.

A   PCR performance

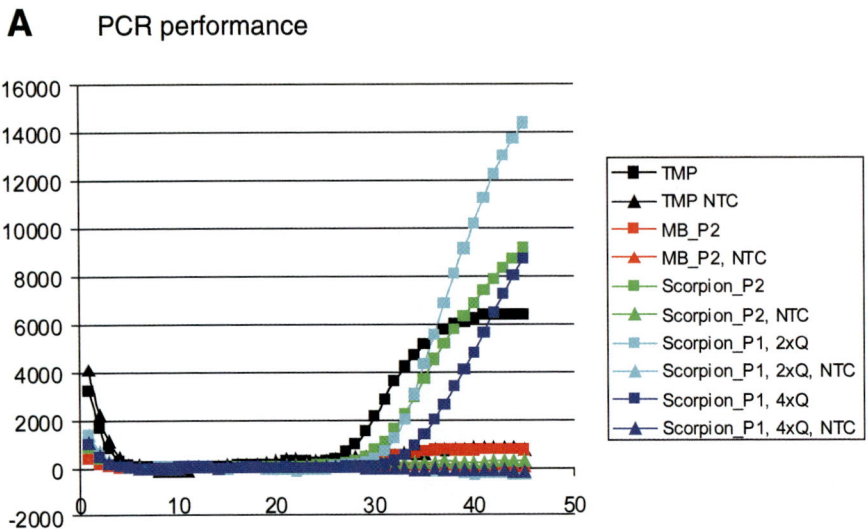

B   Detailed Analysis of Cycle Fluorescence, Cycles 31-36

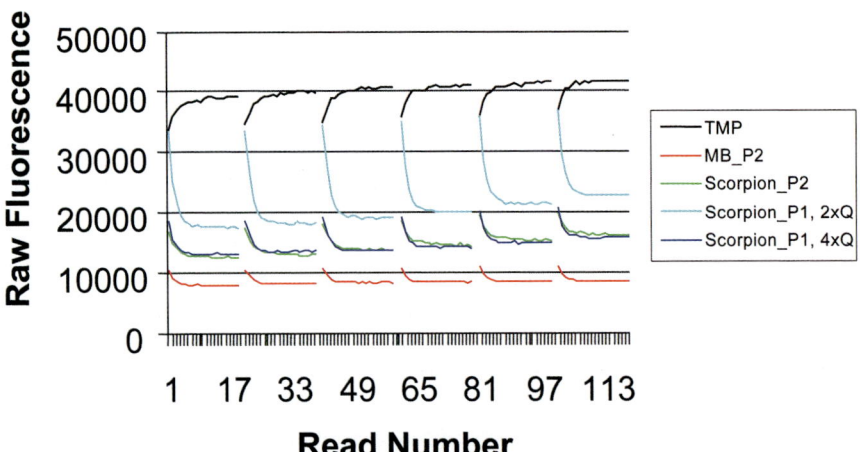

Fig. 10. (A) *PCR performance*. Raw data are processed by the instrument. The baseline/background fluorescence is subtracted. Template positive samples (squares) and NTCs triangles are shown for each assay system. The TaqMan reaction showed the earliest C_t, the stem/loop Scorpion, the Molecular Beacon and the linear Scorpion with twofold excess of Quencher were 1–2 cycles later, while the linear Scorpion with fourfold excess of Quencher was a further 2 cycles later. This latter result may be due to the enzyme sink effect (*see* **Note 4.2**). In terms of fluorescence yield, the order was linear Scorpion with 2×excess>

3.3. Real-time PCR—Comparison of Probe Systems

1. Prepare reaction mixes and add templates as above (steps 1–4).
2. Perform real-time PCR with the following cycling parameters:
   a. Heat at 95°C for 10 min to activate the Hot Start Enzyme (other enzymes may require different activation conditions).
   b. Run 45 cycles of [95°C for 30 s; 60°C for 60 s], collecting fluorescence data at the 60°C anneal/extend step.
3. Compare data to standard curves or other controls.
4. In some instruments, including the Stratagene Mx3000, it is possible to export the individual fluorescence reads from within an anneal/extend step (**Fig. 10**). This allows some elucidation of the mechanism of the various probe reactions.

4. Notes

4.1. Temperature Dependence of Fluorescence

When performing melt curves and other comparative analyses, it is important to note that fluorescence is lower at high temperatures and that an increase in fluorescence at lower temperatures does not necessarily equate to an increase in probe binding. In **Fig. 9A**, for example, the fluorescence from the TaqMan probe increases steadily with decreasing temperature.

4.2. Linear Scorpions—Enzyme Sink

The 250 nM duplex can act as an enzyme sink; ds-DNA binds Taq Polymerase (and probably all polymerases). This can reduce the efficiency of the amplification and is particularly observed in multiplex reactions and at low levels of input template. In order to overcome this issue:

Stem/loop Scorpion>linear Scorpion with 4× excess> TaqMan>> Molecular Beacon. (**B**) Detailed analysis of fluorescence, cycles 31–36. By monitoring fluorescence within the cycle, it is possible to examine the efficiency and mode of action for the different probe systems. For the TaqMan reactions, the fluorescence builds from a low base (due to the temperature effect on fluorescence) and accumulates over a period of 20–30 s; for the stem loop systems, both Scorpion and Molecular Beacons, the fluorescence starts high and decreases to a baseline, as the stem loop reforms or the probe hybridizes to its target. For the linear Scorpions, the cooling/ quenching is kinetically slower because of the bimolecular mechanism, although with a fourfold excess of quencher oligonucleotide, the quenching speed, and efficiency approaches that of the stem/loop Scorpion. With block cyclers such as the Stratagene Mx3000P used here, part of the kinetics that are observed result from the lag time associated with heat transfer through the reaction vessel and the cooling of the reaction liquid. Similar experiments in a capillary-based cycler showed an even more rapid quenching reaction for the stem/loop constructs (data not shown).

1. Use additional enzyme- 2.5 or even 5 U per 25 μL reaction; this overcomes any enzyme limitation.
2. Introduce a three-step cycling protocol incorporating a 72°C step; at this temperature, the duplex is not present and extension occurs with full efficiency. Fluorescence reads must take place at the lower (annealing) temperature.

4.3. NonSpecific Cleavage

Occasionally, one observes a slight but consistent drift in the raw fluorescence throughout the PCR, even in the no template control reactions. This probably results from nonspecific probe cleavage and is more common in the linear format than the stem/loop format. Any double-stranded complex above a certain size (possibly as short as 7 bases in a stem/loop construct), with a T_m 10°C or more above the anneal temperature might be recognized as possible substrates for the cleavase activity of *Taq* Polymerase. Cleavage of the probe region would cause a nonspecific increase in fluorescence.

Several methods exist to overcome this issue:

1. Use an exo⁻ Taq; this resolves the cleavase issue, but may still require the use of higher enzyme levels to counteract the sink effect.
2. Use three-step PCR with a 50–55°C anneal/read step; engineering the quencher duplex appropriately. At these temperatures, the cleavase activity is low. The duplex is not present at the 72°C extension step and PCR is efficient.
3. Use a modified quencher backbone, such as 2′-O-methyl-RNA (2MeRNA). This has several advantages:
   a. The probe cannot be cleaved by standard Taq Polymerase enzymes.
   b. The 2MeRNA probe has a higher T_m on its target than the corresponding DNA probe, permitting shorter probes with higher T_ms.
   c. The 2MeRNA stems have even higher affinities than DNA/2MeRNA hybrids, so stems may be shortened.

The main disadvantages are:
   a. The synthesis of mixed molecules (DNA plus 2MeRNA) is complex and expensive.
   b. The behavior of these probes is less studied and harder to predict.

4.4. Linear Scorpions- Low T_m of Probe/Quencher Hybrid

The probe/quencher hybrid may have a lower T_m than the unimolecular probing event (although the ΔG_0 is more negative). This has the potential to reduce the efficiency of quenching, which in turn may limit the positive signal obtained. It is therefore advisable to ensure that the probe/quencher hybrid has a T_m higher than the anneal/extend temperature (without being too prone to cleavage or the enzyme sink effect). However, more robustly, the use of a three-step PCR with annealing/fluorescence read at 55°C allows the use of probes with T_ms around 60°C, without problem.

4.5. Effects of Labels

The addition of a label at the 5′-end of a molecule typically has also a stabilizing effect on a stem and probably on a duplex. Furthermore, the choice of quencher molecule seems to have an impact on stem stability. In particular, the BHQs from Biosearch Technologies seem to contribute significantly to base-stacking. This is particularly noticeable in the stem/loop format and may lead to significantly elevated T_ms for the Stem. Typical values for these contributions are:

a. 5′-fluorophore adds about −0.4 kcal/mol to the ΔG_0. This can often be modeled in VOMP by including an extra A at the 5′-end of the sequence.
b. The BHQs add approximately −0.8 to −1 kcal/mol to the duplex (particularly strong with ROX). This can be modeled by the addition of a G/C base pair within the stem.
c. The presence of alternative quenchers such as Methyl-Red and DABCYL adds less to the stem strength.

In contrast, the blocking group (C18 or HEG) seems to reduce the T_m of the stem, when compared with the equivalent Molecular Beacon (compare the Molecular Beacon and Scorpion Melt Curves in **Fig. 9**).

References

1. Holland, P. M., Abramson, R. D., Watson, R., and Gelfand, D. H. (1991) Detection of specific polymerase chain reaction product by utilizing the 5′——3′ exonuclease activity of Thermus aquaticus DNA polymerase. *Proc. Natl. Acad. Sci. USA* **88,** 7276–7280.
2. Tyagi, S. and Kramer, F. R. (1996) Molecular beacons: probes that fluoresce upon hybridization. *Nat. Biotechnol.* **14,** 303–308.
3. Whitcombe, D., Theaker, J., Guy, S. P., Brown, T., and Little, S. (1999) Detection of PCR products using self-probing amplicons and fluorescence. *Nat. Biotechnol.* **17,** 804–807.
4. Rychlik, W. and Rhoads, R. E. (1989) A computer program for choosing optimal oligonucleotides for filter hybridization, sequencing and in vitro amplification of DNA. *Nucleic Acids Res.* **17,** 8543–8551.
5. Zuker, M. (2003) Mfold web server for nucleic acid folding and hybridization prediction. *Nucleic Acids Res.* **31,** 3406–3415.
6. SantaLucia, J., Jr. (1998) A unified view of polymer, dumbbell, and oligonucleotide DNA nearest-neighbor thermodynamics. *Proc. Natl. Acad. Sci. USA* **95,** 1460–1465.

9

Hybridization Probe Pairs and Single-Labeled Probes: an Alternative Approach for Genotyping and Quantification

Thomas Froehlich and Oliver Geulen

Abstract

Real-time polymerase chain reaction (PCR) has become a standard tool in both quantitative gene expression and genetic variation analysis. Data collection is performed throughout the PCR process, thus combining amplification and detection into a single step. This can be achieved by combining a variety of different fluorescent chemistries that correlate the concentration of an amplified PCR product to changes in fluorescence intensity. Hybridization probe pairs and single-labeled probes are sequence-specific, dye-labeled oligonucleotides, used in real-time PCR approaches, in particular for genotyping of single nucleotide polymorphisms (SNPs). In that case, a detector probe is designed to cover the polymorphism. Allelic variants are identified and differentiated via post-PCR melting curve analysis. A single melting curve can distinguish different T_ms, and differently labeled probes may be used, theoretically allowing multiplexed genotyping of several SNPs.

Key Words: Hybridization probe pairs; HybProbe chemistry; single-labeled probes; SimpleProbe chemistry; real-time PCR; quantification; melting curve analysis; SNP; genotyping.

1. Introduction

Over the past decade, there has been enormous progress in sequencing the human genome and identifying genes. In the recent years, real-time polymerase chain reaction (PCR) has become a key technology for gene expression analysis and genotyping of single nucleotide polymorphisms (SNPs), both in industrial and in public research with major impact on modern approaches used in the fields of drug discovery, pharmacogenetics, and forensic medicine (1–5). One of the next important steps, e.g., towards understanding of disease mechanisms, is the establishment of genotype–phenotype correlation due to gene-specific variations

From: *Methods in Molecular Biology, Vol. 429: Molecular Beacons: Signalling Nucleic Acid Probes, Methods and Protocols* Edited by: A. Marx and O. Seitz © Humana Press Inc., Totowa, NJ

and alterations in gene products expression levels. Full exploitation of available genomic information requires technologies capable of performing vast numbers of analyses, as large numbers of individuals have to be screened for a large number of markers. Real-time PCR is one of the most promising techniques which can cope with all resulting demands regarding sensitivity, specificity, sample throughput, automation, and cost-effectiveness.

Modern real-time PCR systems are capable of combining simultaneous analysis of at least 384 individual samples with the flexibility of using differently fluorescent-labeled sequence-specific oligonucleotides for detection in different optical channels. A variety of detection chemistries combining fluorescent dyes with special oligonucleotide structures are currently applied for real-time PCR *(6)*.

Hybridization probes are a commonly used format for quantitative PCR and SNP genotyping. They are most frequently used in a four oligonucleotide manner combining two sequence-specific PCR primers with two single-fluorescent dye-labeled sequence-specific probes. In addition, the utilization of just one single-labeled probe is a simple alternative for melting curve-based genotyping experiments *(7)*.

2. Materials

1. All experiments mentioned were run on a LightCycler® 480 instrument, equipped with a 384-well blockcycler module (Roche Diagnostics GmbH, Mannheim, Germany). The instrument was run with ramp rates of 4.8°C/s for heating and 2.5°C/s for cooling, except that cooling down to 40°C was carried out with a slower rate of 2°C/s.
2. White multiwell plates with 384 cavities were used together with an optical transparent seal (LightCycler® 480 multiwell plates 384 and LightCycler® 480 sealing foil; Roche Diagnostics).
3. Primers and hybridization probe pairs were synthesized at TIB MolBiol, Berlin, Germany.
4. SimpleProbe oligonucleotides were in-house synthesized according to standard oligonucleotide synthesis procedures using a 'SimpleProbe 519 Labeling reagent' (Roche Diagnostics).

3. Methods

3.1. Hybridization Probe Pairs for Real-Time PCR

Hybridization probe pairs (HybProbe probes) are two singly labeled oligonucleotides that hybridize side-by-side to an internal sequence of an amplified fragment during the annealing phase of the PCR. The donor probe is labeled at the 3′-end with fluorescein. The second probe is labeled at the 5′-end with an acceptor fluorophore and blocked at the 3′-site to prevent the probe from being extended. During probe hybridization on a reverse complementary PCR strand, these fluorophores are brought into close proximity and fluorescence

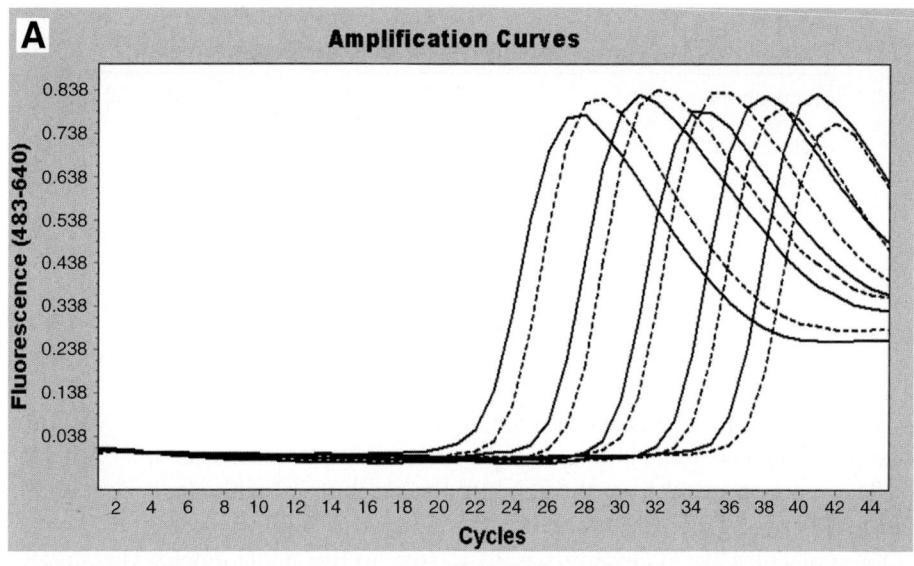

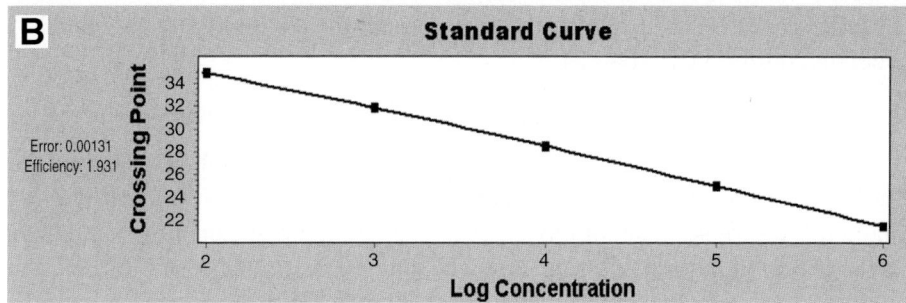

Fig. 1. Absolute quantification using hybridization probe pairs. **(A)** Serial 1:10 dilutions of 10^6-100 copies/reaction (solid line) and 0.5×10^6-50 copies/reaction (dashed line) of a plasmid DNA target detected via FRET at 640 nm. **(B)** Accurate quantification of the target depends on a standard curve that correlates crossing points (cycle numbers) with the concentration of known standard dilutions (10^6-100 copies/reaction). Target sequence: human hypoxanthine-phosphoribosyl-transferase (h-HPRT) gene.

resonance energy transfer (FRET) occurs *(8)*. This allows a pair of probes to be placed anywhere within a primer set. With hybridization probes, an increase in fluorescence resonance energy transfer is observed as a PCR is cooled and probe/target annealing brings donor and acceptor fluorophores into close proximity. On the other hand, upon heating of the reaction mixture, the fluorophores are separated and fluorescence resonance energy transfer decreases. In PCR, once per cycle acquisition during the annealing phase of probes and target provides quantitative information about the initial target copy numbers (**Fig. 1**) *(8)*.

Table 1
Examples of Molecules Used for Hybridization Probe Labeling

Fluorophore	Emission (nm)	Function
LightCycler® 480 Cyan 500	493	Donor
Fluorescein	533	Donor
LightCycler® Red 610	610	Acceptor
LightCycler® Red 640	640	Acceptor
Cy5	670	Acceptor
Cy5.5	705	Acceptor

Depending on instruments emission and detection possibilities, dual hybridization probes can be used either for single- or for multiple-color detection (**Table 1**).

The energy transfer from the donor to the acceptor is highly dependent on the spacing between the two dye molecules. Energy is only transferred efficiently if the molecules are in close proximity (—one to five nucleotides). The amount of fluorescence emitted is directly proportional to the amount of target DNA generated during PCR (**Fig. 2**).

3.2. Melting Curve Analysis Using Hybridization Probe Pairs

Hybridization probes can be successfully used for SNP genotyping via fluorescent melting curve analysis (*9*). After completion of PCR amplification, qualitative information about the target sequence can be obtained by continuous fluorescence monitoring during slow heating (0.05-0.2°C/s) of the reaction mixture. The melting curve analysis starts at low temperature where hybridization probes are mostly annealed to their target sequence. Upon heating, probes drop off the target strand resulting in a decrease of FRET signal (**Fig. 3**). The probe melting temperature (T_m) defined as the point at which 50% of the probe has strand-separated from the target, can be determined from the inflection point of the melting curve or the maximum of the derivative melting curve (*10*). The T_m is characteristic for a particular duplex and depends on factors like length, nucleotide content, primary structure, and Watson-Crick pairing. Base pair mismatches decrease the stability of the duplex, depending on type and position of the mismatch, and sequence neighbors (*11,12*). When a probe hybridizes over a sequence variant, the resulting duplex is destabilized which is reflected by its T_m (**Fig. 4**).

3.2.1. Design of Hybridization Probe Pairs

3.2.1.1 GENERIC DESIGN CONSIDERATIONS

For target selection general rules as with conventional PCR should be applied.

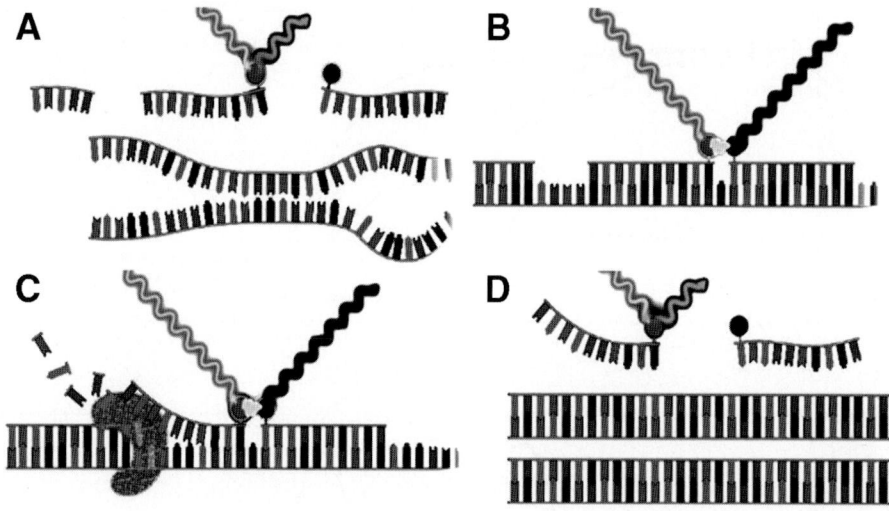

Fig. 2. Principle of real-time PCR monitoring with hybridization probes. (**A**) *Denaturation*: no probe hybridization, no FRET signal. (**B**) *Annealing*: probe hybridization on target. Probes hybridize in a head-to-tail arrangement, thereby bringing the two fluorescent dyes together. Through excitation of the donor dye a FRET is initiated to excite the acceptor dye. Finally, light-emission of the acceptor dye is recorded during each cycle. (**C,D**) *Extension*: After annealing, an increase in temperature leads to elongation and displacement of the probes. At the end of this step, the PCR product is double-stranded; the displaced probes are back in solution and too far apart to allow FRET to occur.

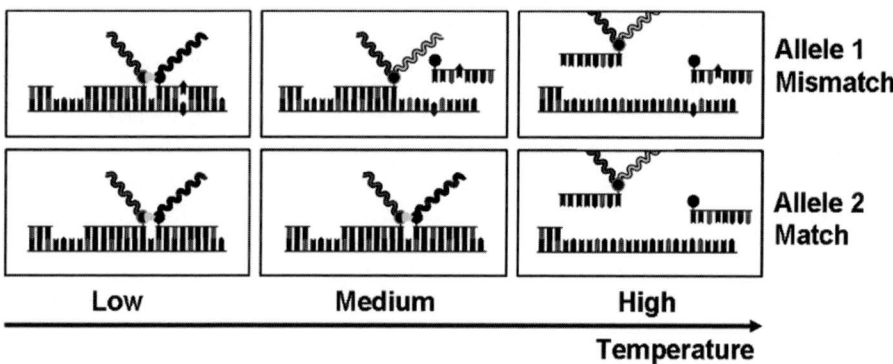

Fig. 3. Principle of probe-based melting curve analysis for genotyping. At low temperatures probes are bound to their targets. With increasing temperature, the mismatched probe leaves its target strand resulting in a decrease of fluorescence. At high temperature the perfectly matched probe backs off its target strand and the fluorescence decreases further more.

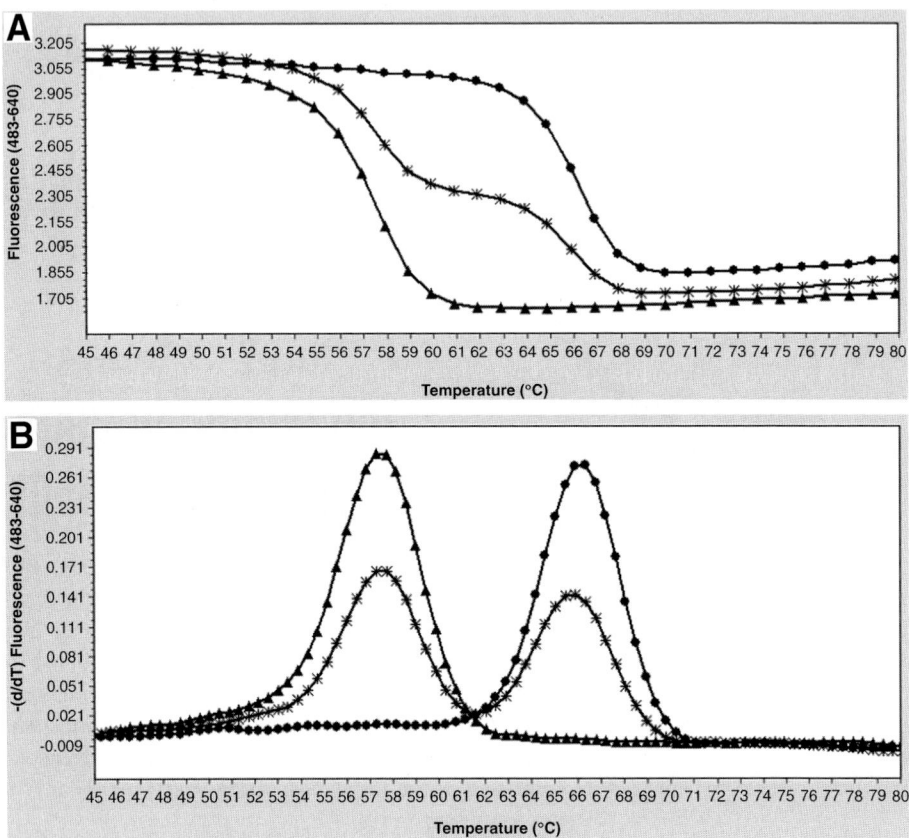

Fig. 4. SNP genotyping of the promoter polymorphism TNF-α-1031 C/T via melting curve analysis. **(A)** Melting curve data (fluorescence versus temperature) generated with LightCycler Red 640 as reporter dye. **(B)** Melting peaks plotted as first negative derivative of the melting curves. Three allelic variants are shown: C/C homozygous (●), T/T homozygous (▲), and C/T heterozygous (*). A single mismatch between the hybridization probes and their target sequences changes the melting temperatures of the bound probes by approximately 8.5°C. *Source*: Data from Ref. *(19)*.

Probes may be designed according to the following considerations *(13)*:

- Avoid placing the probes close to the 5′ primer-binding site on the same target strand.
- Allow a spacing of —one to five bases between 3′-end of the first and 5′-end of the second hybridization probes.
- Probe's T_ms should be approximately 5-10°C higher than the T_ms of the primers. This gives the probes extra time to bind target before being displaced by the polymerase. Usually, the probes have to compete against the elongation reaction of the

Table 2
List of Base Pairing Stabilities *(11)*

Most stable	Least stable
G-C >A-T > G-G >G-T = G-A >T-T = A-A > T-C > A-C > C-C	
Smallest T_m shift	Largest T_m shift

polymerase, which will start regularly after annealing of the primers. To prevent competition or displacement of the probes by the polymerase, a higher annealing temperature will create more stable annealing and proper probes signaling.

- Prevent extension of the probes during PCR by blocking the 3′-end of the acceptor probe, e.g., by phosphorylation. The 3′-end of the donor probe carries a fluorescein label and thus cannot be extended during PCR.

3.2.1.2 DESIGN CONSIDERATIONS FOR GENOTYPING EXPERIMENTS

- The sequence of a SNP-detection probe can be either targeted to the wild-type or mutated sequence on either the sense or the antisense target strand.
- It is possible to position the (predicted) mismatched base either in the middle of the probe or towards one terminus. However, do not let the mismatched base be adjacent to the last three bases of the probe.
- Select the probe sequences that lead to maximum destabilized hybrids in case of a base mismatch, resulting in maximum T_m differences. Either sensor or acceptor probe can be used for covering the polymorphism. The stability of different probe/ target duplexes is given in **Table 2.** Note that the least stable hybrid sequence produces the greatest change in T_m of the hybrid (when compared with the T_m of the perfectly matched probe-target hybrid) *(11)*.

Differences in environment (neighboring bases, pH) can cause slight changes in the stability of the mismatched sequences given above *(14)*.

- Asymmetric PCR conditions may be favorable to gain an excess of target strand for probes binding *(15)*.
- Usually, the simplest form of mutation analysis is to create one amplicon and detect one mutation site within the site of amplification. Combining two amplicons within the same reaction vessel, detection is possible (i) with probes having different detection dyes or (ii) using the same detection dye but different T_m values. Increasing complexity, both types of assays may be combined. Although primer and probe design is supported by software tools (e.g., LightCycler® Primer Probe Design Software 2.0), in particular, the optimization of multiplexed assays may still require several optimization steps.

3.3. Single-Labeled Probes

A single-labeled probe is a special type of simplified hybridization probe, particularly designed for genotyping of SNPs. The so-called 'SimpleProbe format'

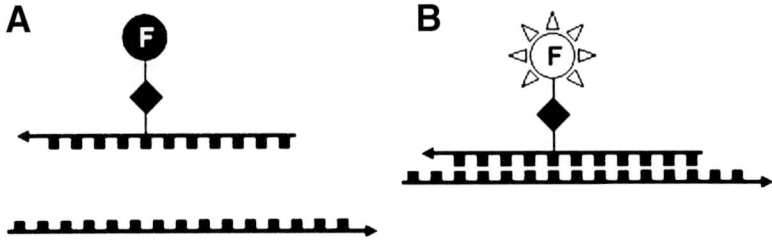

Fig. 5. Principle of SimpleProbe Detection. (**A**) Fluorescence is quenched by an internal linker when the probe is free in solution. (**B**) Fluorescence emission from reporter dye (F) when probe is annealed to its complementary target.

consists of a single oligonucleotide which is internally labeled with Fluorescein through proprietary linker chemistry. Signaling is dependent on the probe's hybridization status: once hybridized to its target sequence, the probe emits more fluorescence light than it does when it is free floating in solution (**Fig. 5**). As a result, a change in fluorescence is based exclusively on the hybridization status of the probe. The probe is designed to hybridize to a target sequence covering one or more SNPs of interest, thereby perfectly matching to a distinct allelic variant. Different allelic variants can be assigned measuring the duplex stability as a result of decreasing fluorescence with increasing temperature during a melting curve experiment. As a practical example simultaneous geno-typing of two SNPs in the promoter region of the TNF-α gene is shown (**Fig. 6**). The designed probe covers both polymorphisms -857C/T and -863C/A *(16)*. Resulting T_m values represent different allelic configurations of the individual SNPs as well as their haplotypic association (**Table 3**).

3.3.1. Design of SimpleProbe Oligonucleotides

Single-labeled SimpleProbe oligonucleotides have been developed for SNP genotyping experiments. Therefore, design considerations regarding probe/ target stability and mismatch instability are exactly the same as mentioned above for paired hybridization probes. Additionally, the following points should be taken into consideration for a successful probe design:

- The sequence that covers the polymorphism must be at least >3 nucleotides apart from either the 3′- or the 5′-terminus of the probe.
- The use of a sequence-internal label provides better signal dynamics when com-pared with simple end labeling. Never use the internal labeling reagent for terminal labeling.
- Position the internal label at least three bases apart from either terminus of the oligonucleotide.

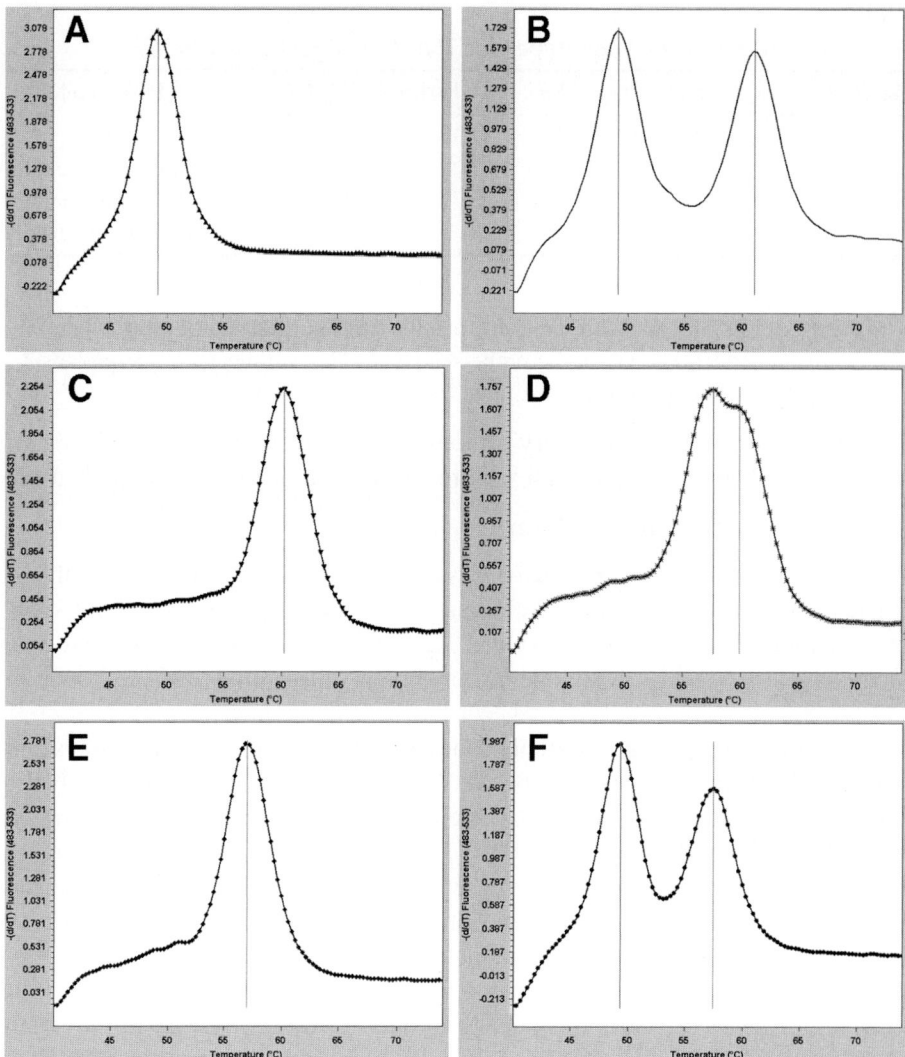

Fig. 6. Duplex-genotyping of TNF-α -857C/T and -863C/A via SimpleProbe melting curve analysis. Different combinations of haplotypes (diplotypes) are illustrated by the derivatives of individual melting curves: **(A)** iv/iv: -863A/A, -857C/C; **(B)** i/iv: -863C/A, -857C/T; **(C)** i/i: -863C/C, -857T/T; **(D)** i/ii: -863C/C; -857C/T; **(E)** ii/ii: -863C/C; -857C/C; and **(F)** ii/iv: -863C/A, -857C/C.

- Do not place the label too close to the polymorphic site. Place the label at least two bases up- or downstream of the potential SNP.
- Avoid reverse complementary dG nucleotides at the reverse complementary target strand in -1, 0, and +1 position, relative to the internal label.

Table 3
T_m Values and Resulting Haplotypes for TNF-α –857C/T and –863C/A

Haplotype	–863	–857	Estimated T_m (°C)	Probe match
I	C	T	60.2	Full match
Ii	C	C	57.3	Single mismatch
Iii	A	T	52.1	Single mismatch
Iv	A	C	49.0	Double mismatch

- Note that the internal fluorescent dye label is used as a substitution for a naturally occurring nucleotide within the probe sequence. This will decrease the probe's T_m value in dependency of the substituted base.

Note: In rare cases, secondary structure may affect proper functioning of single-labeled probes. In such cases, paired hybridization probes should be used.

3.4. Multiplex Genotyping Assays

The power of melting curve analysis is highly expanded by multiplex analysis. There are two principle ways of simultaneously analyzing two or more polymorphisms: multiplexing by color or by melting temperature *(17)*. An example for the analysis of two SNPs with one probe via T_m-discrimination is shown in Section 3.3.

For color multiplexing, different probes are labeled with different fluorescent dyes that have unique emission spectra. One limit for color multiplexing is the broad emission spectra of commonly available dyes. In general, nicely separated emission wavelengths of reporter dyes (e.g., LC-Red610 and Cy5) are favorable for the design of multiplexed reactions. Efficient resonance energy transfer occurs between the hybridization probe donor fluorescein and the acceptors LC-Red610, LC-Red640, and Cy5. Factors influencing the efficiency of resonance energy transfer are reviewed elsewhere *(18)*.

Mathematical correction of optical crosstalk resulting from spectral overlap of fluorescent dyes is necessary to eliminate non-channel specific parts of a measured signal. Color compensation calibration data for each pure fluorescent dye and auto-fluorescence negative control are obtained with fluorescence values acquired in each optical detection channel. These values are used to calculate signal crossover constants which are used to deconvolute measured fluorescence to actual channel-specific signal fluorescence. As fluorescence emission is also temperature dependent, crossover constants have to be calculated and applied over the complete temperature range of a melting curve.

An example for simultaneously genotyping of three SNPs covered by three independent PCR amplicons is shown in **Fig. 7**. Optical signals were recorded and color-compensated for detection channels 610 nm (LC-Red610 probe), 640 nm (LC-Red640 probe), and 670 nm (Cy5 probe).

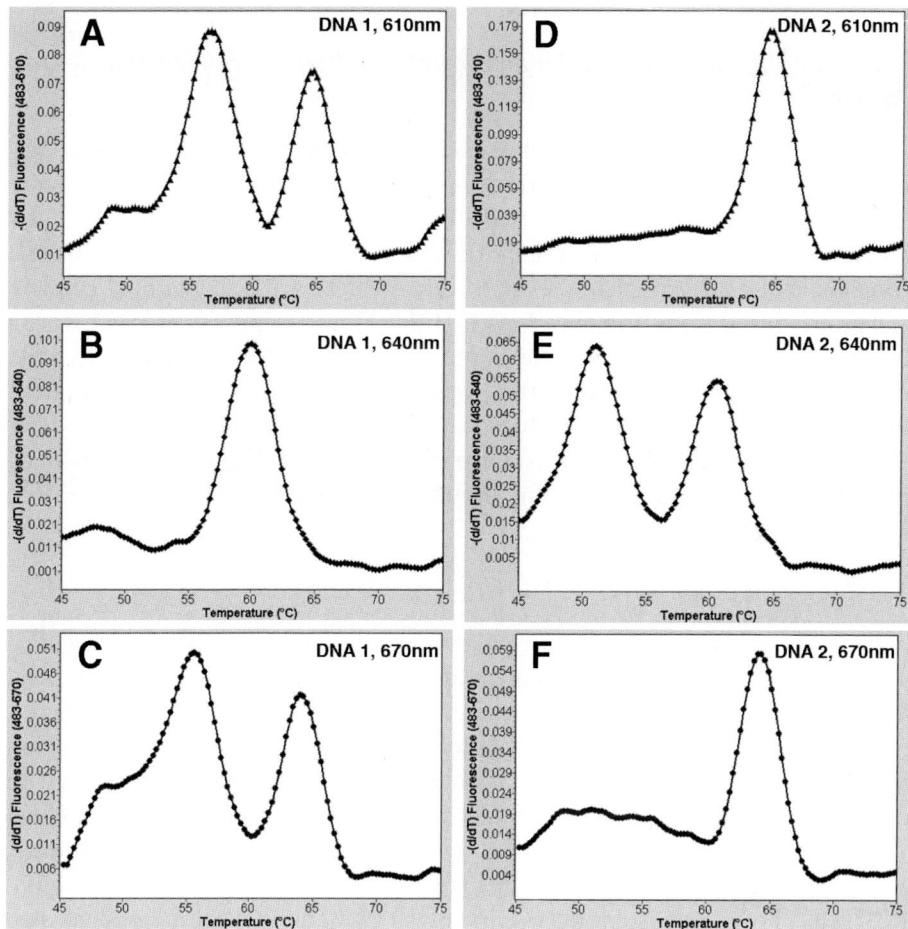

Fig. 7. Genotyping of SNPs in the genes for adiponectin receptor 1 (ADIPOR1) and 2 (ADIPOR2) by color multiplexing. A combination of three optical detection channels and dye-labeled probes is used for simultaneous genotyping of ADIPOR1-55649 C/T (channel 610 nm/LC-Red610), ADIPOR2-89539 C/T (channel 640 nm/LC-Red640), and ADIPOR2-91846 C/T (channel 670 nm/Cy5).

The results for two individual human genomic DNAs are displayed. DNA 1 is C/T heterozygous for ADIPOR1-55649 (A), C/C homozygous for ADIPOR2-89539 (B) and C/T heterozygous for ADIPOR2-91846 (C). DNA 2 is C/C homozygous for ADIPOR1-55649 (D), C/T heterozygous for ADIPOR2-89539 (E) and C/C homozygous for ADIPOR2-91846 (F).

3.5. Description of Protocols

3.5.1. Amplification of Human Hypoxanthine-Phosphoribosyl-Transferase (h-HPRT)

A plasmid construct, containing an inserted sequence of the human HPRT gene was used for serial dilution from 10^6 to 100 and 0.5×10^6-50 copies/ reaction, respectively. PCR was carried out in a total volume of 20 µL. Each reaction contained 10 µL master mix (LightCycler® 480 Probes Master; Roche Diagnostics), 5 µL template dilution, 2 µL detection mix containing primers and probes (used from 'LightCycler® h-HPRT Housekeeping Gene Set'; Roche Diagnostics), and 3 µL PCR grade water. After PCR setup, the multiwell plate was manually sealed. Real-time PCR was performed by an initial denaturation step of 10 min at 95°C followed by 45 cycles of amplification at 95°C for 10 s, 50°C for 20 s and 72°C for 20 s. At the end of each annealing steps fluorescence intensity was measured at 640 nm (single point measurement with automated selection of integration time).

3.5.2. Genotyping of TNF-α-1031 C/T

Genomic sequence information was derived from a public SNP database *(19)*. Ten nanograms of purified anonymous human genomic DNA were used together with salt, enzyme, and buffer concentrations recommended for the kit (LightCycler® 480 Probes Master; Roche Diagnostics). The final concentration of $MgCl_2$ was adjusted to 4 mM. Asymmetric PCR conditions were used to obtain an excess of single-stranded target molecules preferential for probe binding.

PCR primers were 0.5 µM (excess, reverse primer) and 0.025 µM (limiting, forward primer), hybridization probes were used at 0.2 µM final concentration. PCR was carried out in a total volume of 10 µL. The PCR protocol consisted of an initial denaturation step at 94°C for 10 min, followed by 40 cycles of amplification at 94°C for 10 s, 55°C for 15 s, and 72°C for 15 s. Samples were cooled at 40°C for 1 min after a final denaturation at 94°C for 10 s. For melting analysis, samples were slowly heated (1 acquisition/°C; integration time 2 s) from 45°C to 85°C. During heating emission of the LCRed640 detector probe was continuously monitored at 640 nm.

Sequences of primers and probes were as follows:

Oligonucleotide	Sequence/5'→3'
Forward primer	GGTCCTACACACAAATCAGTC
Reverse primer	CTTCTGTCTCGGTTTCTTCTC
Donor probe	AGGATACCCCTCACACTCCCCATC-fluorescein
Acceptor probe	LC Red 640-CCCTGCTC<u>C</u>GATTCCGAG-phosphate*

*SNP position is marked by an underscore.

3.5.3. Simultaneous Genotyping of TNF-α -857 C/T and -863 C/A

Genomic sequence information was derived from a public database *(20)*. The protocol was essentially the same as used for genotyping of TNF-α-1031 C/T (see above), except that primers were used at 0.5 μ*M* (excess, reverse primer) and 0.05 μ*M* (limiting, forward primer) final concentration. Melting curve analysis started from 40°C to 80°C final temperature (5 acquisitions/°C; integration time 0.5 s). During heating emission of the single labeled probe was continuously monitored at 533 nm.

The internally labeled detector probe was synthesized on a standard oligonucleotide synthesizer, using a phosphoramidite of the fluorescent labeled building block (SimpleProbe Labeling Reagent; Roche Diagnostics).

The following primers and probes were used:

Oligonucleotide	Sequence/5′→3′
Forward primer	GGTCCTACACACAAATCAGTC
Reverse primer	CTTCTGTCTCGGTTTCTTCTC
Detector probe	GGACCCCC**C**CT**X**AA**T**GAAGACAGGGC-phosphate*

*SNP positions are marked by underscores; X: internal fluorescent-dye label.

3.5.4. Simultaneous Genotyping of Three SNPs in the Genes for Adiponectin Receptors 1 and 2

As a model for three-color genotyping, one SNP in the gene for adiponection receptor 1 and two SNPs in the gene for adiponection receptor 2 were selected. Reference accession numbers for sequence information are given *(21)*.

For amplification, 20 ng of purified human geno mic DNA were used. Enzyme and buffer concentrations were applied as recommended for the used kit (LightCycler® 480 Master Probes; Roche Diagnostics). The final concentration of MgCl$_2$ was adjusted to 3.5 m*M*. Asymmetric PCR conditions were used to gain an excess of single-stranded targets.

PCR primers were 1.0 μ*M* (excess, forward primer) and 0.02 μ*M* (limiting, reverse primer) for ADIPOR1-55649, 1.0 μ*M* (excess, reverse primer) and 0.02 μ*M* (limiting, forward primer) for ADIPOR2-91846 and ADIPOR2-89539. All hybridization probe pairs were used at 0.2 μ*M* final concentration.

PCR was carried out in a total reaction volume of 20 μL. The PCR protocol consisted of an initial denaturation step at 94°C for 10 min, followed by 40 cycles of amplification at 94°C for 15 s, 55°C for 30 s, and 72°C for 45 s. Samples were cooled at 40°C for 1 min after a final denaturation at 94°C for 10 s. For melting analysis, samples were slowly heated (1 acquisition/°C; automatic integration time selection) from 45°C to 85°C. During heating emission of the detector probes labeled with dyes LC-Red610, LC-Red640, and Cy5 was continuously monitored at 610, 640, and 670 nm. For mathematical crosstalk correction of nonspecific signals in optical detection channels, a color-compensation file was generated. This file was created by single measurement of either single dye (fluorescein, LC-Red610, LC-Red640, Cy5) in all applied optical detection channels. Pure dye signals were recorded from 45°C to 85°C, thereby quantifying the nonspecific signal of dye/channel (e.g., signal of LC-Red610 in optical detection channels 640 and 670 nm). After melting curve analysis, the color-compensation matrix was applied to the non-corrected result of the multiplexed genotyping experiment, finally displaying crosstalk-corrected results valid for genotype assignment.

Sequences of primers and probes are listed in the table below:

SNP	Oligonucleotide	Sequence/5′→3′
ADIPOR1-55649	Forward primer	GAGCCTGACTGACATCTTCT
	Reverse primer	GGCTCTATAATCCCTTCGCA
	Donor probe	CCTTTTCCCCATCCAAATCACTCAC-fluorescein
	Acceptor probe	LCRed610-AGCAGATGGGTCCAGATGTTGC-phosphate
ADIPOR2-89539	Forward primer	AGTTTCGTTTCATGATCGG
	Reverse primer	CCATCCTTTGAGGTATGATTTAT
	Donor probe	CCACGCAGAGCCACTGG-fluorescein
	Acceptor probe	LCRed640-CGTCACAAAGCCCACGAGGC-phosphate
ADIPOR2-91846	Forward primer	GAGTTAATGGTGCTGTGCTG
	Reverse primer	CAATAGGGAGTTAGGACCAG
	Donor probe	GGCCTGAGTGGAATCATTCCTAC-fluorescein
	Acceptor probe	Cy5-TGCACTATGTCATCTCGGAGGGG-phosphate

*SNP position is marked by an underscore.

4. Notes

1. The reliability of the final genotyping result is heavily dependent on all steps of the respective workflow, such as the quality of the input template (integrity and absence of inhibitors) or the PCR assay (specificity and efficiency). The purity, concentration, and sequence of template, primers, and hybridization probes greatly affect the efficiency and specificity of the amplification reaction. Beside the optimized PCR primers amplification, the purity and concentration of template material play an important role.

2. For melting curve-based genotyping experiments asymmetric PCR conditions are often preferred as in later cycles more single-stranded template for probe binding

is produced. This reduces competition between re-annealing of PCR amplicon strands and formation of the probe/target complex. In our hands most genotyping protocols are run with ratios of limiting primer/excess primer of 1:5 up to 1:20.

3. Calculation of primers and probes T_m is depending on sequence and the algorithm used. Consequently, the actual annealing temperature of primers during PCR may be much higher or lower than the T_m calculated. To determine the actual primer annealing temperature, vary the target temperature in the annealing segment of the amplification program. For each trial, we recommend raising or lowering the temperature by 2–3°C and repeating the experiment. If a given annealing temperature produces significant nonspecific background (verified by gel electrophoreses), try a higher annealing temperature.

4. To optimize the detection with paired hybridization probes: for initial experiments, use 0.2 µM for each Hybridization Probe. Try concentrations up to 0.4 µM, alternatively for one or both probes, to improve signal intensity.

5. Hot-start chemistry (e.g., LightCycler® 480 Genotyping Master) for PCR amplification is preferable to prevent nonspecific reactions during PCR set-up.

6. Stock solutions of primers and fluorescent-labeled probes should be aliquoted in advance. Multiple thawing and freezing of the same solution should be avoided.

7. PCR setup can be performed by either manually or pipetting robot. In both cases, microwell plates should be sealed immediately after the last pipetting step by tightly pressing the adhesive seal on top of the plate. No liquid droplets should be on top of the microwells as they might disturb the contact between seal and microwell plate.

8. Before starting the PCR run, the sealed microwell plate should be centrifuged to avoid air bubbles in the liquid phase and/or prevent liquid from sticking to the wall of the microwell plate tubes.

9. If a multicolour experiment is planned, the fluorescent dyes have to be selected carefully to fit into the instruments optical abilities. If the distance between the emission and excitation spectra of the dyes are too small, the bleed over has to be considered.

10. Optical systems containing, e.g., optical filters, lenses, and dichroic mirrors will always show minor variations from instrument to instrument. These hardware variations influence crosstalk values. Therefore, color-compensation files must be used in an instrument-specific manner and not be shared between several instruments.

11. Colour compensation for temperature: the intensity and spectrum of emission from a fluorescent dye is strongly temperature dependent. Thus the temperature chosen for data acquisition during a PCR run significantly influences the crosstalk. This is particularly important during analysis of data measured at different temperatures, e.g., during a melting curve analysis. To provide crosstalk compensation at all temperatures, a color compensation calibration run must include a temperature gradient step (e.g., 40–85°C) in which data from all fluorescent dyes are continuously measured in all channels.

12. Since the spectral distance between channels in the LightCycler 480 system is short, channels might, in rare cases, pick up faint signals from neighboring channels,

even after color compensation activated. These signals are usually very small and just slightly above noise.

13. 'Hook effect': in some experiments there is a decrease in fluorescent signal after the exponential phase of PCR. This is caused by detection phenomenon which is not related to enzymatic or chemical degradation of PCR product. The effect is caused by competition between (i) the reassociation of the single strands of PCR product after denaturation and (ii) the binding of the hybridization probes to target in this late PCR phase. The PCR process amplifies both template strands, so while the amounts of PCR product are high, the two strands reassociate faster than the hybridization probes can bind to their target. The 'Hook effect' does not affect proper quantification or melting curve analysis.

References

1. Little, P. F. (2005) Structure and function of the human genome. *Genome Res.* **15,** 1759–1766.
2. Wong, M. L. and Medrano, J. F. (2005) Real-time PCR for mRNA quantitation. *Biotechniques* **39,** 75–85.
3. Brazeau, D. A. (2004) Combining genome-wide and targeted gene expression profiling in drug discovery: mircroarray and real-time PCR. *Drug Discov. Today* **9,** 838–845.
4. Chen, X. and Sullivan, P. F. (2003) Single nucleotide polymorphism genotyping: biochemistry, protocol, cost and throughput. *Pharmacogenomics J.* **3,** 77–96.
5. Sobrino, B. and Carracedo, A. (2005) SNP typing in forensic genetics: a review. *Methods Mol. Biol.* **297,** 107–126.
6. Roche Applied Science (2005) Assay Formats for Use in real-Time PCR. *Technical Note,* LC 18/20057.
7. Bernard, P. S. and Wittwer, C. T. (2000) Homogenous Amplification and Variant detection by fluorescent hybridization probes. *Clin. Chem.* **46,** 147–148.
8. Wittwer, C. T., Herrmann, M. G., Moss, A. A., and Rasmussen, R. P. (1997) Continuous fluorescence monitoring of rapid cycle DNA amplification. *Biotechniques* **22,** 130–131, 134–138.
9. Lyon, E. (2001) Mutation detection using fluorescent hybridization probes and melting curve analysis. *Expert Rev. Mol. Diagn.* **1,** 92–101.
10. Bernard, P. S., Pritham, G. H., and Wittwer, C. T. (1999) Color multiplexing hybridization probes using the apolipoprotein E locus as a model system for genotyping. *Anal. Biochem.* **273,** 221–228.
11. SantaLucia, J. Jr., Allawi, H. T., and Seneviratne, P. A. (1996) Improved nearest-neighbor parameters for predicting DNA duplex stability. *Biochemistry* **35,** 3555–3562.
12. Guo, Z., Liu, Q., and Smith, L. M. (1997) Enhanced discrimination of single nucleotide polymorphisms by artificial mismatch hybridization. *Nat. Biotechnol.* **15,** 331–335.
13. Roche Applied Science (1999) Selection of Hybridization Probe Sequences for Use with the LightCycler. *Technical Note,* LC 6/99.

14. Peyret, N., Seneviratne, P. A., Allawi, H. T., and SantaLucia, J. Jr. (1999) Nearest-neighbour thermodynamics and NMR of DNA sequences with internal A. A, C. C, G. G, and T. T mismatches. *Biochemistry* **38,** 3468–3477.

15. Szilvasi, A., Andrikovics, H., Kalmar, L., Bors, A., and Tordai, A. (2005) Asymmetric PCR increases efficiency of melting peak analysis on the LightCycler. *Clin. Biochem.* **38,** 727–730.

16. Haga, H., Yamada, R., Ohnishi, Y., Nakamura, Y., and Tanaka, T. (2002) Gene-based SNP discovery as part of the Japanese Millennium Genome Project: identification of 190,562 genetic variations in the human genome. Single-nucleotide polymorphism. *J. Hum. Genet.* **47,** 605–610.

17. Wittwer, C. T., Herrmann M. G., Gundry, C. T., and Elenitoba-Johnson, K. S. (2001) Real-time multiplex PCR assays. *Methods* **25,** 430–442.

18. Wu, P. and Brand, L. (1994) Resonance energy transfer: methods and applications. *Anal. Biochem.* **218,** 1–13.

19. http://www.ncbi.nlm.nih.gov. SNP database accession number: rs1799964.

20. http://www.ncbi.nlm.nih.gov. SNP database accession numbers: rs1799724, rs1800630rs.

21. http://www.ncbi.nlm.nih.gov. Gene bank accession numbers: gil26449126lgbl AC096632.3 (ADIPOR1; SNP at position 55649), gil4165008lgblAC005343.1 (ADIPOR2; SNP at position 91846), gil4165008lgblAC005343.1 (ADIPOR2; SNP at position 89539).

III

NEW TECHNOLOGIES

10

EasyBeacons™ for the Detection of Methylation Status of Single CpG Duplets

Ulf B. Christensen

Abstract

This chapter describes two different fast protocols for optimizing EasyBeacons™ for the challenging detection of the methylation status of a single CpG duplet in bisulfite-treated DNA. EasyBeacons™ can be used in multiplex detections even if they do not have the same affinity for their respective targets giving very specific results for both real-time polymerase chain reaction (real-time PCR) detection and endpoint analysis using inexpensive master mixes. The technology described in this chapter is a very competitive alternative to other real-time PCR technologies.

Key Words: HyNA™; EasyBeacons; real-time PCR; real-time; bisulfite-treated DNA; epignetic; CpG; methylation; qPCR; bisulfite; specific detection; hydrophobic nucleic acid; probe; SNP; expression; alleles; TaqMan; Mx3000; Mx4000; Rotorgene.

1. Introduction

The invention of quantitative real-time polymerase chain reaction (qRT-PCR) has revolutionized the way of determining the expression status of genes in a fast and precise manner. Despite many advances in the technology, there is still a need to find molecular probes that are easy to design and manufacture, highly specific and easy to implement. Some of the most widely used probes today are Molecular beacons, TaqMan® probes,[*] Scorpion primers and dual hybridization FRET probes (recently reviewed by Arya et al. and Stahlberg et al.) (1,2). Each of these technologies has advantages over each other in different applications. Common to all of these technologies is the difficulty of design, often requiring assistance from computer software or specialists, and prolonged optimization to work.[†]

[†]Programs like Primer Express® (www.appliedbiosystems.com), Beacon Designer (www.premierbiosoft.com) and Primer3 (http://frodo.wi.mit.edu/cgi-bin/primer3/primer3_www.cgi).

From: *Methods in Molecular Biology, Vol. 429: Molecular Beacons: Signalling Nucleic Acid Probes, Methods and Protocols* Edited by: A. Marx and O. Seitz © Humana Press Inc., Totowa, NJ

One of the major challenges for the molecular probes is to be able to discriminate between very similar targets, varying in as little as one nucleotide. Generally speaking, the shorter the probe is the larger influence on affinity one mismatch will have and hence the easier it would be to target a fully complementary target over a mismatched one. High-affinity DNA analogues and minor groove binders (MGBs) have previously been used to generate fluorescent probes that can be made shorter than the DNA-based probes. Kutyavin and co-workers conjugated a MGB molecule to the 3′-end of the probe and used it in a 5′-nuclease dependent PCR assay *(3)*. The same laboratory also described a system, called Eclipse, where the MGB is attached to the 5′-end of the probe and does not require nuclease activity of the polymerase *(4)*. In both of these systems, the polymorphic base (point of potential mismatch) should be placed approximately five nucleotides from the attachment site of the MGB, to yield an optimum discrimination between the matched and the mismatched target. However, unmodified DNA probes have previously been shown to give the highest difference in melting temperature and hence the highest specificity, if the polymorphic base is placed in the middle of the probe *(5,6)*. Furthermore, it is often necessary to have a high degree of flexibility in the probe design in order to avoid formation of unwanted secondary structures. Letertre et al. 2003 showed that locked nucleic acid (LNA) could be used in a 5′-nuclease PCR assay, with results comparable to the ones obtained with MGB TaqMan® probes *(7)*. It is, however, difficult to design optimal LNA-containing probes for two reasons. LNA nucleotides are nuclease-resistant, and the probe needs to be hydrolysed during the 5′-nuclease assay and LNA nucleotides have an extremely high self-affinity and hence if not carefully designed tend to form secondary structures. It has previously been shown that a DNA—LNA hybrid with only a few LNA nucleotides can form secondary structures that are so stable that the probe is unable to hybridize with complementary, single-stranded DNA targets *(8)*.

In this chapter, the author will describe a new type of molecular probe, called EasyBeacons™, containing the novel DNA analogue hydrophobic nucleic acid (HyNA™), which are easy to design and very specific probes. Even though some special HyNA™s have inherent fluorescence that previously has been shown to be able to detect single nucleotide variations in a manner that is independent of differences in affinities *(9)*, the author will focus on HyNA™s labeled with fluorophores and quenchers (EasyBeacons™) compatible with all the commercially available RT-PCR instruments. EasyBeacons™ are dual-labeled probes used for the diagnosis of expression, alleles, single nucleotide polymorphisms (SNPs), mutations and methylation status (epigenetic information). Although used routinely for the diagnosis of all these indications at our *in house* facilities, the scope of this chapter has been limited to describe the most challenging situation of them all, where the correct target only differs by a single nucleotide from non-specific targets and where the target is in a highly A-T rich environment, normally requiring longer probes. In other words, this chapter will describe the detection of methylation status of a single CpG duplet using EasyBeacons™.

1.1. Intercalating Nucleic Acids, HyNA™s

The novel features of EasyBeacons™ are introduced by the relatively new DNA analog—HyNA™. HyNA™ is composed of normal DNA nucleotides and hydrophobic stabilising modifications, HSMs (*see* **Fig. 1**). The fact that HyNA™ is mostly composed of normal DNA nucleotides means, that in many respects HyNA™ behave like DNA, making it easier to adapt HyNA™ to assays where DNA has previously been used. This includes the use of standard buffers and reagents and hence reduce costs and optimization efforts. The HSMs are incorporated between nucleotides (not substituting any of the normal DNA nucleotides in the sequence) into the backbone of the growing chain of HyNA™ on a DNA synthesizer, using standard chemistries and conditions *(10)*. The HSMs can be positioned anywhere in the oligonucleotide. Depending on whether the HSMs are placed at the end of a sequence or internally they will, when hybridized to a complementary target, function as either 'lids' on the duplex and/or as bulge insertions 'gluing' the two strands together, respectively (*see* **Fig. 1A**, and **B**).

1.2. EasyBeacons™

An EasyBeacon™ is composed of a fluorophore-like FAM, HEX, TET, VIC, or others and a quencher such as Tamra, Dabcyl, Black hole quencher (BHQ) or others normally linked respectively to the 5′- and 3′-end of an HyNA™. The EasyBeacons™ combine some of the advantageous features associated with conventional probes, like low background fluorescence, ease of design, high stability, high specificity, reduced self-affinity, and fast optimization. In addition, as EasyBeacons™ are not degraded during PCR amplification they can be used at optimal temperatures for the individual probes as well as for the polymerase, in a fast optimization cycle and in endpoint affinity measurements. These features and the usefulness of EasyBeacons™ in real-time detection strategies will be briefly explained below.

1.2.1. Low Background Fluorescence

When hybridized to a complementary target, the fluorophore and quencher of an EasyBeacon™ is separated in spatial distance allowing the fluorophore to fluorescence. The HSMs comprise flat hydrophobic molecules, which, when they cannot intercalate between base pairs in a complementary duplex, like to stack on top of each other having hydrophobic interactions. In this way, they facilitate an interaction between the π–electron clouds of the conjugated systems and minimizes contact with the water in the buffer. Hence, if there are no or fewer complementary targets than EasyBeacons™ in a mixture, the HSMs of the unbound probes will facilitate the two ends of the probe coming

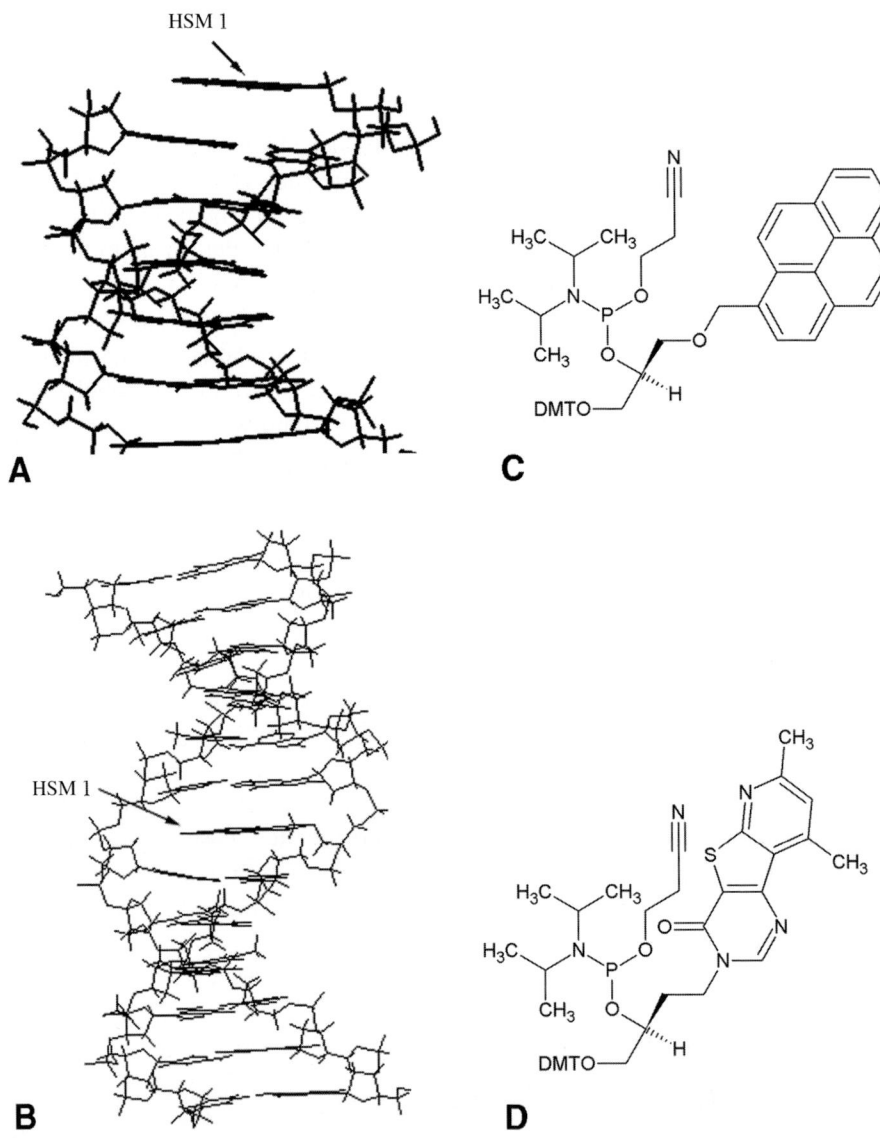

Fig. 1. (A) An HyNA™ oligonucleotide with an end-positioned hydrophobic stabilising modification (HSM). The HSM lies as a lid on top of the formed duplex. **(B)** An HyNA™ oligonucleotide with an HSM inserted in the middle of one strand. The HSM intercalates with neighboring base pairs of both strands "gluing" the strands together. **(C)** One of the two HSMs used in this study. This one is denoted as HSM 1 or "1" when mentioned as part of an HyNA™ sequence. **(D)** One of the two HSMs used in this study. This one is denoted as HSM 2 or "2" when mentioned as part of an HyNA™ sequence.

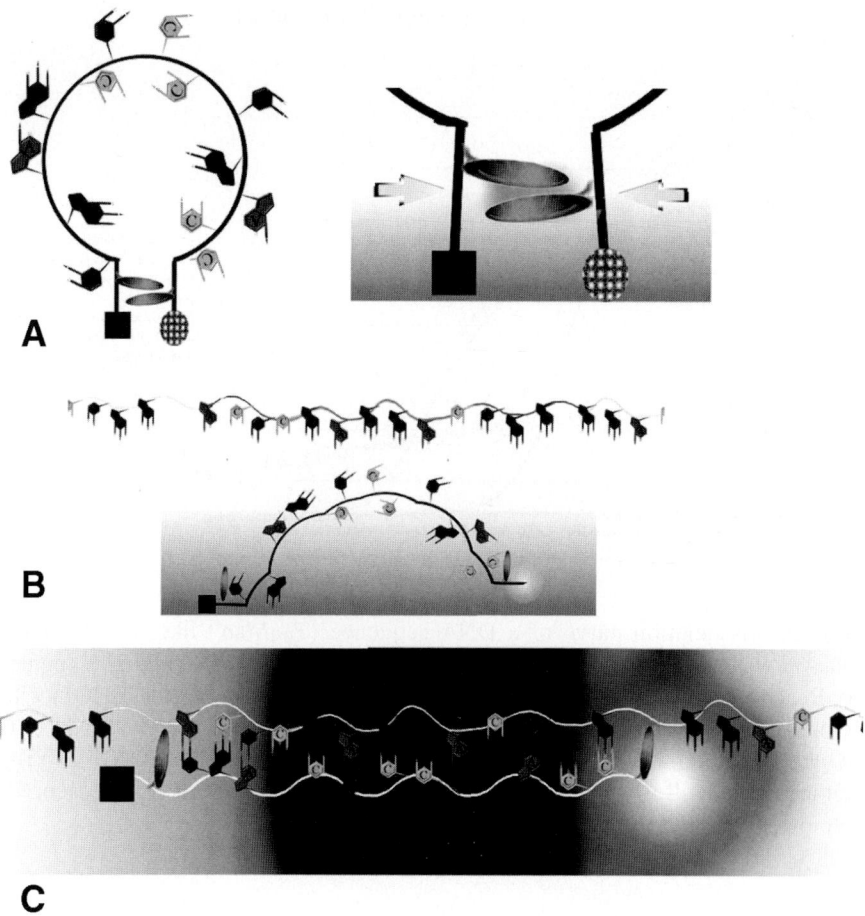

Fig. 2. (A) When no complementary target is present, the HSMs of the EasyBeacon™ like to interact with each other having hydrophobic interactions, bringing the fluorophore (dotted ball) and quencher (black square) in close proximity. Hence the fluorophore is quenched. **(B)** If a complementary target is added or formed for example during an amplification reaction, the EasyBeacon™ will unfold and bind to that. **(C)** When the EasyBeacon™ is binding to its complementary target, the fluorophore, and quencher are no longer in close proximity, hence the fluorophore will fluorescence.

into close proximity to each other, thereby facilitating quenching of the fluorophores (*see* **Fig. 2**). This folding back mechanism is independent of sequence and it is therefore not necessary to design a self-complementary stem structure to obtain a proper quenching of the molecule unlike conventional DNA-based molecular beacons.

Background Fluorescence

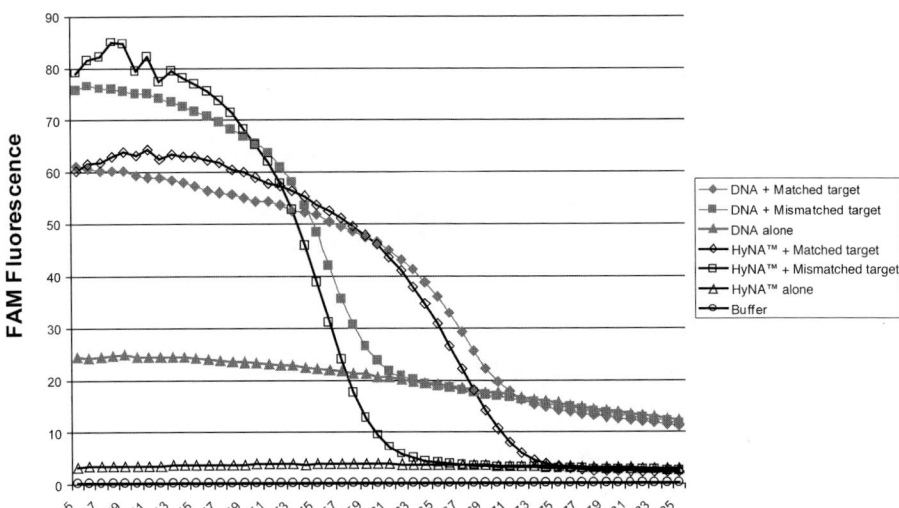

Fig. 3. Dissociation curve of a DNA sequence (TaqMan® like, filled markers, FAM-AAAAATCCCGCGAACTCC-BHQ1) and a comparable HyNA™ sequence (EasyBeacon™, hollow markers, FAM-A2AAA2ATCCCG2CGA2ACT-BHQ1) with 4 IPN insertions (type 2) hybridised to their matched (diamonds, GCGGGAGTTCGC GGGATTTTTTAG) and mismatched (squares, GTGGGAGTTTGTGGGATTTTTTAG) target respectively. It is clear that the background fluorescence of the EasyBeacon™ is much lower than that from the DNA probe.

1.2.1.1. EXAMPLE OF LOW BACKGROUND FLUORESCENCE

As can be seen from **Fig. 3**, the EasyBeacon™ has significantly reduced background fluorescence compared to the DNA probe (fluorescence of unbound probe) with both probes having the same maximum fluorescence (at 35°C). The two probes in the example are designed to distinguish between a methylated and unmethylated target of a bisulfite-treated DNA sample. The preferred reading temperature for that probe during a real-time PCR run is between the melting temperature of the probe against the fully matched target (67°C) and the mismatched target (56°C), let us in this case say midway (62°C). It is important that the probe has a good signal-to-noise ratio at the reading temperature. The signal-to-noise ratio is calculated as follows:

$$\frac{\text{Signal (Probe + Matched target) at annealing - Buffer signal at annealing}}{\text{Signal (Probe + Mismatched target) at annealing - Buffer signal at annealing}}$$

In the example shown in **Fig. 3**, the signal-to-noise ratios at 62°C would be:

$$\text{S/N ratio DNA } \frac{43.2-0.2}{20.7-0.2} = \underline{\underline{2.1}}$$

$$\text{S/N ratio EasyBeacon } \frac{41.1-0.2}{5.9-0.2} = \underline{\underline{7.0}}$$

1.2.2. High Affinity—High Specificity

Due to the high affinity of HyNA™ towards complementary DNA, EasyBeacons™ can be made shorter than their DNA counterparts. Generally speaking, the shorter the EasyBeacon™ is, the better it is to discriminate between single point mismatches. One should keep in mind however that, in order to keep the specificity of the PCR, it is generally not recommended to go below 45°C in annealing temperature. The length of the EasyBeacons™ should, among other considerations, be determined by how hard it is to detect the wanted target over all other possible targets present and the primers annealing strength (if this is restricted or if the primers are already available). A typical EasyBeacon™ comprises between 7 and 18 DNA nucleotides, one fluorophore, one quencher, and 2 to 4 HSMs.

1.2.2.1. EXAMPLE OF HIGH AFFINITY AND SPECIFICITY OF EASYBEACONS™

Figure 4 demonstrates that HyNA™ modified probes (EasyBeacons™) can be designed to achieve a higher specificity than their DNA counterparts. In the example shown, the specificity is 77% higher for the HyNA™ over the DNA, measured as the difference in affinity to a matched versus a mismatched target.

Expression probes do usually not have to meet the same stringency as probes for SNPs, mutations or single CpG methylation detections and hence can be longer. It should be noted that though this chapter is about the detection of methylation of single CpG duplets, it is often possible to design the probe to span several CpG duplets, which are often reported to be co-methylated, improving the probes ability to discriminate between a fully matched target, and targets with several mismatches. However, it is necessary to verify that the CpG duplets spanned are indeed methylated, and not to anticipate that two closely positioned CpGs will be methylated.

1.2.3. Low Self-Affinity—High Specificity

The specificity of the probe is also very dependent on the nature of the mismatch. SantaLucia demonstrated a clear trend in order of decreasing stability: 'G-C' > 'A-T' > 'G·G' > 'G·T' ≥ 'G·A' > 'T·T' ≥ 'A·A' > 'T·C' ≥ 'A·C' ≥ 'C·C', meaning that 'G' is the most promiscuous base, since it forms the strongest base pair and the strongest mismatches *(11)*. On the other hand, 'C' is the most

Specificity

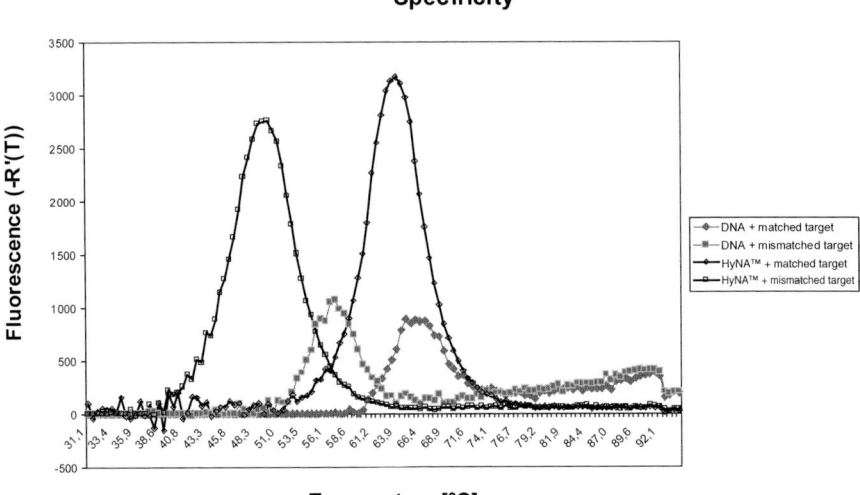

Fig. 4. The negative first derivative to the dissociation curve of an EasyBeacon™ (hollow markers, FAM-T2AGG2GCGT2TTTT2T-BHQ1) compared to a DNA, TaqMan®-like probe (solid markers, FAM-ATTTTAGGGCGTTTTTTTG-BHQ1) annealed to their matched target (diamonds, CCGCAAAAAAACGCCCTAAAATCCC) and mismatched target (squares, CCGCAAAAAAACACCCTAAAATCCC). Even though that there are six more complementary nucleotides in the DNA, the HyNA™ binds almost as strongly to their complementary target but the binding is more specific. The HyNA™ hybridizes at 63.7°C and 49.9°C to its matched and mismatched targets, respectively, giving a 13.8°C difference, while the DNA hybridizes at 64.9°C and 57.1°C to its matched and mismatched target, respectively, giving only 7.8°C difference. The higher signal for the EasyBeacons™ is due to their lower background fluorescence.

discriminating base, since it forms the strongest pair and the three weakest mismatches followed by 'A'. Hence the order of specificity of the single nucleotides is as follows: 'C' > 'A' > 'T' > 'G'. As described below, the methylation status of DNA is translated into a 'C'/ 'T' mismatch in bisulfite-converted DNA on one strand and conversely a 'G'/ 'A' mismatch on the complementary strand after PCR amplification of the converted DNA. It would therefore be most discriminatory if one probe (specific for methylated DNA) comprises the highly selective 'C' base at the polymorphic site, and the other probe (specific for unmethylated DNA) comprises an 'A' at the polymorphic site. Whilst this is most efficient, it means that the two probes will be nearly complementary to each other. This would not be advantageous when using DNA and most DNA analogs, however as HyNA™ binds stronger to its comple-mentary DNA target than to a complementary HyNA™, the EasyBeacons™ are extremely well suited to such a design (*see* **Fig. 5**).

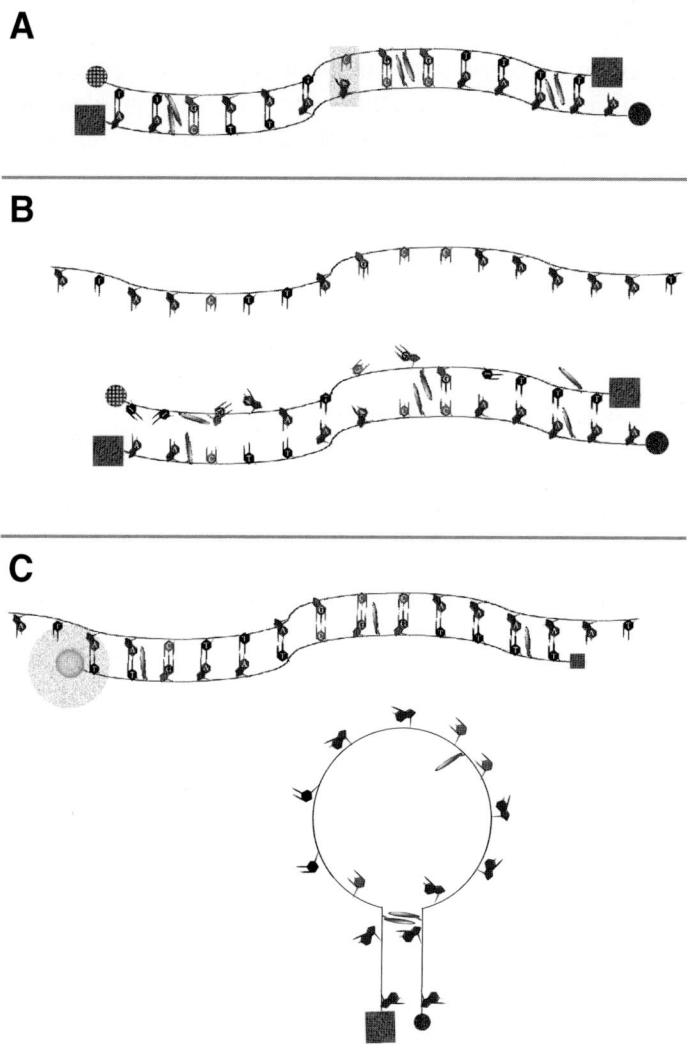

Fig. 5. (A) The partly complementary EasyBeacons™, when there is no or fewer complementary DNA targets present. The two probes binds to each other and quench the fluorescence of the probe sets. The point of methylation detection (the polymorphic site) and the mismatch between the probes is highlighted with a faith grey bar. **(B)** If a complementary DNA target is added or formed during an amplification step, the EasyBeacon™ will preferentially bind to that over the other EasyBeacon™. **(C)** The result will be the complementary EasyBeacon™ binding to its target, the fluorophore and quencher being spatially separated and hence a fluorescent signal generated. This unhybridised EasyBeacon™ folds back on it self resulting in the fluorophore and quencher coming into close contact and the fluorescent signal quenched.

1.2.4. Multiplexing

This chapter only describes the multiplexing of two probes in one tube using one set of primers. The possibility for a higher degree of complexity of multiplexing is possible on many of the newer real-time PCR instruments, where up to five or six different fluorophores can be measured at a time. We believe that multiplexing with HyNA™s will be than with other technologies, mainly because of the shorter length of both primers and probes, reducing the risk of nonspecific hybridization and furthermore because HSMs added internally in a sequence will reduce the risk of the sequence acting as a non-specific template permitting spurious primer extension. Further features of the HyNA™s are described below in the section of designing EasyBeacons™, where these features are put into context.

2. Materials

2.1. Bisulfite Conversion of DNA

- Purified genomic DNA sample(s) of interest.
- MethylEasy (Human Genetic Signatures Pty Ltd, www.methyleasy.com).
- Freshly made 3 M NaOH solution.
- 100% isopropanol (molecular biology grade).
- 70% ethanol (molecular biology grade).
- Water (molecular biology grade).
- Glycogen (Roche cat no. 901 393).
- Cooling or chilled centrifuge (4°C).
- Heated-lid normal or real-time PCR instrument, oven or heating block.

2.2. Real-Time Detection of Methylation Status

- Real-Time PCR instrument.
- Promega mastermix 2× (or any other functional master mix).
- bisulfite-treated template as described in Section 3.2.
- DNA or HyNA™ forward and reverse primers designed as explained in Section 3.3.
- Two EasyBeacons™ per target complementary to methylated and unmethylated DNA respectively, designed as explained in Section 3.4.
- Commercial producers of EasyBeacons™ can be found on the following website: www.EasyBeacons.com and of HyNA™ SuPrimers™ can be found on the following website: www.PentaBase.com.

3. Methods

3.1. Methylation Status Determination—Introduction

Methylation of 'CpG' duplets in the genome is a key component of epigenetic information and has shown to play an important role in imprinting as well as in a range of clinical situations, including mental disorders, cancer biology, and viral infections *(12)*. The commonly used technique for sensitive detection of the methylation status of a gene is: 1) Treating the genomic DNA with sodium bisulfite, converting all unmethylated cytosines 'C' into deoxyribose uracils 'dUs' while methylated cytosines '^{Me}C' are unreactive. 2) Amplify the bisulphite converted DNA. The bisulfite treatment of the genomic DNA converts the DNA into an 'A-T' rich, essentially three-nucleotide genome (with the exception of '^{Me}C' present in methylated 'CpG' duplets) and the difference of a methylated or unmethylated 'CpG' duplet is translated into a sequence difference. The DNA can be analyzed post-amplification, by sequencing, restriction enzyme digestion, or by the presence or absence of bands on a gel after gel electrophoreses when using methylation-specific PCR, (MSP) *(13)*. Alternatively, the DNA can be analyzed real-time using MSP and an unspecific dye or as described herein using primers and methylation-specific probes.

The 'A-T' rich nature of the bisulfite-treated DNA provides a special challenge to primers and probes in the amplification assay due to the relatively low affinity of 'A-T' rich DNAs and the difficult nature of distinguishing between a single 'C' and 'T' nucleotide.

3.2. Target Selection for Methylation Status Determination

Select the DNA region where you want to detect the methylation status and chose either the sense or antisense strand for designing the experiment. Please note that after bisulfite treatment, the sense and antisense strand are no longer complementary, and it is therefore important that either the sense or the antisense strand is used for designing all primers and probes. *In silico* bisulfite treat the sequence to generate one sequence where all the 'CpG' are read as TpG (unmethylated) and one sequence where all the CpG remain (methylated). Hence all 'C' should be converted to 'T' unless they are followed by a 'G' (a 'CpG' duplet, *see* **Fig. 6**). The referenced home page offers a free *in silico* conversion of DNA sequences.‡

3.2.1. Stepwise bisulfite Converting Genomic DNA

The Methyleasy kit by Human Genetic Signatures was used for all bisulfite conversions of the DNA (*see* **Note 1**). Even though the kit is provided with an easy to follow manual, the steps are reported here as well. The main differences between what is written here and the manual are some simplified steps, a common method with two incubation temperatures depending on the amount of starting DNA and different PCR conditions optimized for RT-PCR analysis.

1. Dissolve between 100 pg and 4 μg of purified DNA in 20 μL water.
2. Add 2.2 μL of a freshly produced 3 *M* NaOH (100 mg NaOH in 830 μL water).
3. Incubate at 37°C for 15 min.
4. Add 220 μL combined Reagent 1 and Reagent 2, mix by gentle pipetting.

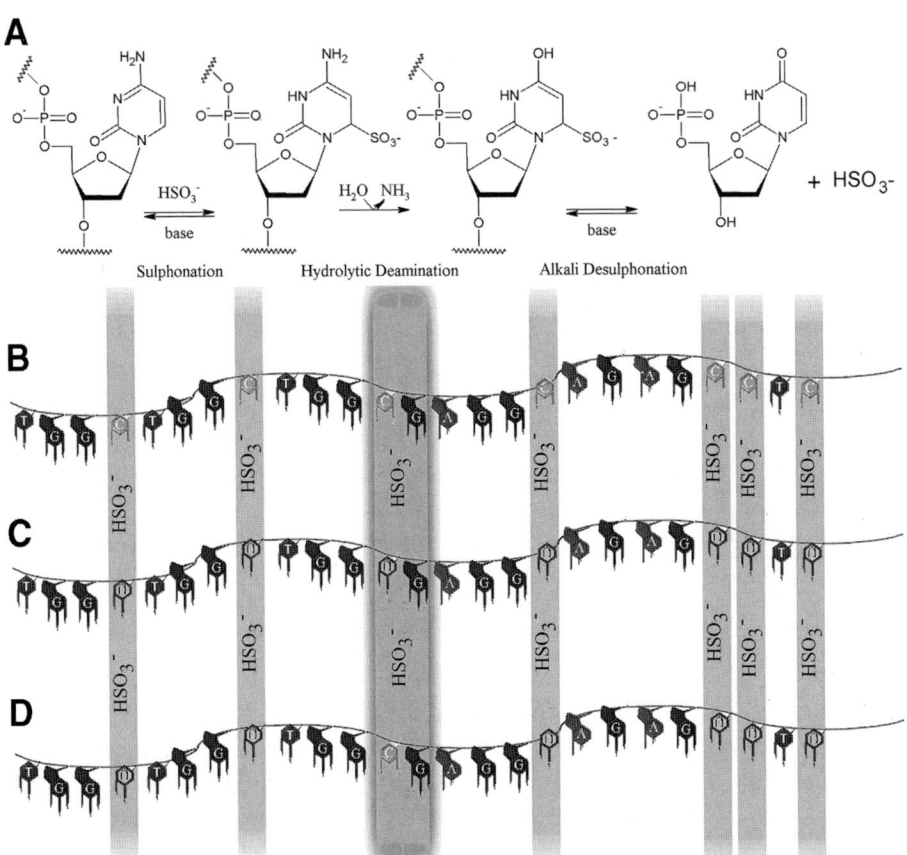

Fig. 6. This illustrates the bisulfite treatment of genomic DNA resulting in all unmethylated deoxyCytosines (dCs) being converted to deoxyUracils (dUs), while leaving the methylated dCs unchanged as dCs. (**A**) The chemical reaction of sodium bisulfite with dC, (**B**) genomic DNA, (**C**) bisulfite-treated unmethylated DNA and (**D**) bisulfite-treated DNA containing a single methylated CpG duplet. The thick bar in the middle highlights the CpG duplet, while all other bars highlight the positions of dCs in the genomic DNA and their resultant conversion to dUs.

5. Incubate in an oven or a PCR machine (with hot lid) at 37°C (for starting amounts up to 50 ng) or at 55°C (for starting amounts from 51 ng to 4 µg) for 4–16 h while shielding the samples from light exposure. Alternatively, the sample can be overlaid with 200 µL of mineral oil and incubated in a heating block and shield from light exposure. Remove mineral oil after incubation.
6. Add and mix 2 µL of glycogen by pipetting.
7. Add 800 µL of Reagent 4 and mix well by pipetting up and down at least 10 times.

8. Add 1 mL of pure isopropanol very slowly, by gradually dispensing and take-up, then Vortex for 5 sec.
9. Incubate at 4°C for 1 h.
10. Centrifuge for 15 min at 15,000 rpm.
11. Very carefully remove the supernatant, making sure not to dislodge the pellet.
12. Add 500 µL of 70% ethanol.
13. Centrifuge for 10 min at 15,000 rpm.
14. Remove the supernatant and allow to air dry for 15 min.
15. Resuspend the pellet in Reagent 3 to make a 2 ng/µL or if starting from less than 2 ng of DNA or needs to run more PCRs make a 100 pg/µL or more concentrated solution. (For long-term storage, *see* **Note 2**).
16. Incubate the sample at 72°C for 60 min in an oven or a PCR machine (with hot lid). Alternatively, use a heating block and spin down at least one during incubation.
17. Use 1 µL for PCR.

Controls are provided with the kit and should be used as described by the manufacturer. It is not recommended to follow the PCR conditions given in the manual, as these are not optimized to real-time detection rather use the conditions described below.

3.3. Designing Primers for Methylation Status Determination

Doing PCR on bisulfite-treated DNA is not a trivial task. Even though it is often possible to use normal DNA primers, we routinely use SuPrimers™ (HyNA™-based primer). SuPrimers™ can be made shorter, have higher sensitivity and are more robust (a broader range of annealing temperatures can be used in the assay) http://www.PentaBase.com. In case of inadequate sensitivity, a nested PCR can be used to pre-amplify the target after the bisulfite conversion (*see* **Note 1**).

The primers should be designed to give an optimal amplification of the target template without yielding unwanted nonspecific products. The primers should have a melting temperature (T_m) higher than the T_m of any of the predicted template secondary structures to ensure that the majority of possible secondary structures have been unfolded before the primer-annealing step. Furthermore, it is important that the two primers have similar T_m values (*see* **Notes 12,15** and **16**).

3.3.1. Stepwise Design of Primers for bisulfite Treated DNA

Issues relevant to the design of primers have been addressed by primer design software.[†] Alternatively, you can follow the rules outlined below:

1. Use one strand of bisulfite-treated DNA as template for designing both primers.
2. Use a sequence identical to part of the bisulfite treated strand as the forward primer and a sequence complementary to the bisulfite-treated strand as the reverse primer.

3. Design either primers with a 20–30% 'G' or 'C' content (dependent on if it is the forward or reverse primer, respectively) to ensure amplification of fully converted DNA.
4. Maintain a T_m for the primers between 48°C and 65°C.
5. Both primers shall have the approximate same T_m.
6. Amplicons should be as short as possible and max. 400 bp.
7. Avoid strong secondary structures like hairpin formation (more than two to three complementary basepairs).
8. Avoid repeats of 'G' or 'C' longer than three bases.
9. Span several bases that are converted by the bisulfite treatment ('C' in the genomic DNA sequence of the forward primer and 'G' in the genomic DNA sequence of the reverse primer) at least one at or close to the 3'-end of the primer (specificity for only converted DNA).
10. Avoid spanning 'CpG' duplets if possible (*see* **Note 4**).
11. To avoid primer—dimers check the two primer sequences against them selves and each other for 3'-complementarities.
12. Avoid long runs of 'A-T' nucleotides in the primers.

3.4. Designing EasyBeacons™ for Methylation Status Determination

When designing EasyBeacons™ for methylation status determination, choose two different fluorophores that are compatible with your RT-PCR instrument and that do not have a spectral overlap and pair them with their respective optimal quenchers. In this study, the following two pairs are used FAM-BHQ1 and HEX-BHQ1, but other combinations can also be used (refer to your instrument manufactures guide). Place the fluorophore in the 5'-end and the quencher in the 3'-end. Avoid having a 'G' nucleotide next to the fluorophore as this can quench the fluorescence and thereby reduce the signal-to-noise ratio.

The first probe complementary to the methylated target should contain a 'C' nucleotide at the potential site of methylation, while the second probe specific to the unmethylated target should be complementary to the first probe, but contain an 'A' nucleotide at the potential site of methylation (*see* **Note 5**).

An EasyBeacons™ is typically between 7 and 18 DNA nucleotides long and contains up to 4 HSMs (*see* **Note 6**). The length of the probe is determined by the 'CG/AT' ratio and the purine/pyrimidine ratio, the lower the ratios the longer the probe has to be. The shorter the probe is the more specific it is, however when used as a real-time probe, it should still be long enough to hybridize to the complementary target at the annealing temperature used in the PCR run. A typical length of the unmethylated specific EasyBeacon™ is 13 DNA nucleotides and 3 HSMs long, while the methylation specific EasyBeacon™ is approximately 12 DNA nucleotides and 3 HSMs long. The reason that the unmethylated-specific probe is longer is that it contains an 'A' base instead of a 'C' base pair in the middle of the probe ('A-T' base pairs are weaker than 'C-G' base pairs). The difference in affinity is also dependent on the sequence surrounding the mismatch. Remember that if the probe is designed in the lower end of the specified lengths then the primers should also be designed to anneal at low temperatures.

In order to avoid interference with the nucleotides specificity at the site of methylation, place the HSMs so that they are not next to any of the potential methylation sites, hence at least one preferably two nucleotide between every HSM and the position(s) of potential methylation. To avoid any quenching of the fluorescent signal from the fluorophore, it is also important to separate the fluorophore and the nearest HSM with at least one, preferable two nucleotides. Furthermore, in order for the HSMs to give maximal affinity for the complementary DNA, position every HSM so that at least three nucleotides separate all and each HSMs. Lastly it is important that as many HSMs as possible are positioned opposite each other or as close as possible to each other, if the two nearly complementary probes are paired, as this ensures that the EasyBeacons™ will have a higher affinity for their complementary DNA targets than for the other probe.

3.4.1. Stepwise Designing of EasyBeacons™

By following the steps described below, you will design two EasyBeacons™ one that is specific for methylated target DNA and the other for unmethylated target DNA.

1. Use one strand of bisulfite-treated DNA as template for designing both EasyBeacons™ (the same as used for designing the primers).
2. Chose a fluorophore and its matching quencher for one probe and another fluorophore (with no spectral overlap) and its matching quencher for the other probe (the fluorophore is incorporated in the 5′-end and the quencher in the 3″-end).
3. Chose the sequence that best fulfils the following rules.
   a. The fluorophores shall not be positioned next to a 'G' nucleotide in the probe.
   b. The polymorphic site shall be close to the centre of the probe.
   c. The length of the probes shall be as short as possible, but still able to hybridize to their fully complementary targets at the used annealing temperature (default length is 13 DNA nucleotides and 3 HSMs).
   d. The probe shall span at least one 'CpG' duplet for which the methylation status is important for the scientist.
4. Add up to 4 HSMs to the sequence using the following rules.
   a. At least have one, preferably two nucleotides between the fluorophore and the first HSM.
   b. Separate all HSMs by at least three nucleotides.
   c. Do not place HSMs immediate next to the polymorphic site.
   d. Place the HSMs so that they will be positioned opposite or close to opposite each other when the two probes are allowed to hybridize.
5. The EasyBeacons™ are ready for synthesis and can be synthesized or ordered at www.EasyBeacons.com.

3.5. Optimization of Primers Annealing Temperature

Depending on the scope of the experiment, it may be sufficient for the primers to work in a non-optimized fashion (if you are looking for the presence of methylation or not) otherwise extensive care will have to be taken in the optimization process (if the level of methylation is important). It is sometimes necessary to optimize the primers before synthesizing or purchasing the probes. A general rule-of-thumb is to use an annealing temperature that is 5°C above the calculated melting temperature for the primers or otherwise obtain an empirical and more precise annealing temperature by optimization using a gradient thermal cycler. Alternatively, multiple PCRs have to be setup and dilution series have to be made.

3.6. Fast Optimization Cycle of Probe Reading Using Gradually Increasing Temperature

The protocol described in this section is for real-time PCR instruments that are capable of making multiple readings (more than four readings in a cycle). Instruments in this category includes Stratagene's real-time PCR instrument Mx3000P, which has been used in this study. A few other real-time PCR machines also support this function (*see* **Note 7**).

Both probes can be optimized in a single run, where the temperature is gradually increased from the annealing temperature measuring the fluorescence at an interval of 2°C. This fast optimization can only be carried out because the EasyBeacons™ are not hydrolysed by a 5′-nuclease activity of the DNA polymerase. Gradually increasing the melting temperature and reading the fluorescence tells exactly at what temperature the probe stops binding to the mismatched target but still binds to its matched target. In this way two annealing intervals are obtained, one for each probe (*see* **Note 8**). One of the properties of EasyBeacons™ that make them so easy to design is that it is not necessary to use a single temperature for annealing and fluorescence measure-ment, but multiple temperatures may be used. As long as the EasyBeacons™ are able to distinguish between its matched target and the mismatched target at a temperature at or higher than the annealing temperature used for the primers the probes will work fine.

The described method in this section can of course be used as a standard method for detecting the methylation status of a DNA sequence, however if the measurement is going to be repeated multiple times, it would be faster and less demanding on the instrument to use the annealing temperature and two points for measuring the fluorescence of the individual probes. At this point the measurements can also be used on other machines able to read at just two stages during a typical thermal cycling protocol.

3.6.1. Stepwise Optimization of the Individual Probes in One Run

By using the procedure described below, it is possible to test if the primers work at the selected annealing temperature with the template whilst also optimizing the reading temperature for both probes individually. The measurement itself can be used for distinguishing between methylated and unmethylated DNA.

1. To an optical PCR-tube add 10 µL of 2× PCR Mastermix, 1 µL of the forward primer (10 pmol/µL), 1 µL of the reverse primer (10 pmol/µL), 1 µL of a the EasyBeacon™ (5 pmol/µL) specific for methylated DNA, 1 µL of a the EasyBeacon™ (5 pmol/µL) specific for unmethylated DNA, 5 µL of H_2O, and 1 µL of bisulfite-treated DNA (100 pG–20 nG/µL) (*see* **Notes 12** and **17**).
2. If available make two tubes for each target of interest as described above, one with methylated bisulfite-treated DNA (methylated control), and one with unmethylated bisulfite-treated DNA (unmethylated control). If these controls are not available, just prepare the tubes with some unknown sample (*see* **Note 9**).
3. Close the tubes with optical lids (if using a RT-PCR machine that measures the fluorescence through the lid, as most models do) or just a standard lid, if using a machine that reads through the side or bottom of the tube.
4. Place the tubes in the RT-PCR machine and run the following program (shown in **Fig. 7**):
   a. If using a hot-start polymerase, activate this according to manufacturer's description, otherwise just heat at 95°C for 1 min to get rid of secondary structures.
   b. Follow the activation by 45 cycles using the following steps (*see* **Note 10**):
      i. Denature at 95°C for 30 sec.
      ii. Anneal the primers and probes (e.g., at 50°C) for 15 sec, endpoint or continuously fluorescence measurement.
      iii. Increase the temperature by 2°C and make endpoint fluorescence measurement. Use as short a time interval as possible (dependent on the number and positions of samples).
      iv. Repeat step ii until approx. 70°C is reached.
      v. Elongate at 72°C for 30 sec.
5. Analyze data—The optimal fluorescence reading temperature for that probe to use in future RT-PCR experiments of that target is where the EasyBeacons™ hybridized to the fully complementary target only and has as high a signal as possible. A separate reading temperature is found for both probes (*see* **Note 18**).

An example of the difference in fluorescence at different temperature stages and how this protocol is used for optimizing the future RT-PCR runs of that particular target can be seen in **Fig. 8**.

3.3. Fast Optimization Cycle Using Endpoint Affinity Test

This section describes how an endpoint affinity measurement can be used for optimizing EasyBeacons™. This method is preferred when multiple readings

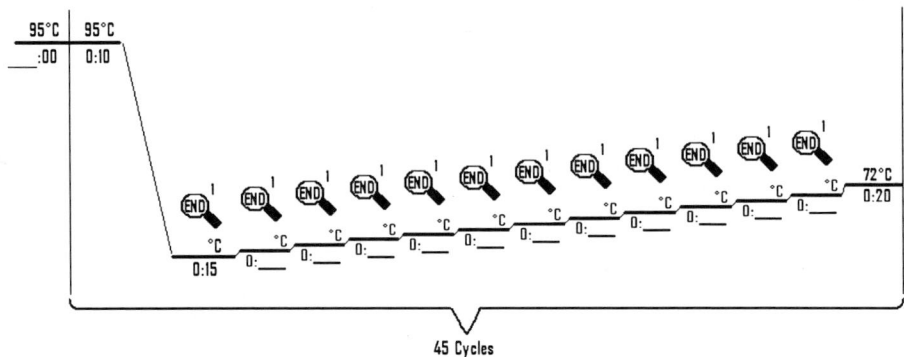

Fig. 7. The amplification profile with gradual increase of temperature incorporating fluorescence measurements at each stage. Initially, the DNA polymerase is activated according to the manufacturer's instructions. Next 45 cycles of PCR amplification is performed. The cycles comprise the annealing and first measurement (e.g., 50°C), after which the temperature is gradually increased in 2° increments measuring the fluorescence at each stage. This is repeated until the normal extension temperature of 72°C is reached. The duration of each measurement stage should be as short as possible. The magnifying glass with "END" written in it and a '1' at the border, illustrates that the fluorescence is measured one (1) time at the end of the stage.

during the amplification cycle are not possible, and is especially suitable on RT-PCR instruments with high ramp rates such as e.g., Corbett Life Science's Rotorgene™ 3000, Corbett Life Science's Rotorgene™ 6000, Applied Biosystems 7500 Fast Real-Time PCR System, and other fast ramping RT-PCR instruments with a ramp (cooling) rate of ≥ 3°C/sec, but can also be carried out on standard RT-PCR instruments. Measurements based on this procedure in this chapter were carried out on a Corbett Research's Rotorgene™ 3000.

1. To an optical PCR-tube is added 10 µL of 2× PCR Mastermix, 1 µL of the forward primer (10 pmol/µL), 1 µL of the reverse primer (10 pmol/µL), 1 µL of a the EasyBeacons™ (5 pmol/µL) specific for methylated DNA, 1 µL of a the EasyBeacons™ (5 pmol/µL) specific for unmethylated DNA, 5 µL of H_2O, and 1 µL of bisulfite-treated DNA (100 pG–20 nG/µL) (*see* **Notes 12 and 17**).
2. If available make two tubes for each target of interest as described above, one with methylated bisulfite-treated DNA (methylated control), and one with unmethylated bisulfite-treated DNA (unmethylated control). If these controls are not available, just prepare the tubes with some unknown sample (*see* **Note 9**).
3. Close the tubes with optical lids (if using a RT-PCR machine that measures the fluorescence through the lid, as most models do) or just a standard lid, if using a machine that reads through the side of the tube as the Rotorgene™, Corbett Life Science, does.

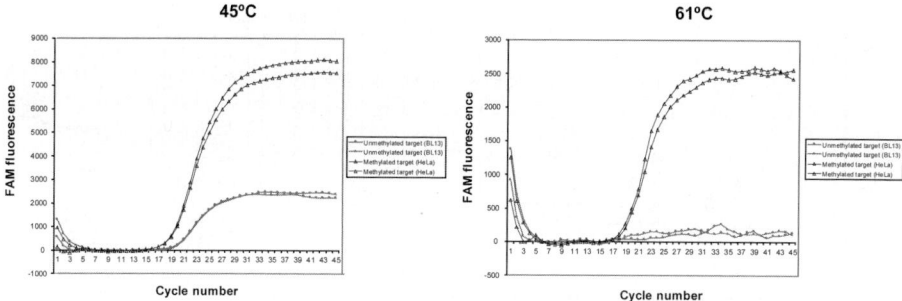

Fig. 8. An example showing how it is possible to differentiate fluorescence from the fully complementary (methylated DNA) and the mismatched target (unmethylated DNA). At 45°C up to 59°C the methylation-specific probe (FAM-A2AAA2ATCCCG2 CGA2ACT-BHQ1) also binds to a small extent to the mismatched target, but at 61°C and up to 71°C the binding of the probe is specific for the fully complementary target. In future measurements, a reading point of 61°C could be used for this probe. The polymorphic sites are highlighted in grey.

4. Place the tubes in the RT-PCR machine and run the following program (shown in **Fig. 9**):

   a. If using a hot-start polymerase, activate this according to manufacturer's description, otherwise just heat at 95°C for 1 min to get rid of secondary structures.

   b. Follow the activation by 45 cycles of amplification using the following steps (*see* **Note 11**):

   i. Denature at 95°C for 10 sec.

   ii. Anneal the primers and probes (e.g., at 50°C) for 15 sec, read endpoint fluorescence.

   iii. Optionally read fluorescence at additional one or two temperatures.

   iv. Elongate at 72°C for 20 sec.

   c. Make an endpoint affinity test the following way (*see* **Note 13**):

   i. Denature at 95°C for 15 sec.

   ii. 15 sec annealing at 1°C lower than the previous annealing, starting at 72°C. Read endpoint fluorescence.

   iii. Repeat above two steps 27 times (72°C to 45°C).

5. Analyze data—The best fluorescence reading temperatures to use in future RT-PCR experiment of that target is just above the temperature where the fluorescence of the EasyBeacon™ hybridized to the mismatched target starts to take off (*see* **Notes 14** and **18**). The found temperatures can be used as reading points in future measurements using the same setup. It is recommended that the endpoint affinity measurement is always carried out, as this is a very sensitive way to distinguish between similar targets.

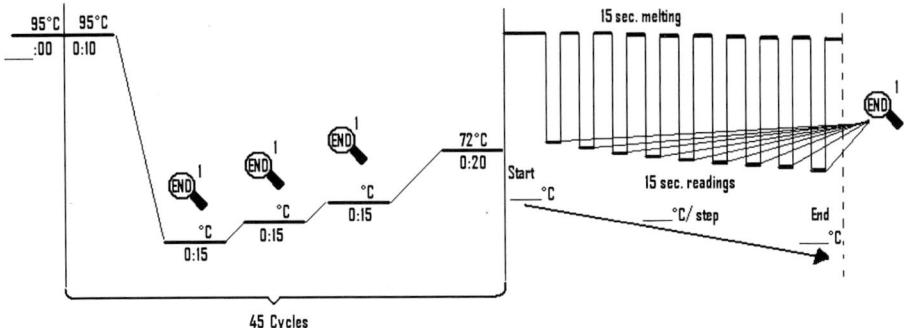

Fig. 9. Amplification profile using endpoint affinity readings. Firstly, the DNA polymerase is activated according to the manufacturer's instructions. The DNA region is then amplified using an optimal annealing temperature (e.g., 50°C) and two additional reading points (e.g., 55°C and 60°C). This profile is repeated for 45 cycles. After amplification an endpoint affinity study is made where the two strands are denatured at up to 95°C and then the reaction cooled to 72°C in the first step gradually lowering the temperature at each step (for example with 1°C per step). This is repeated until the desired temperature is reached at least to the level where both probes bind to their complementary target (e.g., 45°C). The magnifying glass with '**END**' written in it and a '**1**' at the border, illustrates that the fluorescence is measured one (1) time at the end of the stage.

An example of how the endpoint affinity measurement can be used for optimization of future RT-PCR runs for the same target and how the affinity measurement is very powerful in distinguishing between two very similar targets can be seen in **Fig. 10**.

4. Notes

1. The Methyleasy™ kit was chosen for two main reasons. There is next to no degradation of sample, meaning that a correct representation of methylation is obtained (the standard procedure for bisulfite conversion destroys up to 96% of the DNA) *(14)* and the bisulfite-converted DNA is extremely stable (I have used bisulfite-converted samples that is over 2 yr old) meaning that I can return to old samples for further studies.
2. For a longer storage time, the concentration of the mixture should be as high as possible (e.g., 20 nG/μL).
3. If it is necessary to perform a nested PCR, design the outer primers in accordance to the rules described herein for designing primers for methylation status determination. The amplicon shall be as short as possible, no longer than 600 bp. Reduce the number of first round cycles.
4. If spanning a 'CpG' duplet cannot be avoided and the methylation status of that 'CpG' duplet is unknown, insert the degenerate base 'Y' (mix of 'C' and 'T') or 'R'

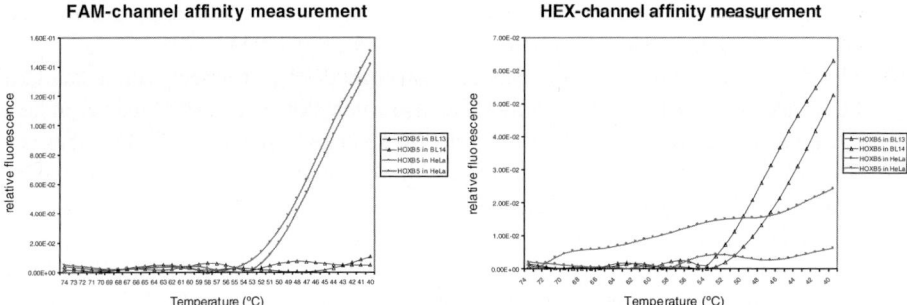

Fig. 10. An example showing how it is possible to distinguish the fluorescent signal from the fully complementary (methylated DNA) and the mismatched non-target (unmethylated DNA) using either FAM (FAM-TT1GAATCG1GTTT1T-BHQ1) or HEX-labeled probes (HEX-TT1AAAAAC1CAATTC1AAT1A-BHQ1). Both probes anneal specifically between 45°C and 50°C. In future measurements a common reading point for the two differently labeled probes should be at 46°C. It is clear from the different profiles, that it is simple to distinguish methylated from unmethylated DNA using the described protocol.

(mix of 'A' and 'G') depending on if the complementary template at that position is a purine or pyrimidine, respectively. The degenerate base should be as far from the 3'-end of the primer as possible, to minimize priming efficiency difference between the two targets.

5. It depends on the sequence surrounding the point of methylation and the length of the probes, if it is better to design nearly complementary probes with an 'A' nucleotide in one probe and a 'C' nucleotide in the other or probes that are nearly identical with for example a 'C' nucleotide in one probe and a 'T' nucleotide in the other probe. This chapter deals mostly with the case that the probes are nearly identical but using nearly complementary probes have also been shown to work fine—especially when using short probes.

6. If the EasyBeacon™ comprises 7 or less DNA nucleotides insert two HSMs, if it comprise between 8 and 14 nucleotides insert 3 HSMs and if it comprises 15 or more nucleotides insert 4 HSMs.

7. The website www.EasyBeacons.com will update the instruments that can be used with the different protocols, as soon as the instruments have been tested.

8. If the ramp rate is too slow or there is too many measurements, there can pass so much time that long amplicons might start to reanneal and displace the probe. If this happens either start the measurement closer to the melting temperature of the probe or reduce the time or measuring points in the annealing step prior to reaching the melting temperature.

9. The optimization is easiest if you have control samples of known methylation status available. But even without these control samples it should be easy to determine

the methylation status of the DNA target, by using differently labelled methylated and unmethylated specific probes combined in the same tube.

10. The described protocol is for a standard real-time PCR instrument with a standard PCR mixture. If using fast ramping PCR instruments or special PCR master mixes, the annealing, extension, and/or denaturing time intervals might be different (shorter).

11. The described protocol is only tested on the Rotorgene™ 3000 using a standard (Promega) PCR master mix. It could therefore be that other time intervals are optimal on other instruments or using other master mixes.

12. Often when you receive probes from oligonucleotide manufactures, they have determined the concentration by using the absorption of the probe at 260 nm and converting that into a concentration. However not all these measurements are precise, often done in high-throughput machines without proper quality control and do not take into account if the fluorophore is not coupled, partially degraded or in the presence of impurities. If the curve is flat or have several bumps, it might therefore be necessary to adjust the concentration of the probe.

13. The way that each real-time PCR machine is programmed is different from machine to machine, and the author can only refer to the instructions of the individual machines. However on most machines, you are able to easily perform an endpoint affinity measurement. The author has programmed the endpoint affinity measurement as a touch-down cycle, with a denaturing stage (at 95°C) and an annealing/reading stage where the temperature is gradually reduced by 1°C for each cycle.

14. It can be necessary to smooth the results if 'bumps' occur on the curve. The T_m is defined as the temperature where half of the formed duplexes have separated. It is found as a peak on the negative first derivative to the affinity measurement. The probes should still be intact after the affinity measurement, so if the measurement for some reasons went wrong, it can be repeated without having to remake the solutions (within 8 h of the preparation of the first experiment). Be aware that the fluorophores are light sensitive and will be degraded by prolonged exposure to light or multiple repeated measurements.

15. Make sure that the primers used for the real-time PCR is optimized to work at the annealing temperature found for the EasyBeacons™ and vice versa. The primers can be tested by running a normal PCR using the optimal annealing temperature and analysing the result by gel-electrophoreses after the PCR, preferable using the same PCR buffer and enzymes as used in the real-time PCR. Alternatively a real-time PCR using the nonspecific dye SYBR Green can be performed, with an end-point melting temperature measurement to check for the formation of primer—dimers.

16. The efficiency of the PCR can be determined by performing a serial dilution experiment, using 100,000, 10,000, 1000, 100, and 10 copies of template, and constructing a standard curve. Refer to the manual of the used thermal cycler to automatically determine the C_t for the standards and to calculate a standard curve. PCR efficiency can be calculated using the following formula: $E = e^{\ln 10/-s} - 1$, where E is the efficiency, s the slope of the standard curve, $e = 2.718$ and $\ln 10 = 2.303$. A slope of –3.322 represents a PCR efficiency of 100%.

17. One cell contains approximately 6 pG of DNA. In order to avoid a false negative due to the possibility of no DNA being present in large dilutions (the Poisson distribution) use at least 60 pG of DNA.

18. If no signal is produced by one of the probes, try hybridizing it to a complementary oligonucleotide and do a melt measurement. If no signal is produced still, it could be a sign of either no fluorophore in the probe, degraded fluorophore, no quencher (a small signal change should still be observable), or formation of a strong intra-strand, secondary structure making the probe unable to hybridize to its target.

Acknowledgments

All graphic illustrations in this chapter were done with great competence by Mr Preben Firkloever at Scientific Graphics, Denmark. Furthermore, the author would like to thank Corbett Life Science Pty Ltd (www.corbettlifescience.com) for making one of their fabulous, fast ramping Rotorgene™ 3000 available for me to verify the EasyBeacon™ technology on. I would also like to thank Integrated Science, Australia (www.integratedsci. com.au) for making the flexible and user friendly Stratagene Mx3000p available for me as well as for offering the best service one could wish for. Additionally, I would like to thank Mrs Christina Nielsen for help with the HyNA™ synthesis and Dr Kiranjit Kaur for cellular work and finally Drs Esben Flindt, John Melki, and Douglas Millar for scientific discussions and review of this chapter.

References

1. Arya, M. et al. (2005) Basic principles of real-time quantitative PCR. *Expert. Rev. Mol. Diagn.* **5,** 209–219.

2. Stahlberg, A., Zoric, N., Aman, P., and Kubista, M. (2005) Quantitative real-time PCR for cancer detection: the lymphoma case. *Expert. Rev. Mol. Diagn.* **5,** 221–230.

3. Kutyavin, I. V. et al. (2000) 3′-Minor groove binder-DNA probes increase sequence specificity at PCR extension temperatures. *Nucleic Acids Res.* **28,** 655–661.

4. Afonina, I. et al. (1997) Efficient priming of PCR with short oligonucleotides conjugated to a minor groove binder. *Nucleic Acids Res.* **25,** 2657–2660.

5. Dorris, D. R. et al. (2003) Oligodeoxyribonucleotide probe accessibility on a three-dimensional DNA microarray surface and the effect of hybridization time on the accuracy of expression ratios. *BMC. Biotechnol.* **3,** 6.

6. Letowski, J., Brousseau, R., and Masson, L. (2004) Designing better probes: effect of probe size, mismatch position and number on hybridization in DNA oligo-nucleotide microarrays. *J. Microbiol. Methods* **57,** 269–278.

7. Letertre, C., Perelle, S., Dilasser, F., Arar, K., and Fach, P. (2003) Evaluation of the performance of LNA and MGB probes in 5′-nuclease PCR assays. *Mol. Cell Probes* **17,** 307–311.

8. Filichev, V. V., Christensen, U. B., Pedersen, E. B., Babu, B. R., and Wengel, J. (2004) Locked nucleic acids and intercalating nucleic acids in the design of easily denaturing nucleic acids: thermal stability studies. *Chembiochem.* **5**, 1673–1679.

9. Christensen Ulf, B. and Pedersen, E. B. (2003) Intercalating Nucleic Acids with pyrene nucleotide analogues as next-nearest neighbors for excimer fluorescence detection of single-point mutations under nonstringent hybridization conditions. *Helv. Chim. Acta* **86**, 2090–2097.

10. Christensen, U. B. and Pedersen, E. B. (2002) Intercalating nucleic acids containing insertions of 1-O-(1-pyrenylmethyl)glycerol: stabilisation of dsDNA and discrimination of DNA over RNA. *Nucleic Acids Res.* **30**, 4918–4925.

11. SantaLucia, J., Jr. and Hicks, D. (2004) The thermodynamics of DNA structural motifs. *Annu. Rev. Biophys. Biomol. Struct.* **33**, 415–440.

12. Das, P. M. and Singal, R. (2004) DNA Methylation and Cancer. *J. Clin. Oncol.* **22**, 4632–4642.

13. Herman, J. G., Graff, J. R., Myohanen, S., Nelkin, B. D., and Baylin, S. B. (1996) Methylation-specific PCR: a novel PCR assay for methylation status of CpG islands. *Proc. Natl. Acad. Sci. U.S.A* **93**, 9821–9826.

14. Grunau, C., Clark, S. J., and Rosenthal, A. (2001) Bisulfite genomic sequencing: systematic investigation of critical experimental parameters. *Nucleic Acids Res.* **29**, E65.

11

Quenched Autoligation Probes

Adam P. Silverman, Hiroshi Abe, and Eric T. Kool

Abstract

Methods are described for preparation and use of quenched autoligation (QUAL) probes. These modified oligonucleotide fluorescent probes can be used to detect DNA and RNA in solution, on solid surfaces, and in fixed and living bacterial and human cells. They are quenched probes, and thus provide a "lighting up" signal in a single step, without removing unbound or unreacted probes from the analyte. QUAL probe signals can be detected by fluorescence spectrometer, fluorescence microscope, or flow cytometry. These probes can distinguish between very small variations, including single nucleotide differences, in nucleic acid targets. The described method includes a description of how to prepare the needed dabsyl quencher linker, how to prepare the QUAL probes by DNA synthesizer, and how to employ them in detecting nucleic acids in solution and in detecting RNAs in bacterial and human cells.

Key Words: Autoligation; quencher; dabsyl; rRNA; mRNA; bacteria; amplification.

1. Introduction

Methods for identifying and distinguishing nucleic acid sequences are extremely important for applications in molecular diagnostics. Fluorescence *in situ* hybridization methods targeting cellular RNAs are particularly useful because they are typically rapid, facile, and can be easily monitored directly in the cells. However, discrimination of small sequence differences such as single nucleotide polymorphisms *in situ* has been problematic. Quenched autoligation (QUAL) probes are a relatively new class of hybridization probes that are designed to give highly specific signal for cellular assays, even at single nucleotide resolution.

The source of fluorescence in QUAL probes is a nucleophilic displacement reaction which releases a quencher *(1)*. QUAL probes consist of a probe pair: an electrophilic "quenched probe" which has an internal fluorescein tethered to

From: *Methods in Molecular Biology, Vol. 429: Molecular Beacons: Signalling Nucleic Acid Probes, Methods and Protocols* Edited by: A. Marx and O. Seitz © Humana Press Inc., Totowa, NJ

a dT nucleotide as well as a "dabsyl" (dimethylaminoazobenzenesulfonyl) quencher at the 5′-terminus, and a nucleophilic "phosphorothioate probe" which has a phosphorothioate group at its 3′-terminus. When the quenched probe and phosphorothioate probe bind side-by-side, the nucleophilic phosphorothioate displaces the quencher, joining the two strands, and resulting in a substantial increase in fluorescence signal (**Fig. 1**). In many cases, the product can dissociate from the target, allowing a new probe pair to react, thus generating amplified quantities of signal by turnover. Reaction turnover numbers up to 96 have been reported *(2)*. A high degree of selectivity is observed for single nucleotide mismatches in solution, solid-state, and cellular assays using QUAL probes *(1,4,5)*.

QUAL probes can be used in solution or on solid supports. However, the most highly developed application of QUAL probes is for *in situ* imaging of rRNA *(5,6)*. Notably, they have been used for identification of sequences in bacteria, and for discrimination of closely related species of bacteria by single nucleotide differences in 16S RNA *(5)*. QUAL probes have also been used for identifying rRNA sequences in human cell lines, and recent results from our lab suggest that such probes offer promise for *in situ* detection of abundant mRNAs as well *(3)*.

2. Materials

2.1. Preparation of Quencher Phosphoramidite

1. *Dry solvents*: freshly distilled tetrahydrofuran, dichloromethane, and acetonitrile. Reagent grade hexanes, ethyl acetate, and dichloromethane for chromatography. Plates for thin layer chromatography (TLC), silica gel, and column for flash chromatography.
2. *Reagents*: triethylamine, 1,4-butanediol (*see* **Note 1**), *N*, *N*-dimethylaminopyridine (DMAP), 4-(dimethylamino)azobenzene-4′-sulfonyl chloride (dabsyl chloride), diisopropylamine and *N*, *N*-diisopropylmethyl phosphonamidic chloride.

2.2. Design and Synthesis of rRNA-Targeted Probes

1. DNA synthesizer with all standard reagents.
2. *Phosphorothioate probe*: standard DNA (or 2′-*O*-Me RNA) phosphoramidites and sulfurizing reagent (*see* **Note 2**). Concentrated aqueous ammonium hydroxide.
3. *Quencher probe*: ultramild DNA (or 2′-*O*-Me RNA) phosphoramidites, fluorescein-dT phosphoramidite, and quencher linker phosphoramidite from Section 3.1. Potassium carbonate in methanol (0.05 *M* solution).
4. *HPLC solvents*: 0.1 *M* triethylammonium acetate solution filtered through 0.2 µm filter paper and HPLC-grade acetonitrile.

2.3. Detection of QUAL Probes in Solution

1. Fluorometer and quartz cuvette.
2. An appropriate hybridization buffer, e.g., 6× SSC, or 10 m*M* Na•PIPES (pH 7.0) with 1.0 *M* NaCl.

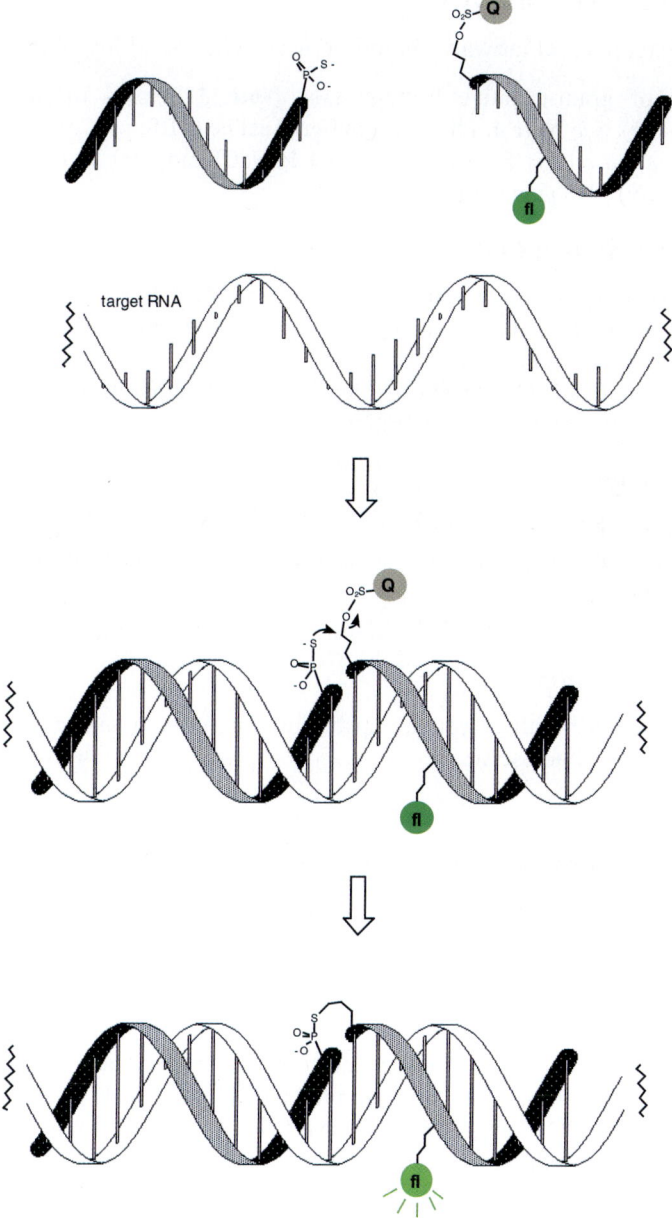

Fig. 1. Mechanism of QUAL probe signal generation. When a pair of QUAL probes binds side-by-side on a DNA or RNA template, the templated autoligation reaction causes the displacement of a dabsyl group by a phosphorothioate, thereby unquenching the fluorophore.

2.4. Staining Bacterial rRNA

Sterile reagents and labware should be used when working with live cells.

1. Culture of gram-negative bacteria (*see* **Note 3**) grown to mid-log phase (OD_{600}~0.5). *See* **Note 4**. Phosphate buffered saline (PBS; pH 7.0).
2. *Hybridization buffer*: 6× SSC buffer (0.9 *M* NaCl and 0.09 *M* sodium citrate) + 0.05% (w/v) SDS (*see* **Note 5**).

2.5. Staining Human Cells

1. HL-60 or other cells, grown in Dulbeco's-modified Eagle medium (DMEM) without phenol red, and containing 10% fetal bovine serum, 50 U/mL penicillin, and 50 µg/mL streptomycin.
2. Streptolysin O (SLO), Mg^{2+}/Ca^{2+}-free PBS, dithiothreitol (DTT), and bovine serum albumin (BSA), for transfection.

2.6. Microscopy

1. Microscope slides and cover slips.
2. Epifluorescence microscope with excitation at 488 nm. Digital camera if photographs are desired.
3. Confocal microscope if desired.

2.7. Flow Cytometry

1. Flow cytometer with detection capabilities for forward scattering, orthogonal scattering, and fluorescein.

3. Methods

3.1. Preparation of Quencher Linker Phosphoramidite

The quencher phosphoramidite is not currently commercially available, and at present must be synthesized in two steps from commercially available reagents (**Fig. 2**). Complete characterization of the intermediates and product has been reported *(2)*.

1. In a dry round-bottom flask with a magnetic stirbar, butanediol (1.3 mL, 14.8 mmol) is dissolved in 20 mL of dry tetrahydrofuran/dichloromethane (1:10).
2. Triethylamine (0.413 mL, 2.96 mmol) and DMAP (50 mg, 0.4 mmol) are added to the round-bottom flask, followed by dabsyl chloride (500 mg, 1.48 mmol).
3. The reaction is stirred for 2–4 h while monitoring progress by silica TLC using pentane/ethyl acetate (4:1) as eluent.
4. When TLC reveals that all starting material has been converted, the reaction solution is evaporated *in vacuo* to a small volume (<5 mL) and purified by flash column chromatography on silica gel using pentane/ethyl acetate (5:1) as the eluent. The purified fractions are combined, and the solvent is removed *in vacuo*, leaving a red solid (*see* **Note 6**).

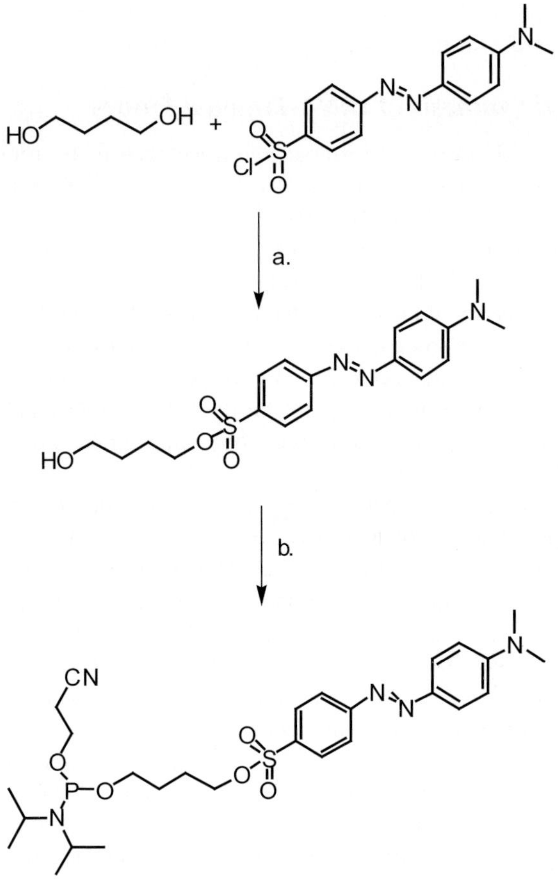

Fig. 2. Two-step preparation of the dabsyl butyl-linker. Reagents: (a) NEt$_3$; (b) *N*, *N*-diisopropylmethyl phosphonamidic chloride, diisopropylethylamine.

5. The solid is dissolved in freshly distilled dichloromethane (~2 mL) and diisopropylethylamine (0.7 mL, 4.0 mmol) and *N*, *N*-diisopropylmethyl phosphonamidic chloride (0.5 mL, 2.6 mmol) are added. The reaction is stirred for 30–60 min, with monitoring by TLC using 1:1 pentane/ethyl acetate + 1% triethylamine as eluent (*see* **Note 7**).

6. When the TLC reveals that all starting material has been converted, the reaction mixture is loaded directly onto a short silica gel column for purification. The column is packed with pentane:ethyl acetate:triethylamine (2:1:0.005), and pentane: ethyl acetate (2:1) is used as eluant. The solvent from the combined product fractions is removed *in vacuo*, leaving a red solid (*see* **Note 8**), which is dried overnight and stored as a solid under argon at −80°C.

7. For use on the DNA synthesizer, the quencher amidite is dissolved in 10:1 acetonitrile:tetrahydrofuran to make a 0.8 *M* solution.

3.2. Design and Synthesis of rRNA-Targeted Probes

1. A region of the rRNA that is to be targeted is selected, usually around 30 nucleotides long. Four probes are prepared, the "quencher probe," the "phosphorothioate probe," and two "helper probes." (*see* **Note 9**).

2. The quencher and phosphorothioate probes are typically 8–15 and 12–18 nucleotides long, respectively (*see* **Note 10**). The probes must be designed such that the 5′-end of the quencher probe should line up with the 3′-end of the phosphorothioate probe. In addition, the quencher probe must have a T base <5 nucleotides from its 5′-terminus (for incorporation of fluorescein-dT). The helper probes should be 18 nucleotides long; one should bind adjacently to the quencher probe and the other should bind adjacently to the phosphorothioate probe. An example of QUAL probe design is shown in **Fig. 3**.

3. For ribosomal targets, "helper" probes are recommended to be used along with QUAL probes. Helper probes are prepared with standard DNA backbones, and may be purchased commercially if desired. Helper probes are oligonucleotides that bind adjacently to the quencher and phosphorothioate probes, are also commonly used for other *in situ* applications, as they help disrupt secondary structure and improve the accessibility of many sites on a folded RNA target *(7)*. If one wishes to synthesize the helper probes, normal DNA phosphoramidites are used with the standard protocol on the DNA synthesizer. The DNA is cleaved from the solid support and deprotected by heating to 55°C in 1 mL concentrated aqueous ammonium hydroxide for 16–24 h. After cooling the solution, the solvent is removed using a speed-vac. The resulting solid is dissolved in 1 mL distilled deionized water, and any insoluble material is removed by centrifugation and decanting. No purification is necessary.

4. For synthesis of the phosphorothioate probe, phosphoramidites with standard protecting groups may be used (DNA or 2′-*O*-Me RNA). (Note that phosphorathioate probe synthesis uses only commercial reagents, so these probes may be purchased commercially if desired.) A 3′-phosphate CPG column is employed. The first base is added using a "sulfurization" procedure, in which the standard oxidation reagent is replaced with a solution of sulfurizing reagent *(8)*. Subsequent bases are added using the normal coupling cycle. The DNA is cleaved from the solid support and deprotected by heating to 55°C in 1 mL concentrated aqueous ammonium hydroxide for 16–24 h. After cooling the solution, the solvent is removed using a speed-vac. The resulting solid is dissolved in 1 mL distilled deionized water, and any insoluble material is removed by centrifugation and decanting. No purification is necessary.

5. For synthesis of the quencher probe, Ultramild phosphoramidites must be used (DNA or 2′-*O*-Me RNA). At a position two to three nucleotides from the 5′-terminus, fluorescein-dT phosphoramidite is used in place of the standard T amidite. The quencher phosphoramidite is coupled at the 5′-terminus (*see* **Note 11**). The DNA

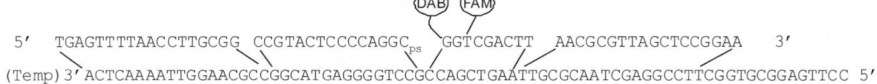

```
                                            (DAB) (FAM)
5'   TGAGTTTTAACCTTGCGG  CCGTACTCCCCAGGCₚₛ  GGTCGACTT  AACGCGTTAGCTCCGGAA   3'
(Temp) 3' ACTCAAAATTGGAACGCCGGCATGAGGGGTCCGCCAGCTGAATTGCGCAATCGAGGCCTTCGGTGCGGAGTTCC  5'
```

Quencher Probe: 5' LGG T'CG ACT T 3'
Phosphorothioate Probe: 5' CCG TAC TCC CCA GGCps 3'
Helper 1: 5' TGA GTT TTA ACC TTG CGG 3'
Helper 2: 5' AAC GCG TTA GCT CCG GAA 3'

L = Dabsyl Linker
T^f = Fluorescein-dT
ps = 3'-phosphorothioate

Fig. 3. Example of QUAL probe design with specific target and probe sequences shown.

is cleaved from the CPG beads and deprotected using 1 mL of 0.05 M K_2CO_3 in methanol at room temperature for 4–6 h. The probe is then purified by HPLC using a gradient of acetonitrile in 0.1 M TEAA buffer (*see* **Note 12**).

6. Concentration of the probes is determined on a UV/Vis spectrometer. Stock solutions of each probe are prepared at the following concentrations: phosphorothioate probe 100 μM, quencher probe 10 μM, helper probes 300 μM (in both helpers).

3.3. Detection of QUAL Probes in Solution

1. A solution containing appropriate amounts of quencher probe, phosphorothioate probe, and analyte DNA or RNA strand (typically 200 nM each) is prepared in hybridization buffer and transferred to a quartz cuvette.
2. Typical fluorometer settings are: excitation = 495 nm, emission = 515 nm, integration time = 1 s. Fluorescence is measured as a function of time; significant increase should be observed only for probes that are fully complementary to the analyte sequence.

3.4. Staining Bacterial rRNA

1. Aliquots of bacterial culture (1 mL) are centrifuged at 10,000 rpm for 5 min. The supernatant is removed, and the pellet is washed with 0.5 mL PBS.
2. Each aliquot of bacteria is resuspended in 1 mL hybridization buffer. Aliquots of 100 μL are distributed in 0.2 mL Eppendorf tubes.
3. To the sample is added: 1 μL quencher probe, 2 μL phosphorothioate probe, and 1 μL helper probes stock solutions. The suspension is vortexed thoroughly and incubated at 37°C (*see* **Note 13**).
4. After 2–3-h incubation, the bacteria are analyzed by microscopy or flow cytometry.

3.5. Staining RNAs in Human Cells

1. Cells are permeabilized using SLO for the transfection of QUAL probes. SLO (1000 U/mL) is incubated in Mg^{2+}/Ca^{2+}-free PBS buffer solution containing 10 mM

DTT and 0.05% BSA at 37°C for 2 h *(9)*. Small aliquots of activated SLO can be stored at −20°C.

2. 10 mL aliquots of cells (10^5 cells/mL) are centrifuged at 1000 rpm (9300 *g*) for 5 min. The supernatant is removed, and the pellet is washed twice with 5 mL of Mg^{2+}/Ca^{2+}-free PBS.

3. The cells are suspended in 300 µL of Mg^{2+}/Ca^{2+}-free PBS buffer containing dabsyl- FAM-labeled probe (200 nM), phosphorothioate probe (200 nM), calf thymus DNA (1 µg/mL) and, if desired, Hoechst 33342 stain (0.2 µg/mL). SLO is then added (30 U/mL final concentration). The suspension is incubated for 30 min.

4. Cells are resealed by the addition of 1 mL DMEM containing $CaCl_2$ (0.2 g/L) and incubated for 1 h at 37°C.

5. The cell suspensions are concentrated to 20 µL volumes.

3.6. Analysis of Stained Cells

3.6.1. Epifluorescence Microscopy

1. Cell suspensions are spotted directly onto a glass slide without any washing steps. The spots are covered with glass cover slips.

2. Cells are imaged by fluorescence excitation (488 nm excitation and 520 low-pass filter). For negative controls, bacteria can be imaged with light.

3.6.2. Laser Confocal Microscopy

1. Cells are directly spotted onto a glass slide after reaction with QUAL probes, without any washing step, and covered with a glass cover slide.

2. Microscope settings were as follows: frame size, 1024×1024; data depth, 12 bits; scan speed, 1.76 µs $\times$ 1.76 µs; scan mode, three channels (bright field, DAPI, FAM). DAPI imaging: excitation by Ti:sapphire laser, 780–800 nm (two photon system); emission, 435–485 nm BP filter; pinhole, 852 µm. FAM imaging: excitation by argon laser, 488 nm; emission, 520 LP filter; pinhole, 886 µm. The resulting raw data were analyzed by LSM image browser (Zeiss).

3.7. Flow Cytometry

1. It may be necessary to dilute samples to 0.2–0.3 mL volume prior to running flow cytometry, depending on the instrument.

2. Events for bacteria should be triggered by orthogonal scattering, events for human cells should be triggered by forward scattering, at appropriate log amp settings.

4. Notes

1. 1,4-Butanediol is a regulated chemical in the United States and may be difficult for some labs to obtain. Pentanediol or propanediol may be used alternatively.

2. 2′-*O*-Me RNA is more stable and less prone to intracellular degradation than standard DNA probes, but probe specificity is diminished, so standard DNA probes are recommended in cases where single nucleotide discrimination is desired.

3. We have thus far been unable to deliver the probes into gram-positive bacteria. Although these organisms should be applicable to this methodology, permeabilization remains a challenge.

4. The procedure will also work for fixed bacteria. Fixation media should be thoroughly removed prior to treating the bacteria with QUAL probes. Subsequent steps apply for fixed or live bacteria.

5. Maximum signal-to-background was obtained for *E. coli* using 6× SSC buffer. Other buffers also work well, such as 20 mM Tris–HCl or HEPES pH 7.0 + 0.9 M NaCl. If lower-salt conditions are desired, 10 mM MgCl$_2$ may be used instead of NaCl, but signal-to-background tends to be lower under these conditions. Regardless of specific buffer conditions, at least 0.05% SDS must be used for permeabilization.

6. The product of the first reaction may be characterized by NMR, but is unstable and will decompose if it is not carried on to the second step immediately.

7. If after 1 h starting material is still present, another 0.5–1 equiv. of *N, N*-diisopropylmethyl phosphoramidic chloride should be added and the reaction stirred for an additional 0.5 h.

8. If not highly pure, the product may be a red oil. Often it is still sufficiently pure to use for DNA coupling; overall purity can be determined by NMR characterization. The product should be stored dry with dessicant at −80°C until use. It should not be stored as a solution.

9. Helpers may not be necessary for signal in accessible regions of RNA, but we recommend preparing a set and empirically testing to see if they improve signal.

10. If high specificity is desired (for discrimination of one to two mismatches), the quenched probe should be shorter, ideally seven to nine nucleotides. Check the melting temperature of the sequence, however, since in some cases the T_m may be too low if 7-mers are used.

11. Use 15 min coupling times for fluorescein-dT and the quencher amidite. Be sure that the procedure is "DMT-On," as trichloroacetic acid treatment will cleave the quencher. The CPG beads should be deep orange to red in color after washing with dichloromethane and drying.

12. After purification, the solution should be stored at −80°C. The solution may be lyophilized, but this often results in some decomposition of the probe (loss of quencher), so it is not recommended. If purification is not done immediately after deprotection (recommended), the solution may be stored at −80°C for a few days, but lower yields may result.

13. Depending on the length of probe, target site accessibility, etc., optimal temperature for incubation may differ, though we have found 37°C to be ideal for most cases. It is not recommended that temperatures above 37°C are used, as higher temperatures speed up the hydrolysis of the quencher, which leads to nonspecific fluorescent signal.

References

1. Sando, S. and Kool, E. T. (2002) Quencher as leaving group: efficient detection of DNA-joining reactions. *J. Am. Chem. Soc.* **124,** 2096–2097.

2. Abe, H. and Kool, E. T. (2004) Destabilizing universal linkers for signal amplification in self-ligating probes for RNA. *J. Am. Chem. Soc.* **126,** 13,980–13,986.
3. Abe, H. and Kool, E. T. (2006) Flow cytometric detection of specific RNAs in native human cells with quenched autoligating FRET probes. *Proc. Natl. Acad. Sci.* **103,** 263–268.
4. Sando, S., Abe, H., and Kool, E. T. (2004) Quenched auto-ligating DNAs: multicolor identification of nucleic acids at single nucleotide resolution. *J. Am. Chem. Soc.* **126,** 1081–1087.
5. Silverman, A. P. and Kool, E. T. (2005) Quenched autoligation probes allow discrimination of live bacterial by single nucleotide differences in rRNA. *Nucleic Acids Res.* **33,** 4978–4986.
6. Silverman, A. P., Baron, E. J., and Kool, E. T. (2006) RNA-templated chemistry in cells: discrimination of escherichia, shigella, and salmonella bacterial strains with a new two-color FRET strategy. *Chem. Bio. Chem.* **7,** 1890–1894.
7. Fuchs, B. M., Glockner, F. O., Wulf, J., and Amann, R. (2000) Unlabeled helper oligonucleotides increase the in situ accessibility to 16S rRNA of fluorescently labeled oligonucleotide probes. *Appl. Environ. Microbiol.* **66,** 3603–3607.
8. Xu, Y. and Kool, E. T. (1999) High sequence fidelity in a non-enzymatic DNA autoligation reaction *Nucleic Acids Res.* **27,** 875–881.
9. Faria, M., Spiller, D. G., Dubertret, C. et al. (2001) Phosphoramidate oligonucleotide as potent antisense molecules in cells and in vivo. *Nat. Biotechnol.* **19,** 40–44.

12

HyBeacon® Probes for Rapid DNA Sequence Detection and Allele Discrimination

David J. French, David G. McDowell, Paul Debenham, Nittaya Gale, and Tom Brown

Abstract

HyBeacon probes are single-stranded oligonucleotides with one or more internal base(s) labeled with a fluorescent dye. When a probe forms a duplex with its target sequence, the level of fluorescence emission increases considerably. HyBeacons have been developed as new tools for rapid sequence detection and discrimination and have been employed in a wide variety of applications including infectious diagnostics and analysis of human polymorphisms. Single-labeled (FVG1) and dual-labeled (FVG11) probes were designed to analyze the factor V Leiden (R506Q) polymorphism which causes an increased risk of deep vein thrombosis and pulmonary embolism. Detection and identification of factor V alleles is performed by melting curve analysis and determination of probe melting temperature (T_m). HyBeacon hybridization to the glutamine allele (Q) causes the formation of mismatched DNA duplexes that are detected through decreases in T_m. HyBeacon probes are included in homogeneous PCR assays to genotype samples with respect to the factor V polymorphism within 20 min, using purified DNAs and unpurified saliva/blood samples. This paper describes the preparation of homogeneous PCR assays, LightCycler target amplification, and subsequent melting curve analysis. This chapter also describes the use of homologous oligonucleotides and melting curve analysis as a method for probe evaluation.

Key Words: Fluorescent probes; HyBeacon; LightCycler; single nucleotide polymorphisms.

1. Introduction

In the fields of clinical diagnostics, forensic science, food analysis, and pathogen detection, there is a considerable interest in technologies that permit rapid interrogation of DNA sequences. As an example, analysis of single nucleotide polymorphisms (SNPs) is an important method for the investigation of human genetic variation, aiding research into areas such as population

From: *Methods in Molecular Biology, Vol. 429: Molecular Beacons: Signalling Nucleic Acid Probes,
Methods and Protocols* Edited by: A. Marx and O. Seitz © Humana Press Inc., Totowa, NJ

genetics, genetic disease, and drug development (pharmacogenomics) *(1–3)*. SNPs are positions within a DNA sequence that vary by a single nucleotide (substitution, insertion, or deletion) from person to person and are the most abundant form of sequence variation in the human genome. SNPs located in coding and regulatory regions may influence phenotypic characteristics directly by changing amino acid sequences and affecting protein production respectively. SNPs have been identified as possible causative agents of certain genetic diseases (e.g., cystic fibrosis gene defects and susceptibility to Alzheimer's disease) *(4,5)* and have been associated with variations in patient response to therapeutic medication (e.g., codeine) *(6)*. SNPs may also be employed indirectly as genetic markers to identify complex traits through linkage dissociation studies. The importance of SNPs has driven the discovery of novel polymorphisms (e.g., through the human genome project) for use as genetic markers and many techniques have been developed to screen for known polymorphisms in rapid diagnostic and high throughput formats.

Homogeneous PCR methods of DNA sequence analysis, such as those employing fluorescent oligonucleotide probes, possess benefits over traditional heterogeneous techniques. Homogeneous methods do not require post-amplification manipulation (e.g., restriction digestion and gel analysis) to generate sequence data, thereby reducing hands-on-time, assay duration and the potential for cross-contamination.

Homogeneous assay methodologies may be specific or nonspecific. Nonspecific methods commonly employ fluorescent molecules, such as double-strand specific DNA-binding dyes (e.g., SYBR-green I) and Amplifluor primers (formerly sunrise primers) *(7–10)*, to detect the presence of particular DNA sequences. However, these techniques also detect the formation of spurious amplification products and primer-dimers during PCR. In contrast, specific methods of homogeneous analysis frequently utilize fluorescently labeled oligonucleotide probes that hybridize exclusively to amplified target sequences, such that spurious products and primer-dimers are not detected. Examples of fluorescent oligonucleotide probes include TaqMan probes, molecular beacons, hybridization probes, Scorpion primers, and light-up probes *(11–21)*. Fluorescent probes typically emit lower quantities of fluorescence when in isolation (i.e., single-stranded) than when hybridized to complementary target sequences (i.e., double-stranded). Therefore, measuring the amount of fluorescence emitted at each cycle of PCR allows the real-time accumulation of amplicon to be monitored and the presence or absence of specific target sequences to be assessed. Furthermore, the structures of oligonucleotide probes permit the discrimination of closely related sequences, when used in conjunction with optimized reaction conditions, such that only those sequences that are fully complementary to the probes cause elevated levels of fluorescent signal to be generated during real-time PCR analysis *(22–24)*.

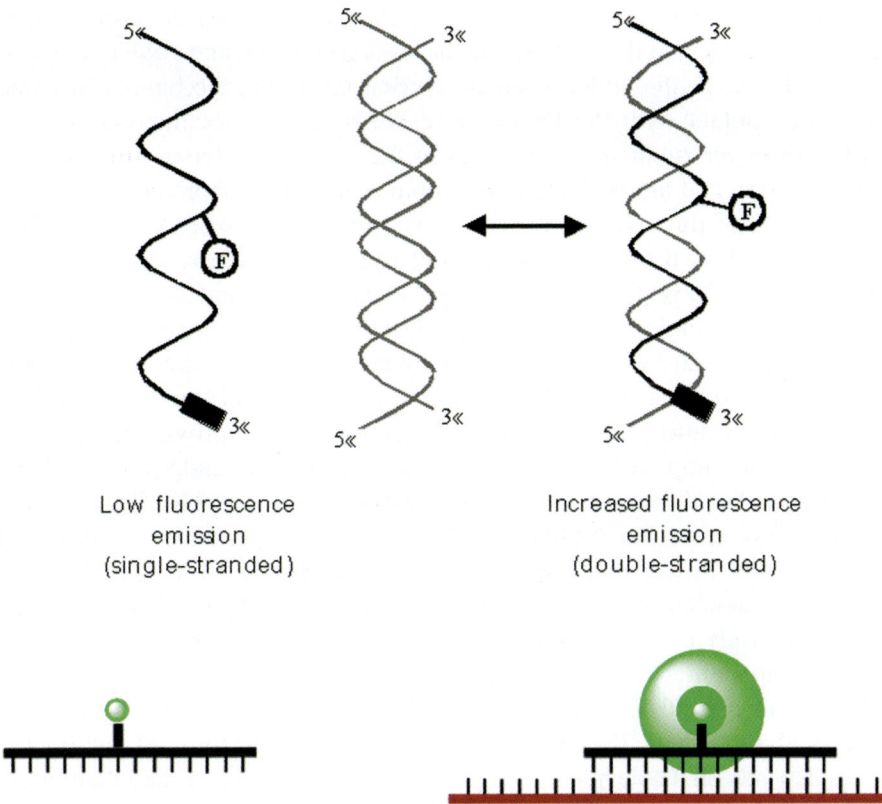

Fig. 1. Structure of a HyBeacon probe and the method of target detection. HyBeacons consist of oligonucleotide sequences with fluorophore moieties "F" attached to internal nucleotides and 3′-end blockers "■" to prevent PCR extension from probes. The amount of fluorescence emitted from hybridized HyBeacons is considerably greater than that of single-stranded probes, permitting detection of target sequences.

We have developed a novel fluorescent probe technology, called HyBeacons® which provides a homogeneous method for fluorescence-based sequence detection, allele discrimination and DNA quantification *(25–27)*. HyBeacons comprise a sequence-recognising oligonucleotide, which lacks significant secondary structure, with fluorescent dye labels attached to internal nucleotides and a 3′ phosphate or similar blocker to prevent PCR extension (**Fig. 1**). The HyBeacon probe is included in PCR assays and emits a greater fluorescent signal when hybridized to complementary target sequences than when single-stranded. In the absence of a specific organic quencher dye, the inherent quenching properties of DNA lead to the observation of an alteration in fluorescence emission upon probe hybridization. This is a consequence of the change in the degree of interaction of the fluorophore and the heterocyclic bases in the single- and double-stranded forms of DNA. In the single-stranded HyBeacon probe, the bases can come into close contact with the fluorophore, leading to molecular interactions and fluorescence quenching, whereas in the duplex the bases are involved in base pairing and are no longer accessible to the fluorophore (**Fig. 1**, right). Enhancement of fluorescence is particularly pronounced when the HyBeacon is labeled with two or more fluorescent dyes separated by three or more base pairs, as fluorophore—fluorophore quenching interactions, which are prevalent in the single-strand are not possible in the rigid duplex. HyBeacons offer a specific advantage over other sequence-specific probe technologies. They have no internal secondary structure (unlike Molecular Beacons) and are not cleaved during PCR (unlike TaqMan probes), so provide a means of analyzing the amplicon by melting analysis at the endpoint of PCR. Furthermore, the HyBeacon mode of action does not require FRET to a second fluorophore or quencher (unlike hybridization probes), making probe design uncomplicated. In this context, HyBeacons have been used to identify short genomic sequences and associated SNPs, since closely related sequences differing by only a single nucleotide may be discriminated by measuring the melting temperatures (T_m) of various probe/target duplexes.

The presence and identity of target sequences is determined by melting curve analysis, where the stability and melting temperature of hybridized probes depends on the degree of homology between HyBeacons and their target sequences. The interaction that occurs between a HyBeacon probe and target DNA sequence may be fully complementary or contain positions of mismatch, depending on the origin of the polymorphic sequence. Increasing the reaction temperature above the T_m of the HyBeacon causes probe/target duplexes to dissociate and the amount of fluorescence emission to decrease. Probes that are hybridized to mismatched sequences possess significantly reduced T_ms compared with fully complementary probe/target duplexes.

HyBeacon probes may be employed to genotype individuals for known polymorphisms using high-quality genomic DNA extracted from samples such as blood and buccal swabs. However, DNA extraction protocols are time-consuming, laborious, and expensive. Therefore, to reduce the duration and cost of analyses, direct PCR amplification from saliva, buccal swabs, and blood samples may be performed, removing the requirement for genomic DNA extraction. Direct saliva analysis, in combination with the rapid thermal cycling conditions of the LightCycler™, permits samples to be genotyped within 20 min. The development of a rapid, cheap, and non-invasive diagnostic system may permit "point-of-care" and genetic testing to be performed in hospitals, doctors' surgeries, and pharmacies.

HyBeacon probe assays were developed to genotype a variety of sample types with respect to the factor V Leiden SNP. Activated protein C (APC) is a serine protease and anticoagulant that is responsible for the inactivation of factor V. Normally, factor V is inactivated by cleavage of the protein at Arg506 and subsequent cleavage at Arg306. A SNP within the factor V gene (R506Q; G to A polymorphism) causes inactivation to occur only by cleavage at R306. Without prior cleavage at R506, inactivation of factor V occurs at an approximately 10 times slower rate. Therefore, the factor V Leiden polymorphism affects the anticoagulation potential of APC, such that individuals that possess the SNP have an increased risk of deep vein thrombosis and pulmonary embolism (5–10 fold and 50–100 fold increased risk in heterozygous and homozygous individuals, respectively). The factor V Leiden SNP is present in approximately 6% of the general population and has been linked to the fatal thromboses ("economy class syndrome") that occur in individuals during, or shortly after, long flights.

Single-labeled *(25–27)* and dual-labeled HyBeacon probes were designed to analyze the R506Q polymorphism. Both probes yielded high quality melting peaks and enabled sensitive, robust and accurate analysis of purified DNA and unpurified saliva samples. The dual-labeled probe exhibited considerably larger signal-to-noise ratios and melting peaks compared with the single-labeled probe.

2. Materials

2.1. General Equipment

1. 0.5 and 1.5 mL flip-cap tubes (Alpha laboratories).
2. P2, P10, P20, P200 and P1000 pipettes and tips. Pipettes and working areas are regularly decontaminated using 1% sodium hypochlorite (*see* **Note 1**).
3. Vortex (Whirlimixer™).
4. Ice bucket with lid.
5. Microcentrifuge.
6. LightCycler capillaries. Stored at room temperature.
7. LightCycler block and adaptors. Stored at 4°C ± 2°C in a sterile plastic bag within the template addition area. Sterilize under U. V. between runs to prevent cross-contamination between assays.
8. LightCycler instrument (Roche Diagnostics).

2.2. HyBeacons and Synthetic Target for Probe Evaluation

1. Single-labeled HyBeacon probe FVG1: 5′-CTGTAFTCCTCGCCTGTCCP-3′ and dual-labeled HyBeacon probe FVG11: 5′-CTGTAFTCCTCGCCFGTCCP-3′. Both probes were synthesized by ATDBio Ltd, Southampton, UK and purified by reversed-phase HPLC (*see* **Note 2**). "P" indicates the 3′ phosphate and "F"

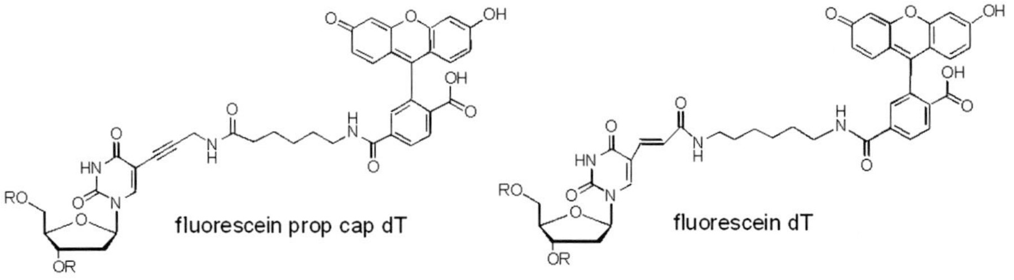

Fig. 2. Structures of fluorescein prop cap dT[28] and fluorescein dT.

are the fluorophore-labeled bases, fluorescein prop cap dT or fluorescein dT (Glen Research, **Fig. 2**). Probes were designed to be fully complementary to the arginine variant of the factor V gene. Probes were stored as 50 µL aliquots in 0.5 mL flip-cap tubes, in the dark at −20°C ± 2°C.

2. Complementary oligonucleotide FVGRC: 5′-GGACAGGC<u>G</u>AGGAATACAG-3′ (Sigma-Genosys, Cambridge, UK). Prepare 100 µ*M* stock solutions by adding the volume of tissue culture water recommended by the oligonucleotide company. Then prepare 10 µ*M* working stock solutions by adding 10 µL of the 100 µ*M* stock to 90 µL of tissue culture water in 0.5 mL flip-cap tubes, vortex and store at −20°C ± 2°C. The underlined "G" base indicates the position of the R506Q polymorphism, which would be replaced with "A" in the glutamine variant.
3. 10× TaKaRa PCR buffer containing 15 m*M* MgCl$_2$ (TaKaRa Bio Inc.). Stored at −20°C ± 2°C.
4. 25 m*M* MgCl$_2$ (GE Healthcare). Stored at −20°C ± 2°C.
5. Tissue culture water (Sigma-Aldrich Co. Ltd, catalogue number: W3500).

2.3. Polymerase Chain Reaction

1. DNA Taq polymerase (GE Healthcare, formerly Amersham Biosciences UK Ltd, catalogue number: 27-0799-05). Stored at −20°C ± 2°C.
2. 10× TaKaRa PCR buffer containing 15 m*M* MgCl$_2$ (TaKaRa Bio, Inc.). Stored at −20°C ± 2°C.
3. 25 m*M* MgCl$_2$ (GE Healthcare). Stored at −20°C ± 2°C.
4. dNTPs (GE Healthcare, catalogue number: 27-2035-01), 100 m*M* dATP, dCTP, dGTP, and dTTP. Prepare 5 m*M* working stocks by adding 125 µL dATP, 125 µL dTTP, 125 µL dGTP, and 125 µL dCTP to 9.5 mL of tissue culture water in 15 mL sterile apex centrifuge tube (Alpha laboratories, catalogue number LW 3076). Mix using a vortex, aliquot 500 µL into 1.5 mL flip-cap tubes and store at −20°C ± 2°C.
5. Forward primer FVF1.3: 5′-GGACTACTTCTAATCTGTAAGAGCAGATC-3′ and reverse primer FVFR3.5: 5′-GCCCCATTATTTAGCCAGGAGACCTAA-CAT-3′ (Sigma-Genosys). Prepare 100 µ*M* stock solutions by adding the volume of tissue culture water recommended by the oligonucleotide company. Then

prepare 10 μM working stock solutions by adding 20 μL of the 100 μM stock to 180 μL of tissue culture water in 0.5 mL flip-cap tubes, vortex and store at $-20°C \pm 2°C$.

6. Bovine serum albumin (BSA) 20 mg/mL (Roche Diagnostics, catalogue number: 711454). Prepare 1 mg/mL stocks of BSA by diluting 5 μL of 20 mg/mL stock in 95 μL tissue culture water. Store at $-20°C \pm 2°C$.
7. HyBeacon probes FVG1 (ATDBio Ltd) or FVG11 (ATDBio Ltd), prepared as in Section 2.2.1.
8. Tissue culture water (Sigma-Aldrich Co. Ltd, catalogue number: W3500).

3. Methods

The quality of HyBeacon melting peaks is dependent on the efficiency of the PCR process. As with any PCR test, development of HyBeacon assays may require optimization of reagents (e.g., $MgCl_2$ concentration), amplification strategy (e.g., symmetric versus asymmetric PCR), and thermal protocol (e.g., anneal temperature and hold-times). As PCR is a very sensitive method for DNA amplification, preventing cross-contamination is essential. Comprehensive quality procedures for laboratory design and operation are therefore required (*see* **Note 1**).

3.1. Evaluation of HyBeacon Probes Using Synthetic Oligonucleotide Targets

3.1.1. Reagent Preparation

1. Defrost all reagents at room temperature, ensuring that the HyBeacon probe is kept in the dark.
2. Once defrosted, vortex reagents to mix and centrifuge briefly.
3. Prepare two mastermixes in 0.5 mL flip-cap tubes. FVG1 (mastermix 1) is composed of 12 μL of TaKaRa PCR buffer, 7.2 μL of 25 mM $MgCl_2$ (total concentration 3 mM), and 0.81 μL of 22.16 μM FVG1 (final concentration 150 nM). Top up with 88 μL of sterile water to total volume of 108 μL and vortex. FVG11 (mastermix 2) is composed of 12 μL of TaKaRa PCR buffer, 7.2 μL of 25 mM $MgCl_2$ (total concentration 3 mM), and 1.00 μL of 17.87 μM FVG11 (final concentration 150 nM). Top up with 88 μL of sterile water to total volume of 108 μL and vortex.
4. Dilute the 10 μM stock solution of FVGRC oligonucleotide to 1.5 μM in a 0.5 mL flip-cap tube by adding 3 μL to 17 μL of sterile water. Vortex well and briefly centrifuge.
5. Place 10 capillaries in a LightCycler adapter block.
6. Add 2 μL of 1.5 μM FVGRC onto the side walls of the LightCycler capillaries in positions 1–4 and 6–9.
7. Add 2 μL of tissue culture water onto the side walls of the capillaries 5 and 10 as negative controls.

8. Add 18 µL of mastermix 1 directly to the 2 µL droplets in the capillaries 1–5 to enable mixing. Seal the capillaries with LightCycler caps.
9. Add 18 µL of mastermix 2 directly to the 2 µL droplets in the capillaries 6–10 to enable mixing. Seal the capillaries with LightCycler caps.

NOTE: It is better to add mastermix into the 2 µL droplet of sample than to add sample into the mastermix, as this ens ures better mixing and target amplification.

10. Centrifuge the capillaries briefly (~5 s) at 3345×g using the LightCycler adaptors to spin the reaction mixtures to the bottom of capillaries. Centrifugation at high speed and/or long periods can cause breakage of the capillaries.
11. Add the capillaries to positions 1–10 of a LightCycler rotor. Place the rotor in the LightCycler instrument.

3.1.2. Thermal Protocol and Melting Peak Analysis

1. Edit parameters for melting curve analysis as follows:
   - Denature by heating to 95°C (20°C/s) for 5 s.
   - Cool to 35°C (20°C/s) and incubate for 30 s to anneal probe and target.
   - Melting curve analysis from 35°C to 95°C with a 0.2°C/s transition rate, acquiring fluorescence continuously.
2. Process melting peaks using the LightCycler software by plotting the negative derivative of fluorescence with respect to temperature (-dF/dT on the *y*-axis) against temperature (*x*-axis).
3. Example results are presented in **Fig. 3**. The melting curves of 150 n*M* single-labeled probe (FVG1) and dual-labeled probe (FVG11) hybridized to 150 n*M* of complementary oligonucleotide FVGRC are presented in **Fig. 3A**. The dual-labeled probe exhibits approximately double the amount of background signal compared with the single-labeled probe, when single-stranded at 90°C. However, the LightCycler software demonstrates that the dual-labeled peak is approximately five times greater in height than the equivalent single-labeled probe peak (**Fig. 3B,C**). The signal-to-noise ratios of single- and dual-labeled factor V probes are calculated as 1.33 and 1.82, respectively (*see* **Note 3**). The improvement to probe signal-to-noise ratio is significantly greater than would be expected if the effect of the additional fluorophore was simply additive. Despite the elevated background fluorescence noise, dual-labeled probes consistently yield considerably greater signal-to-noise ratios and melting peak heights.

Fig. 3. Comparison of single- and dual-labeled factor V probes. (**A**) The dual-labeled probe FVG11 exhibits approximately twice the background fluorescence when dissociated (e.g., at 90°C) compared with the single-labeled FVG1 HyBeacon. (**B**) The dual-labeled probe generates melting peaks and derivatives which are more than five times as intense as those of the equivalent single-labeled HyBeacon. (**C**) Direct comparison of single- and dual-labeled probes.

(*Continued on opposite page*)

A

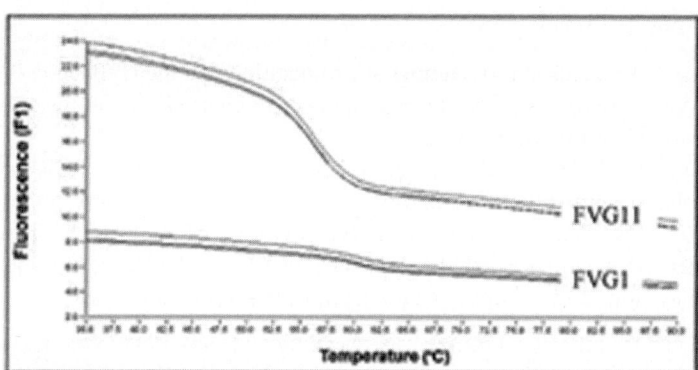

B

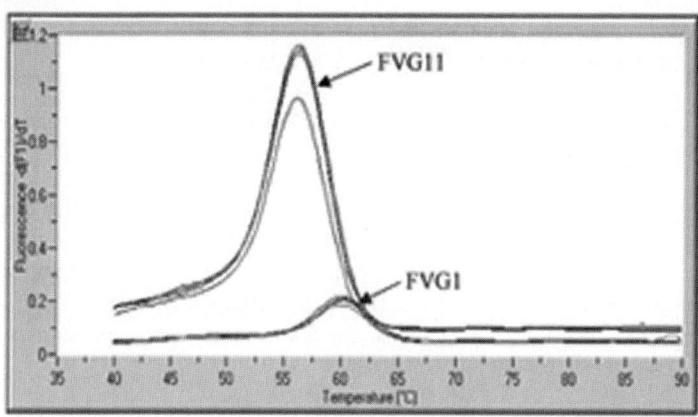

C

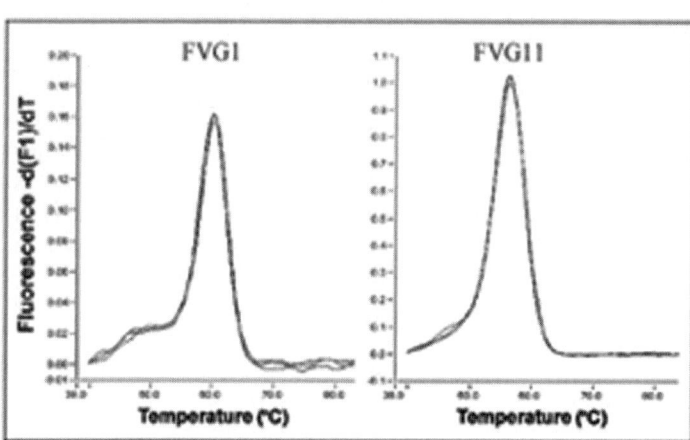

Fig. 3.

3.2. Polymerase Chain Reaction

3.2.1. Reagent Preparation

1. Defrost all reagents at room temperature, ensuring that the HyBeacon is kept in the dark. Vortex reagents for 5 s to mix and centrifuge briefly to let the solution settle down to the bottom of the tubes.
2. Prepare 35× PCR mastermix using 1.5 mL flip-cap tubes. PCR mastermixes comprise 1× TaKaRa PCR buffer, 0.5 μM FVF1.3 forward primer, 0.5 μM FVFR3.5 reverse primer, 1 U *Taq* polymerase, 3 mM total MgCl$_2$, 1 mM dNTPs, 5 ng/μL BSA, and 150 nM of dual-labeled FVG11 HyBeacon probe. The final concentrations and volumes of all reagents are presented below:

Reagents and concentration:	working concentration	volume (μL) to add
TaKaRa PCR buffer	1×	70.0
5 mM dNTPs	1 mM	140.0
25 mM MgCl$_2$	3 mM	42.0
10 μM forward primer FVF1.3	0.5 μM	35.0
10 μM reverse primer FVFR3.5	0.5 μM	35.0
1 mg/1 mL BSA	5 ng/μL	3.50
Taq polymerase	1 U	7.00
17.87 μM HyBeacon FVG11	150 nM	5.88
Tissue culture water		291.6
Total		630

3. Vortex reagents for 5 s to mix and store on ice in the dark until required.
4. Saliva samples are prepared by dilution with sterile water in a ratio of 1:1 in 1.5 mL flip-cap tubes. Samples are vortexed for 5 s to mix but are not centrifuged as this causes buccal epithelial cells to be pelletted.
5. Place 32 capillaries in a LightCycler adaptor block.
6. Add 2 μL of each sample to be analyzed (purified DNA or saliva) onto the side walls of LightCycler capillaries in positions 1–28.
7. Add 2 μL of known genotype samples (purified DNA or saliva) onto the side walls of LightCycler capillaries in positions 29 and 30 as positive controls.
8. Add 2 μL of tissue culture water to positions 31 and 32 for negative controls.
9. Add 18 μL of PCR mastermix directly into the 2 μL sample droplet to enable mixing. High quality procedures are required to prevent cross-contamination between samples.
10. Seal capillaries with LightCycler caps and centrifuge briefly (~5 s) at 3345×g using the LightCycler adaptors to spin the reactions mixtures to the bottom of the capillaries. Centrifugation at high speed and/or long periods can cause breakage of the capillaries.
11. Add capillaries to the appropriate positions of a LightCycler rotor. Place the rotor in a LightCycler instrument.

3.2.2. Thermal Protocol and Melting Analysis

1. Edit parameters for PCR amplification and melting curve as follows:
   - Initial denaturation stage for 1 min at 95°C.

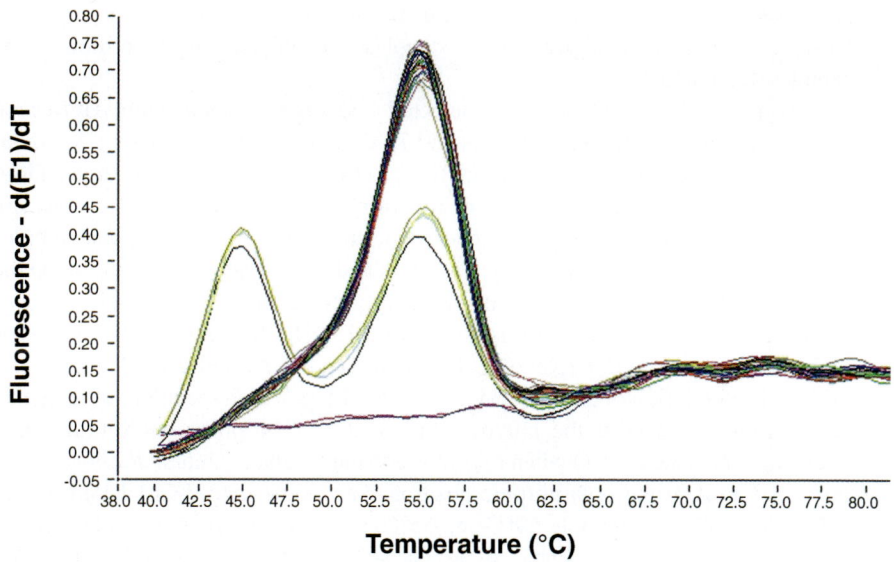

Fig. 4. Melting peak analysis of purified DNA samples using the dual-labeled factor V HyBeacon FVG11. Homozygous "wild-type" samples generate a single matched melting peak with a T_m of approximately 55°C. Whereas heterozygous samples generate melting traces possessing both matched and mismatched melting peaks, exhibiting T_ms of 55°C and 45°C, respectively. Multiple melting peaks for homozygous and heterozygous samples are presented along with no template controls.

- Amplification stage using 50 cycles comprising 0 s denaturation at 95°C; 10 s at 50°C for primer annealing and 10 s at 72°C for extension of products. Fluorescence is acquired during the annealing stage of cycling such that target amplification may be monitored in real-time through fluorescence increases.
- Melting curve analysis of HyBeacon probes comprises of 0 s denaturation at 95°C, cooling to 35°C for 30 s and a slow heat from 35°C to 75°C using a 0.1°C/s temperature transition rate and continuous fluorescence acquisition.
- The speed of analysis may be increased using modified thermal protocols and alternative polymerase enzymes (*see* **Note 4**).

2. Process melting peaks using the LightCycler software by plotting the negative derivative of fluorescence with respect to temperature (-dF/dT on the *y*-axis) against temperature (*x*-axis). Example melting peak data for the factor V assay is presented in **Fig. 4**.

3. Investigate melting curves and peaks to ensure that targets have been amplified and detected with all samples and positive controls. Also ensure that negative controls do not contain any melting peaks. The presence of melting peaks in negative controls will require that all samples be reanalyzed. The absence of peaks

from positive controls may also require that the whole sample set be repeated. However, the absence of peaks from samples will only require repetition of the samples that failed.

4. Polymorphic DNA sequences are detected and discriminated using HyBeacon probes through melting peak analysis and T_m determination. The melting temperatures derived from melting curve analyses permits identification of target DNA sequences, where T_ms derived from the fully complementary HyBeacons are considerably greater than mismatched probe interactions. Homozygous samples generate single melting peaks, which are matched or mismatched depending on the identity of the target sequence, whilst heterozygous samples produce melting traces possessing both matched and mismatched peaks. The factor V Leiden genotype is caused by a G to A SNP (R506Q). The single- and dual-labeled HyBeacons were designed to be fully complementary to the "wild-type" arginine allele of the gene, which possesses a G at the polymorphic position. The probe exhibits a single nucleotide mismatch (C/A) when hybridized to the "mutant" glutamine allele, which possesses an A at the polymorphic position. The single- and dual-labeled probes exhibit T_ms of approximately 59°C and 55°C, respectively, when hybridized to the fully complementary "wild-type" allele. The stability of the single- and dual-labeled probes are reduced when hybridized to the mismatched "mutant" allele, exhibiting T_ms of approximately 49.5°C and 45°C, respectively (**Fig. 4**; *see* **Note 5**).

4. Notes

1. It is highly recommended that sample preparation, PCR set-up, template addition and post-PCR areas of the laboratory are physically separated to prevent contamination. Transfer of suitable materials between laboratory areas should involve decontamination using 1% sodium hypochlorite and/or UV treatment. Pipettes cleaned with sodium hypochlorite should be wiped with sterile water to remove the bleach and dried to prevent rusting. Maintenance of high quality procedures precludes the requirement for decontamination of reaction tubes, pipette tips and LightCycler capillaries, e.g., through autoclaving and UV treatment.

2. The single-labeled probe FVG1: 5′-CTGTAFTCCTCGCCTGTCCP-3′ was labeled with fluorescein prop cap dT, where F and P are the fluorophore-labeled base and 3′-phosphate, respectively. Carboxyfluorescein was attached to the 5-position of a uracil nucleotide through a novel linkage chemistry and incorporated into probes as a phosphoramidite during solid-phase oligonucleotide synthesis *(28)*. The dual-labeled probe FVG11: 5′-CTGTAFTCCTCGCCFGTCCP-3′ was labeled with fluorescein dT (**Fig. 2**).

3. Fluorescence values were determined from HyBeacon melting curves at temperatures of probe $T_m \pm 10$°C for dissociated and hybridized states, respectively. Fluorescence measurements were made over this defined temperature range to remove some of the effects of temperature on fluorescence emission. The hybridized and dissociated fluorescence values provide a measure of signal-to-noise ratio and an indication of the effect of probe hybridization on emission, where signal-to-noise ratio is calculated as hybridized signal divided by dissociated signal.

The single-labeled FVG1 probe possesses a T_m of approximately 59°C when hybridized to fully complementary target sequence and exhibits fluorescence levels of 8 and 6 (arbitrary LightCycler fluorescence units) at 49°C and 69°C, respectively. The dual-labeled FVG11 probe possesses a T_m of approximately 55°C when hybridized to the fully complementary target sequence and exhibits fluorescence levels of 21.8 and 12 at 45°C and 65°C, respectively. The signal-to-noise ratios of single and dual-labeled factor V probes are calculated as 1.33 and 1.82, respectively. Probe signal-to-noise ratios have been found to be highly reproducible and exhibit low run-to-run and instrument-to-instrument variation. Determination of signal-to-noise ratios and T_ms using complementary and mismatched oligonucleotide targets enables evaluation and comparison of probe designs.

4. Faster thermal cycling protocols may be employed to reduce the duration of factor V tests. Employing alternate polymerase enzymes may also permit test duration to be reduced. For example, employing TaKaRa Z-Taq™ DNA polymerase, which is reported to possess a five-fold greater processivity compared with *Taq* polymerase, enables extremely rapid target amplification completing 50 cycles of PCR and melting analysis within 16 min. A three-step LightCycler PCR protocol may comprise 0 s denaturation at 95°C, 1 s for primer annealing at 50°C and 1 s extension at 72°C. A two-step LightCycler protocol may comprise 1 s denaturation at 95°C and 5 s at 65°C for primer annealing and target extension. The temperature transition rate during melting curve analysis may also be increased to reduce assay duration. Transition rates of 0.1 to 0.4°C/s may be employed depending on the number of samples in the analysis batch.

5. Over 500 purified DNA samples have been analyzed with the single-and dual-labeled HyBeacon probes. Melting peak analyses generated the correct genotype for each sample tested, confirmed by either RFLP or the Roche factor V Leiden hybridisation probe kit. Homozygous "wild-type" samples generated a single matched melting peak, whilst heterozygous samples yielded both matched and mismatched peaks. Homozygous "mutant" samples were not encountered during evaluation of the factor V tests. A series of saliva samples were analyzed with the single-and dual-labeled probes, without DNA purification. Mouthwashes, which comprise of approximately 50% saliva in water, were included directly in the assays. Buccal swabs expressed in water were also analyzed directly.

6. HyBeacon assays may also be performed in a high-throughput format, for example, simultaneously analyzing 384 samples using a block thermal cycler and a Light Typer/LightCycler 480 instrument (Roche Diagnostics).

7. The use of HyBeacon technology for the detection of the Factor V Leiden mutation has been licensed to Osmetech Molecular Diagnostics (Rockland, MA) for the production of an Analyte Specific Reagent (ASR) test.

Acknowledgments

Part of this work was supported by a grant from the Department of Trade and Industry (DTI) in the UK. Nittaya Gale was funded by a research grant from the EPSRC.

References

1. McCarthy, J. J. and Hilfiker, R. (2000) The use of single-nucleotide polymorphism maps in pharmacogenomics. *Nat. Biotechnol.* **18,** 505–508.
2. Kwok, P. Y. and Gu, Z. (1999) Single nucleotide polymorphism libraries: why and how are we building them? *Mol. Med. Today* **5,** 538543.
3. Brookes, A. J. (1999) The essence of SNPs. *Gene* **234,** 177–186.
4. Rommens, J. M., Iannuzzi, M. C., Kerem, B., *et al.* (1989) Identification of the cystic fibrosis gene: chromosome walking and jumping. *Science* **245,** 1059–1065.
5. Kerem, B., Rommens, J. M., Buchanan, J. A., *et al.* (1989) Identification of the cystic fibrosis gene: genetic analysis. *Science* **245,** 1073–1080.
6. Eichelbaum, M., Kroemer, H. K., and Mikus, G. (1992) Genetically determined differences in drug metabolism as a risk factor in drug toxicity. *Toxicol. Lett.* **64–65,** 115–122.
7. Wittwer, C. T., Herrmann, M. G., Moss, A. A., and Rasmussen, R. P. (1997) Continuous fluorescence monitoring of rapid cycle DNA amplification. *Biotechniques* **22,** 130–131, 134–138.
8. Ririe, K. M., Rasmussen, R. P., and Wittwer, C. T. (1997) Product differentiation by analysis of DNA melting curves during the polymerase chain reaction. *Anal. Biochem.* **245,** 154–160.
9. Winn-Deen, E. S. (1998) Direct Fluorescence Detection of Allele-Specific PCR Products Using Novel Energy-Transfer Labeled Primers. *Mol. Diagn.* **3,** 217–221.
10. Uehara, H., Nardone, G., Nazarenko, I., and Hohman, R. J. (1999) Detection of telomerase activity utilizing energy transfer primers: comparison with gel- and ELISA-based detection. *Biotechniques* **26,** 552–558.
11. Livak, K. J., Flood, S. J., Marmaro, J., Giusti, W., and Deetz, K. (1995) Oligonucleotides with fluorescent dyes at opposite ends provide a quenched probe system useful for detecting PCR product and nucleic acid hybridization. *PCR Methods Appl.* **4,** 357–362.
12. Lay, M. J. and Wittwer, C. T. (1997) Real-time fluorescence genotyping of factor V Leiden during rapid-cycle PCR. *Clin. Chem.* **43,** 2262–2267.
13. Happich, D., Schwaab, R., Hanfland, P., and Hoernschemeyer, D. (1999) Allelic discrimination of factor V Leiden using a 5′ nuclease assay. *Thromb. Haemost.* **82,** 1294–1296.
14. Tyagi, S. and Kramer, F. R. (1996) Molecular beacons: probes that fluoresce upon hybridization. *Nat. Biotechnol.* **14,** 303–308.
15. Bernard, P. S., Lay, M. J., and Wittwer, C. T. (1998) Integrated amplification and detection of the C677T point mutation in the methylenetetrahydrofolate reductase gene by fluorescence resonance energy transfer and probe melting curves. *Anal. Biochem.* **255,** 101–107.
16. Isacsson, J., Cao, H., Ohlsson, L., et al. (2000) Rapid and specific detection of PCR products using light-up probes. *Mol. Cell. Probes* **14,** 321-8, doi: 10.1006/mcpr/2000.0321.
17. Livak, K. J., Marmaro, J., and Todd, J. A. (1995) Towards fully automated genome-wide polymorphism screening. *Nat. Genet.* **9,** 341–342.

18. Svanvik, N., Westman, G., Wang, D., and Kubista, M. (2000) Light-up probes: thiazole orange-conjugated peptide nucleic acid for detection of target nucleic acid in homogeneous solution. *Anal. Biochem.* **281,** 26–35.

19. Tyagi, S., Marras, S. A., and Kramer, F. R. (2000) Wavelength-shifting molecular beacons. *Nat. Biotechnol.* **18,** 1191–1196.

20. Whitcombe, D., Theaker, J., Guy, S. P., Brown, T., and Little, S. (1999) Detection of PCR products using self-probing amplicons and fluorescence. *Nat. Biotechnol.* **17,** 804–807.

21. Crockett and Wittwer. (2001) Fluorescein-Labeled Oligonucleotides for Real-Time PCR: Using the Inherent Quenching of Deoxyguanosine Nucleotides. *Anal. Biochem.,* **290,** 89–97.

22. Bonnet, G., Tyagi, S., Libchaber, A., and Kramer, F. R. (1999) Thermodynamic basis of the enhanced specificity of structured DNA probes. *Proc. Nat. Acad. Sci. USA* **96,** 6171–6176.

23. Livak, K. J. (1999) Allelic discrimination using fluorogenic probes and the 5′ nuclease assay. *Genet. Anal.* **14,** 143–149.

24. Marras, S. A., Kramer, F. R., and Tyagi, S. (1999) Multiplex detection of single-nucleotide variations using molecular beacons. *Genet. Anal.* **14,** 151–156.

25. French, D. J., Archard, C. L., Brown, T., McDowell, D. G. (2001) HyBeacon probes: a new tool for DNA sequence detection and allele discrimination. *Mol. Cell. Probes.* **15,** 363–374.

26. French D. J., Archard, C. L., Andersen, M. T., and McDowell, D. G. (2002) Ultra-rapid DNA analysis using HyBeacon probes and direct PCR amplification direct from saliva. *Mol. Cell. Probes.* **16,** 319–326.

27 Dobson, N., McDowell, D. G., French, D. J., Brown, L. J., Mellor, J. M., and Brown, T. (2003) Synthesis of HyBeacons and dual-labelled probes containing 2′-fluorescent groups for use in genetic analysis. *Chem. Commun.* 1234–1235.

28 McKeen, C. M., Brown, L. J., Nicol, J. T. G., Mellor, J. M., and Brown, T. (2003) Synthesis of fluorophore and quencher monomers for use in Scorpion primers and nucleic acid structural probes. *Org. Biomol. Chem.* **1,** 2267–2275.

13

FIT-Probes in Real-Time PCR

Elke Socher and Oliver Seitz

Abstract

Forced intercalation probes (FIT-probes) are peptide nucleic acid-based probes in which the thiazole orange dye replaces a canonical nucleobase. FIT-probes are used in homogenous DNA detection. The analysis is based on sequence-specific binding of the FIT-probe with DNA. Binding of the FIT-probe places thiazole orange in the interior of the formed duplex. The intercalation of thiazole orange between nucleobases of the formed probe–target duplex restricts the torsional flexibility of the two heterocyclic ring systems. As a result, FIT probes show strong enhancements of fluorescence upon hybridization. A remarkable attenuation of fluorescence is observed when forcing thiazole orange to intercalate next to a mismatched base pair. This base specificity of fluorescence signaling, which adds to the specificity of probe–target recognition, allows the detection of single base mutations even at non-stringent hybridization conditions.

The performance of FIT-probes in real-time PCR is demonstrated in an assay for the SNPtyping of human H-*ras*. FIT-probe was added at the start of a real-time amplification containing the *wild-type* (G,G)-allele, *mutant* (T,T)-allele or *heterozygous* (G,T)-allele of the human H-*ras* gene. The identity of the target DNA is determined in real time due to significant differences in signal intensities.

Key Words: Cyanine dye; FIT probe; fluorogenic probe; genotyping; hybridization; peptide nucleic acid; real-time PCR; SNP; thiazole orange.

1. Introduction

The real-time monitoring of PCR amplification is an extremely powerful investigational and diagnostic tool (*1*) that has found almost ubiquitous use in a range of experimental research fields (*2*) as well as medical applications (*3*). The development of reliable, fast, and inexpensive, and most importantly, accurate methods for the detection of nucleotide mutation and polymorphisms is central to the modern science of molecular genetics. Environmentally sensitive dyes that experience increases of fluorescence when bound to double-stranded DNA

From: *Methods in Molecular Biology, Vol. 429: Molecular Beacons: Signalling Nucleic Acid Probes,
Methods and Protocols* Edited by: A. Marx and O. Seitz © Humana Press Inc., Totowa, NJ

represent a simple way of reporting the production of amplicon during the PCR. These dyes typically bind to the minor groove *(4)* or by intercalating *(5)* to detect accumulation of both specific and nonspecific PCR products. The latter may generate false positive signals. Thus, post-PCR sequence analyses or melting analyses are required to confirm the identity of the PCR product. The use of fluorogenic nucleic acid-based probes allow for improvements of target specificity *(6)*. Since only specific amplification products are detected there is no need to perform post-PCR analyses.

Most of the current sequence-specific DNA-detection probes such as adjacent probes *(7,8)*, Molecular Beacons *(9)* or TaqMan probes *(10,11)* draw on the energy transfer between two interacting chromophores. These methods distinguish between the bound and the unbound state of probe molecules and report the formation of the probe–analyte duplex. The sequence specificity of DNA-detection is governed by the selectivity of probe hybridization. To increase target specificity, one must add another sequence discriminating event or improve the hybridization selectivity of the probe molecule itself. DNA analogous probes such as peptide nucleic acids (PNAs) bind complementary DNA stretches with higher affinity than DNA-based probes *(12)*. As a result PNA probes can be shortened to a degree that enables very high sequence specificity of probe–target binding.

We designed a fluorogenic probe with an aim to combine the excellent hybridization properties of PNA with another level of sequence discrimination provided by the fluorophore itself *(13,14)*. The so-called FIT (forced intercalation of thiazole orange)-PNA probes *(15)* contain a thiazole orange dye that has been introduced by replacing a canonical nucleobase (**Fig. 1**). FIT–PNA probes signal hybridization by enhancement of fluorescence which occurs due to binding of the perfectly complementary target. It is a particular feature that FIT–PNA probes show attenuated fluorescence in case undesired binding to single-mismatched targets had occurred.

A homogeneous three step PCR genotyping strategy was developed which does not require processing after PCR. **Fig. 2** shows how FIT probes can be used for allelic discrimination. Fluorescence will be measured during the annealing step when the FIT probe binds to the target. Hybridization of the FIT–PNA probes to a perfectly complementary sequence results in strong enhancements of thiazole orange emission because intercalation between the base pairs of the newly formed target–probe duplex restricts the torsional flexibility around the central methine bridge which closes an important non-radiative decay channel. The degree of fluorescence enhancement depends on the probe sequence. A mismatched base pair greatly reduces the efficiency of probe hybridization which leads to lower melting temperature of mismatched probe–target duplexes. PNA probes (11–16 bp) exhibit greater differences in T_m values between matched and mismatched probes compared to longer DNA probes (20–30 bp). The larger

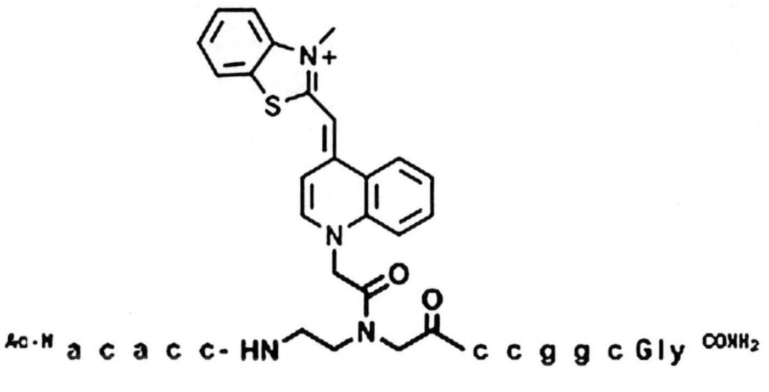

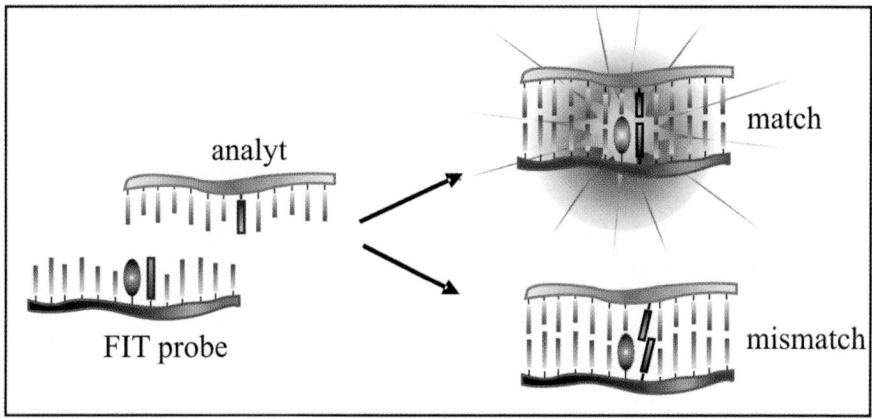

Fig. 1. A typical FIT–PNA probe.

Fig. 2. Concept of forced intercalation. Fluorescence occurs only with the perfectly matched target.

T_m window (8–20°C) achieved with PNA probes provides more accurate allelic discrimination. This broad T_m window makes it easy to design probes that have an optimal T_m (65–70°C) for the matched probe–target duplex above the annealing temperature of PCR (normally 60–65°C) and a T_m below this temperature for the mismatched probe–target duplex (**Fig. 2**). In addition to hybridization specificity, FIT probes offer an additional level of sequence discrimination that has its origin in the responsiveness of the thiazole orange base surrogate. Mismatched base pairs are characterized by a rapid equilibrium between innerhelical and extrahelical conformations. The high flexibility of mismatched base pairs provides a low viscosity environment to thiazole orange. Adjacent base mismatches are hence expected to allow torsions around the methine** bridge between the two ring systems of thiazole orange. Thus, thiazole orange fluorescence remains low even when a mismatched probe–target duplex had formed. As a result, the mismatch fluorescence is always lower than match fluorescence even at temperatures below the T_m of the mismatched probe–target duplexes (*see* **Fig. 3A**).

The FIT probe does not interfere with the PCR and can be included in the sample mixture. The fluorescence signal is equal to the amount of PCR product present. In the heterozygous sample, only half of the PCR product is complementary to the PNA probe which affords lower signal intensity. After mixing the sample and reaction components, the user runs the assay in a closed-tube format. According to significant different signal intensity, it is possible to determine the identity of the target DNA in real time (**Fig. 4**). As model system a 275 bp fragment of the human H-*ras* gene including a carcinogenic single base mutation was chosen for amplification.

2. Materials

2.1. Synthesis of FIT Probes (2 μM)

1. 1-ml reactors for the synthesizer
2. NovaSyn®TGR resin loaded with Fmoc-glycine (190 μmol/g, Novabiochem)
3. Fmoc cleavage: DMF/piperidine (4:1) [v/v]
4. Coupling of PNA-building blocks: 4 eq. Fmoc/Bhoc-protected PNA monomer (Applied Biosystems) in *N*-methylpyrolidine (NMP), 8 eq. *N*-methylmorpholine (NMM) in DMF, 3.6 eq. 1H-benzotriazolium 1-[bis(dimethylamino)methylene]-5chloro,-hexafluorophosphate (1-),3-oxide (HCTU) in NMP
5. Coupling of Fmoc-Aeg(TO)-COOH: 4 eq. Fmoc-protected TO-PNA monomer in NMP, 8 eq. NMM in DMF, 3.6 eq. HCTU in NMP, 4 eq. Pyridinium *p*-toluenesulfonate (PPTS) (*see* **Note 1**)
6. Capping: acetic ester anhydride/2,6-lutidine/DMF (5:6:89) [v/v/v]
7. Final washing: 600 μL ethanol
8. Final cleavage: cysteine methyl ester hydrochloride (5 mg, 29 μmol) in TFA/*m*-cresol/H_2O (37:2:1, 1 mL); 500 μL TFA
9. Purification: diethyl ether; water-equilibrated SepPak® C18 cartridge (gradient solution Acetonitril/H_2O: 20:80, 0.1% TFA; 20:80, 0.1% TFA; 40:60, 0.1% TFA; 60:40, 0.1% TFA; 80:20, 0.1% TFA; 2 mL each); semipreparative HPLC using RP-C18 A polaris (5 μ, PN A 2000-250 × 100)-columns (Varian)

2.2. Characterization of FIT Probes

1. FIT probe is dissolved in degassed water (*see* **Note 2**) at 8 μM concentration, stored in aliquots at −20°C (*see* **Note 3**)
2. *Ras* G,G-allele FIT probe (8 μM): AcNHacacc Aeg(TO) ccggc-GlyCONH2
3. Short synthetic DNA (purchased from BioTeZ-Buch GmbH), sequence corresponds to the *ras* G,G-allele or to the *ras* T,T-allele PCR target. DNA is dissolved in degassed water (*see* **Note 2**) at 8 μ*M* concentration, stored in aliquots at −20°C

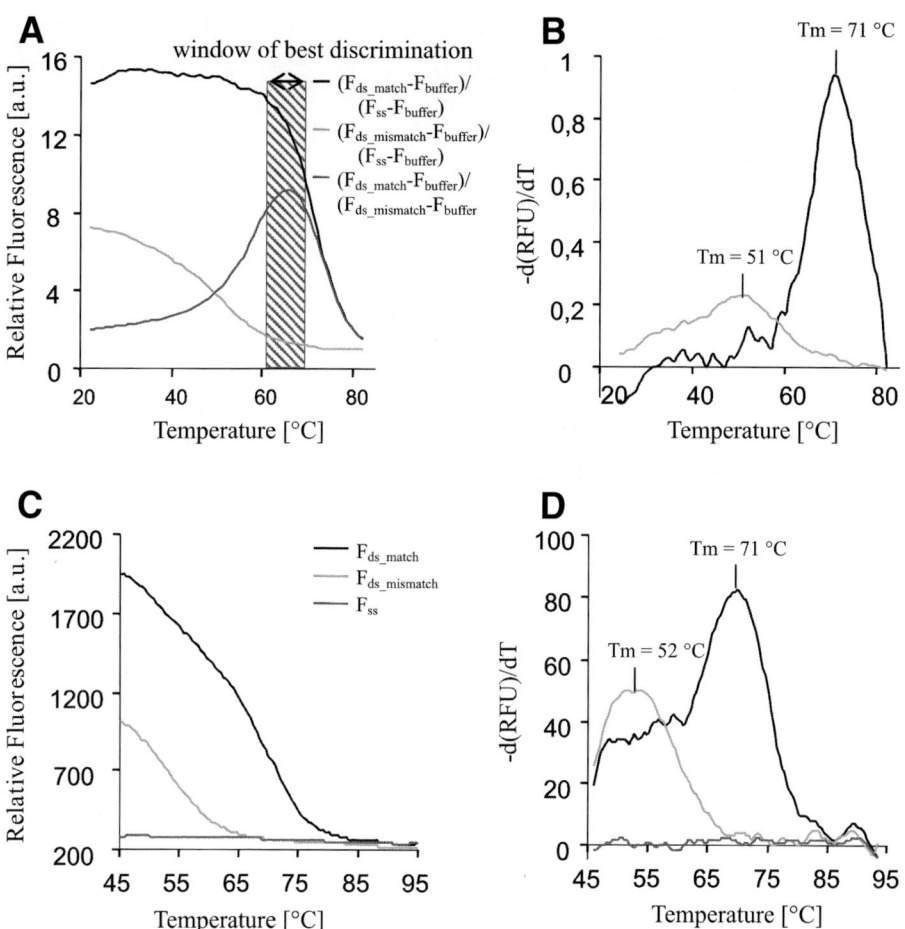

Fig. 3. Melting analysis of FIT-probes and FIT-probe–target complexes. (**A**) Temperature dependence of fluorescence increase upon matched hybridization $(F_{ds\_match}-F_{buffer})/(F_{ss}-F_{buffer})$ and single mismatched hybridization $(F_{ds\_mismatch}-F_{buffer})/(F_{ss}-F_{buffer})$ and match/mismatch discrimination $(F_{ds\_match}-F_{buffer})/(F_{ds\_mismatch}-F_{buffer})$. Performed on the Cary Eclipse Fluorescence spectrometer (Varian). (**B**) Negative first derivative to fluorescence increases measured for matched and mismatched hybridization. (**C**) Melting curve of single stranded FIT-probe and FIT-probes hybridized to matched DNA or mismatched DNA as measured on the iCycler (Bio-Rad). (**D**) The negative first derivative shows the differences in T_m of matched and mismatched duplexes.

matched DNA (G, G-allele): 3′-TGT GGC **G**GC CG-5′
mismatched DNA (T,T-allele): 3′-TGT GGC **T**GC CG-5′
4. Buffer (Qiagen), 10× PCR buffer (50 mM KCl, 20 mM Tris–HCl, 1.5 mM MgCl$_2$, pH 8.7 at 20°C)

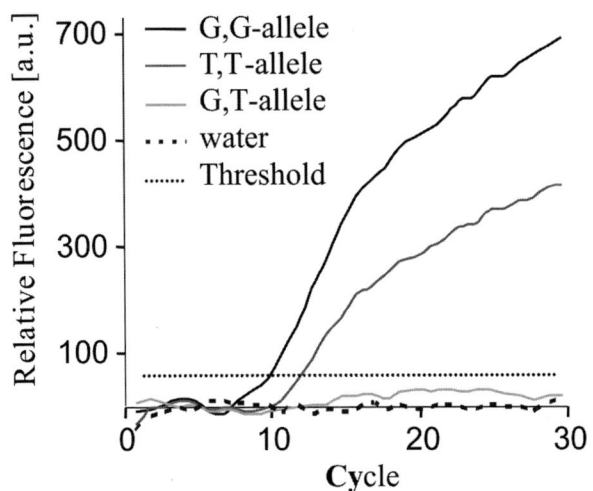

Fig. 4. Real-time PCR amplification plot of H-*ras* (G,G)-allele DNA, H-*ras* (T,T)-allele DNA and heterozygous DNA using T,T-allele specific FIT-probe ^{AcNH}acacc-Aeg (TO)-ccggc-Gly^{CONH2} performed on the iCycler (Bio-Rad).

2.3. Real-Time PCR

1. Taq DNA polymerase 5 U/μL (Qiagen)
2. 10× PCR buffer (50 mM KCl, 20 mM Tris–HCl, 1.5 mM MgCl$_2$, pH 8.7 at 20°C), (Qiagen)
3. BSA 1 mg/mL (Fluka, Biochemika)
4. dNTPs: 10 mM each of dATP, dCTP, dGTP, and dTTP (Qiagen)
5. 10× primers (BioTeZ, Berlin-Buch GmbH)
   ras forward primer (6 μM):   5′-GATAAACGACGGCCAGTGAATTG-3′
   ras reverse primer (6 μM):   5′-CAGGAAACAGCTATGACCATGAT-3′
6. 20× FIT probe (*see* **Note 4**)
   ras G,G-allele specific FIT probe (8 μM): ^{AcNH}acacc Aeg(TO) ccggc-Gly^{CONH2}
7. DNA Template:
   A 68-bp fragment of synthetic DNA was ligated into the pCR2.1 plasmid vector, transformed in *E. coli*, and dilutions of the cloned plasmid were used in PCR as template.

2.4. Instrumentation

1. ResPep parallel synthesizer (Intavis AG)
2. 1100 HPLC system (Agilent)
3. Cary Eclipse Fluorescence spectrometer (Varian)
4. NanoDrop® ND-1000 UV-Vis Spectrophotometer
5. Amplification, detection, and data analysis were performed with the iQ5 real-time PCR detection system (Bio-Rad).

3. Methods

3.1. Design and Synthesis of FIT Probe

1. Thiazole orange is placed adjacent to the expected mutation site (*see* **Note 5**).
2. Probes were designed to fulfill the following criterion: primers had an annealing temperature of around 65°C, the T_m of the PNA should preferably be 3°C higher, T_m between 65°C and 70°C (*see* **Note 6**).
3. T_m for mismatch hybridization of the PNA probe should preferably be 5°C lower (or even lower) than the annealing temperature of the PCR primers (*see* **Note 7**).
4. Sequence length has to be adjusted to fulfill the criteria 3.1.2 and 3.1.3 (*see* **Note 8**).
5. FIT probes were synthesized by using automated solid-phase synthesis and purified by high pressure liquid chromatography:

   a. Resin (11 mg) was allowed to swell in DMF (2 mL) for 30 min in 1 mL reactors, than transferred to the synthesizer and washed with DMF ($2 \times 180 \mu L$)
   b. *Fmoc cleavage*: $2 \times 100 \mu L$ cleavage solution was added to the resin for 2 min, then resin was washed with DMF ($1 \times 180 \mu L$, $3 \times 100 \mu L$, $1 \times 180 \mu L$)
   c. *Coupling of Standard PNA-building blocks*: A preactivation vessel was charged for 8 min with a 0.6 *M* HCTU solution in NMP (12 μL), a 4 *M* NMM solution in DMF (4 μL), and a 0.6 *M* PNA monomer solution in NMP (27 μL). Forty microliters of the preactivation solution was transferred to the resin. After 30 min, the resin was washed with DMF ($2 \times 180 \mu L$) (**Note 9**).
   d. *Coupling of Fmoc-Aeg(TO)-COOH*: A preactivation vessel was charged for 8 min with a 0.6 *M* HCTU solution in NMP (12 μL), a 4 *M* NMM solution in DMF (4 μL), and a 0.6 *M* Fmoc protected TO-PNA monomer solution in NMP (0.6 *M* TO-PNA monomer, 0.6 *M* PPTS, 27 μL). Forty microliters of the preactivation solution was transferred to the resin. After 30 min, the resin was washed with DMF ($2 \times 180 \mu L$). The procedure was repeated once.
   e. *Capping*: The capping solution was added to the resin for 3 min. Resin was washed with DMF ($3 \times 200 \mu L$, $3 \times 100 \mu L$).
   f. *Final washing*: $2 \times 200 \mu L$ ethanol
   g. *Cleavage*: The cleavage solution was passed through dried resin in 40 min ($5 \times 200 \mu L$). The resin was washed with TFA (500 μL). The combined filtrates were concentrated *in vacuo*.
   h. *Purification*: To the concentrated cleavage solution was added cold diethyl ether. The precipitate was collected by centrifugation and disposal of the supernatant. The residue was dissolved in water and precleaned by using a water-equilibrated SepPak® C18 cartridge. Colored eluates obtained upon gradient elution ($1 \times$ 20:80 acetonitril:H_2O: 0.1% TFA; $1 \times$ 40:60 acetonitril:H_2O: 0.1% TFA; $1 \times$ 80:20 AcN:H_2O: 0.1% TFA; $1 \times$ 80:20 acetonitril:H_2O: 0.1% TFA; 2 mL each) were analyzed by HPLC and MALDI-TOF/MS and purified by semipreparative HPLC (Eluent A: 0.1% TFA in water + 1% acetonitril, Eluent B: 0.1% TFA in acetonitril + 1% water, gradient: 3–40% B in 30 min)
   i. *Determination of yields*: Purified PNA was dissolved in 500 μL of water. An aliquot of 5 μL was diluted to 1 mL and the optical density was measured

at 260 nm by using a quartz cuvette with a 10-mm path length. For estimating the absorption coefficient at 260 nm, the calculation program provided by Gensetoligos *(16)* was used (**Note 10**).

3.2. Characterization of FIT Probes

1. To ensure that the selected FIT probe provides maximum sequence discrimination at the PCR-annealing temperature, perform melting analyses with synthetic oligonucleotides. Melting curves were generated by measuring fluorescence of the FIT–PNA probe as a function of temperature in the presence of oligonucleotide target as well as in the absence of target (*see* **Fig. 3**). These measurements can be performed with a fluorescence spectrometer equipped with a thermostated cuvette block or in a real-time PCR cycler (**Fig. 3A,C**, respectively)
   Melting curve program: 95°C for 1 min, 45°C for 1 min, followed by a temperature ramp from 55 to 95°C with a slope of 0.1°C/s.
2. The concentration of FIT probe and DNA target (if added) is 1 μM. To determine the best annealing temperature, that means the temperature window of highest match/mismatch discrimination, divide the buffer corrected fluorescence intensity of matched duplex by the buffer corrected fluorescence intensity of mismatched duplex $((F_{\text{double strand\_match}}-F_{\text{buffer}})/(F_{\text{double strand\_mismatch}}-F_{\text{buffer}}))$ (*see* **Fig. 3A**). For assessing the fluorescence increase expected during formation of matched duplexes divide the buffer corrected fluorescence intensity of matched double strand by the buffer corrected fluorescence intensity of single-strand probe $((F_{\text{double strand\_match}}-F_{\text{buffer}})/(F_{\text{single strand}}-F_{\text{buffer}}))$.
3. *Analysis*: The discrimination curve $((F_{\text{double strand\_match}}-F_{\text{buffer}})/(F_{\text{double strand\_mismatch}}-F_{\text{buffer}}))$ shows that FIT probe ^{AcNH}acacc Aeg(TO) ccggc-Gly$^{\text{CONH2}}$ fluoresces on matched target DNA up to ninefold brighter than on mismatched DNA (**Fig. 3A**). The window of best discrimination (8.2–9.1-fold) is between 61°C and 69°C. Note that matched hybridization of the FIT probe always gives higher fluorescence enhancements than mismatched hybridization at ANY temperature (due to the fluorophor's responsiveness). Thus, mismatched targets can be discriminated at non-stringent conditions, which means at temperatures below the T_{m} of mismatched probe—target duplexes.
4. *Melting analysis*: The maximum of the negative first derivative to the fluorescence-detected dissociation curve gives the T_{m} of the probe–target duplex. **Fig. 3B** shows a $T_{\text{m}} = 71°C$ for the matched probe–target duplex and a $T_{\text{m}} = 51°C$ for the mismatched probe–target duplex. The high $\Delta T_{\text{m}} = 20°C$ attests to the high target specificity of PNA hybridization.

3.3. PCR Preparation

1. Measure the absorbance of DNA (plasmid) at 260 nm and adjust to 0.1 ng/μL.
2. *Master mix*: 2 μL PCR buffer, 1.5 m*M* MgCl$_2$, 500 n*M* of forward and reverse primer, 150 μ*M* dNTPs, 400 n*M* FIT probe (*see* **Note 11**), 3 μL BSA (1 mg/mL), 0.1 μL (0.5 U) Taq DNA Polymerase (*see* **Note 12**).
3. Aliquot the above master mix into 12 wells (*see* **Note 13**).

4. To obtain three repeats add 3 × 2 µL wild-type target (2 ng plasmid), 3 × 2 µL heterozygous target, 3 ×2 µL mutant target to the prepared wells.

3.4. Real-Time PCR Amplification in 96 Plates

1. Briefly spin the plate in a centrifuge.
2. PCR thermal profile consisted of an initial hold of 94°C, 1 min followed by activation of 94°C, 3 min, followed by 30 cycles of PCR program: of 95°C for 15 s, 65°C for 15 s, and 72°C for 20 s.
3. Fluorescence was recorded during the annealing phase of each PCR cycle with the FAM filter pair (excitation: 485 nm/20×, emission: 530 nm/30 *M*).

3.5. Analysis of Real-Time PCR Data

1. The obtained data are analyzed by means of amplification plots (see **Fig. 4**) produced by iQ™5 Optical System Software Version 1.0 (Bio-Rad).
2. Baseline value was the average fluorescence value of PCR cycles 2 to around 10 (*see* **Note 14**).
3. The threshold value will be defined on the basis of the matched amplification plot (*see* **Note 15**).
4. Analysis: The no-template control shows that the relative fluorescence remains on the baseline level. Strong fluorescence of G,G-allele-specific FIT probe occurs when matched *ras* G,G-allele has been subjected to PCR. In contrast fluorescence is low through all cycles in case of PCR of the ras T,T-allele. Heterozygous DNA-target also gives a clear fluorescence signal. The results demonstrate that FIT probe fluorescence allows the single nucleotide-specific genotyping.

4. Notes

1. Add to the solution of Fmoc protected TO-PNA monomer 8 eq. of PPTS for better solubility of monomer.
2. Water was purified with a Milli-Q® Ultrapure Water Purification System (Millipore Corp.)
3. Avoid repeat freeze/thaw cycles of fluorogenic probe, store working stock in water in 50 µL aliquots at −20°C protected from light to prevent degradation of probe.
4. Vortex FIT probe before use for 1 min.
5. Our previous measurements suggested that more than 20-fold fluorescence enhancements can be observed when TO is placed next to an adenine-thymine base pair; however, the amplification plots shown in **Fig. 4** show that real-time geno-typing can also be performed if the polymorphic site in question is embedded in a gc-rich segment (like in the *ras*-G,G-allele specific probe studied). Avoid self-complementary sequence stretches in the FIT–PNA, consider parallel and antiparallel intra- and intermolecular hybridization.
6. Annealing temperatures of the primers were 65°C with a 258 bp product. In order to minimize primer–dimer formation, the maximum 5′-self-complementarity score was 4 and the maximum 3′-self-complementarity score was 2. Check interaction between primers and probes.

7. To ensure an adequate discrimination between the two alleles it may prove necessary to shift the polymorphic site along the probe sequence such that probe length can be minimized.
8. The length of FIT-probe depends on the sequence especially on the gc content. Purine-rich PNA sequences have higher T_m's than pyrimidine-rich sequences. Please note that TO confers no destabilization.
9. Double coupling of PNA monomer following the TO-PNA monomer coupling and of each PNA monomer after the 10th coupling.
10. The sample concentration was calculated by approximating for thiazole orange $\varepsilon = 9400$ L mol^{-1} [ε_{260}(TO) $\approx \varepsilon_{260}$(thymine)].
11. The optimal concentration depends on concentration of primer and amplication efficiency. A good staring range for the probe concentration is $0.2–0.5$ μM. Optimal annealing temperature, primer concentration, and probe concentration must be determined in respective order.
12. Total reaction volume is 20 μL. Add first water, buffer, MgCl$_2$, forward and reverse primer together with FIT-probe and vortex them briefly and centrifuge the mixture for 15 s. Than add dNTP's and Taq DNA polymerase and centrifuge shortly again.
13. The usual total reaction volume is 20 μL in plates or tubes.
14. Baseline value will be fixed for every PCR sample individually.
15. Under optimal conditions the intensity of mismatched amplification plot should be below the threshold value.

References

1. Shi, M. M. (2001) Enabling large-scale pharmacogenetic studies by high-throughput mutation detection and genotyping technologies. *Clin. Chem.* **47**, 164–172.
2. Lantz, P. G., Abu al-Soud, W., Knutsson, R., Hahn-Hagerdal, B., and Radstrom, P. (2000) Biotechnical use of polymerase chain reaction for microbiological analysis of biological samples. *Biotechnol. Annu. Rev.* **5**, 87–130.
3. Bustin, S. A. and Dorudi, S. (1998) Molecular assessment of tumour stage and disease recurrence using PCR-based assays. *Mol. Med. Today* **4**, 389–396.
4. Morrison, T. B., Weis, J. J., and Wittwer, C. T. (1998) Quantification of low-copy transcripts by continuous SYBR Green I monitoring during amplification. *Biotechniques* **24**, 954–958, 960, 962.
5. Lee, L. G., Chen, C. H., and Chiu, L. A. (1986) Thiazole orange: a new dye for reticulocyte analysis. *Cytometry* **7**, 508–517.
6. Walker, N. J. (2002) Tech.Sight. A technique whose time has come. *Science* **296**, 557–559.
7. Cardullo, R. A., Agrawal, S., Flores, C., Zamecnik, P. C., and Wolf, D. E. (1988) Detection of nucleic acid hybridization by nonradiative fluorescence resonance energy transfer. *Proc. Natl. Acad. Sci. USA* **85**, 8790–8794.
8. Heller, M. J. and Morrision, L. E. (1985) In Rapid Detection and Identification of Infectious Agents (Kingsbury, D. T., and Falkow, S., eds.), Academic Press, New York, pp. 245–256.

9. Tyagi, S. and Kramer, F. R. (1996) Molecular beacons: probes that fluoresce upon hybridization. *Nat. Biotechnol.* **14,** 303–308.

10. Holland, P. M., Abramson, R. D., Watson, R., and Gelfand, D. H. (1991) Detection of specific polymerase chain reaction product by utilizing the 5′–3′ exonuclease activity of Thermus aquaticus DNA polymerase. *Proc. Natl. Acad. Sci. USA* **88,** 7276–7280.

11. Livak, K. J., F., S. J., Marmaro, J., Giusti, W., and Deetz, K. (1995) Oligonucleotides with fluorescent dyes at oppostie ends provide a quenchend probe system useful for detectiong PCR product and nucleic acid hybridization. *PCR Meth. Appl.* **4,** 357–562.

12. Egholm, M., Buchardt, O., Christensen, L., et al. (1993) PNA hybridizes to complementary oligonucleotides obeying the Watson-Crick hydrogen-bonding rules. *Nature* **365,** 566–568.

13. Köhler, O. and Seitz, O. (2003) Thiazole orange as fluorescent universal base in peptide nucleic acids. *Chem. Commun. (Camb)*, 2938–2939.

14. Seitz, O., Bergmann, F., and Heindl, D. (1999) A Convergent Strategy for the Modification of Peptide Nucleic Acids: Novel Mismatch-Specific PNA-Hybridization Probes. *Angew. Chem. Int. Ed.* **38,** 2203–2206.

15. Köhler, O., Jarikote, D. V., and Seitz, O. (2005) Forced intercalation probes (FIT Probes): thiazole orange as a fluorescent base in peptide nucleic acids for homogeneous single-nucleotide-polymorphism detection. *Chembiochem* **6,** 69–77.

16. www.gensetoligos.com

14

SNP Genotyping by Unlabeled Probe Melting Analysis

Maria Erali, Robert Palais, and Carl Wittwer

Abstract

Fluorescent nucleic acid detection in polymerase chain reaction (PCR) generally uses oligonucleotide probes labeled with covalently attached dyes. However, unlabeled oligonucleotides in the presence of saturating DNA dyes can also serve as hybridization probes. The DNA dye, LCGreen® Plus, and a 3′-blocked unlabeled probe are added before amplification, and asymmetric PCR is performed at a 1:5 to 1:10 primer ratio. After PCR is complete, fluorescent melting curves reveal both probe melting at low temperature and amplicon melting at high temperature. After background removal, the melting temperature(s) of the probe/target duplex specific to the allele(s) amplified are revealed. Probes between 20 and 40 bp with T_ms between 50 and 85°C are effective. The method requires only three standard oligonucleotides and endpoint fluorescence melting. No real-time PCR or allele-specific amplification is needed. Unlabeled probes are inexpensive, provide the sequence specificity of probes, and allow simultaneous identification of multiple alleles by melting analysis.

Key Words: Unlabeled probes; genotyping; LCGreen Plus; asymmetric PCR; melting analysis.

1. Introduction

Dye methods remain very popular in real-time polymerase chain reaction (PCR), even without the specificity of an internal probe. The reason is simple. Fluorescent probes with covalently attached labels are expensive irrespective of the specific design. In addition to the expense, they require more time to obtain from commercial suppliers, quality control is problematic, and a new probe is needed for each target of interest. In contrast, a DNA dye can be used for any target. Furthermore, product melting analysis can identify the amplified product with an accuracy that depends on the resolution of the melting instrument (*1*). Even so, probes remain preferred by many, especially in clinical diagnostic

From: *Methods in Molecular Biology, Vol. 429: Molecular Beacons: Signalling Nucleic Acid Probes, Methods and Protocols* Edited by: A. Marx and O. Seitz © Humana Press Inc., Totowa, NJ

applications. A method with the specificity of a probe and the cost and simplicity of closed-tube dye analysis would be desirable.

The dye, SYBR® Green I, was first used in real-time PCR in 1997 *(2,3)*. SYBR Green I is commonly used to detect duplex PCR products. In addition, SYBR Green I can successfully detect probe/product melting if single-stranded product is isolated and immobilized *(4,5)*. Although this requires processing, genotyping by melting is possible without labeled probes. However, closed-tube analysis in solution with SYBR Green I is problematic when multiple duplexes are present in solution. Higher T_m products are preferentially detected and heteroduplexes are not observed well at dye concentrations compatible with PCR *(6)*. Indeed, genotyping in solution with unlabeled probes after asymmetric PCR was not successful when SYBR Green I was used *(7)*. At PCR compatible concentrations, SYBR Green I does not saturate all DNA duplexes present and the dye appears to redistribute to higher T_m duplexes during melting.

The LCGreen® family of dyes detects all duplexes present in solution because saturating concentrations can be used that do not inhibit PCR *(8)*. Unlabeled probes are included in PCR that are not extended by polymerase because they are 3′-blocked. After asymmetric PCR, the probes anneal to single-stranded product. Melting curves show regions of both probe/product and product/product melting *(9)*. Different alleles result in different probe/product melting transitions based on the stability of the mismatches present *(7)*. It is easiest to see the transitions by plotting the negative derivative (-dF/dT) of fluorescence (F) versus temperature (T). **Fig. 1** shows the method. Often, background fluorescence is high, resulting in an elevated background, especially at low temperatures.

Background fluorescence can be subtracted using a technique which has been found to be superior to the traditional method of estimating the melting curve as fluorescence percentage between two baselines. The new technique fits a decreasing exponential to the slope of the raw fluorescence versus temperature curve in regions where no melting occurs. Although exponentials are usually fit to values rather than slopes, both approaches require two equations with two unknowns. Since the contribution of background to total fluorescence is not known, raw fluorescence values cannot be used directly to determine background. However, the slope of fluorescence versus temperature in any two regions where no melting occurs is entirely attributable to background (because the linearly additive slope contribution from melting is zero) and these are sufficient to fit the background. This exponential approach is superior to linear baseline normalization *(10)* in removing background fluorescence from unlabeled probe and combined unlabeled probe/amplicon melting curves that cannot be normalized using baselines. It also provides better background removal when multiple small amplicons are analyzed *(11,12)*.

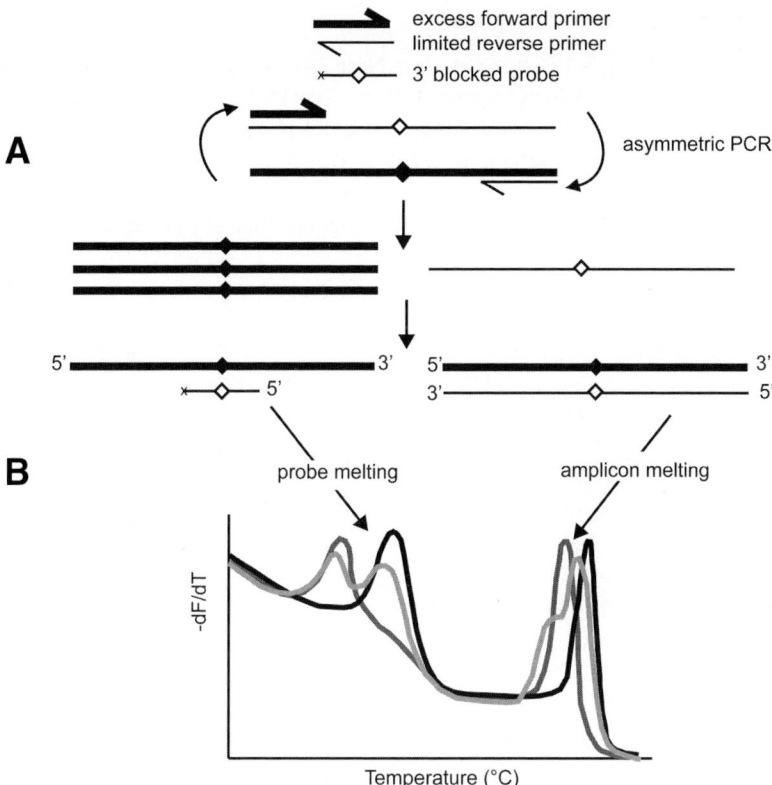

Fig. 1. Genotyping by unlabeled probe melting analysis. (**A**) Asymmetric PCR in the presence of LCGreen dye and an unlabeled probe produces both probe/product and product/product duplexes. (**B**) Melting analysis reveals both low temperature (probe/product) and high temperature (product/product) melting transitions. Complete single nucleotide polymorphism genotyping is possible by analysis of either region. However, probe genotyping does not require high-resolution instrumentation or analysis techniques. Data from Ref. *(9)*.

The following procedure is for unlabeled probe genotyping of the common clinical target, factor V Leiden, a risk factor for coagulation. Any single nucleotide polymorphism or small deletion/insertion can be genotyped by similar means.

2. Materials

2.1. DNA Isolation (see Note 1)

1. Sample: EDTA/ACD/sodium citrate/sodium heparin anti-coagulated human blood.
2. QIAamp DNA Blood Mini Kit (Qiagen).
3. Ethanol (96–100%).

2.2. Polymerase Chain Reaction

1. 10× thermostable DNA Polymerase (*see* **Note 2**), 0.4 U/µL, diluted from concentrated stock in Taq dilution buffer (2.5 mg/mL BSA, 10 m*M* Tris (pH 8.3), Idaho Technology). For example, Taq (Roche Applied Science) or KlenTaq1 (AB Peptides) can be used
2. 10× PCR buffer, 500 m*M* Tris–HCl (pH 8.3), 2.5 mg/mL BSA, and 30 m*M* MgCl$_2$ (Idaho Technology).
3. 10× dNTPs: 2 m*M* each of dATP, dCTP, dGTP, and dTTP (Idaho Technology) (*see* **Note 3**).
4. 10× LCGreen Plus (Idaho Technology) (*see* **Note 4**).
5. 10× primers/probe (*see* **Notes 5/6**):
   Factor V forward primer (5 µ*M*): CTGAAAGGTTACTTCAAGGAC
   Factor V reverse primer (1 µ*M*): GACATCGCCTCTGGG
   Factor V probe (4 µ*M*): TGGACAGGCGAGGAATACAGGTT-P

2.3. Instrumentation (see Note 7)

1. GeneAmp® PCR System 9700 (Applied Biosystems).
2. LightCycler® (Roche).
3. LightScanner (Idaho Technology).
4. HR-1™ Instrument (Idaho Technology).

3. Methods

3.1. DNA Isolation

1. Whole blood obtained in EDTA, ACD, sodium citrate, or sodium heparin tubes is processed according to the QIAamp DNA Blood Mini Kit Handbook.
2. Measure the absorbance at 260 nm and adjust to an absorbance of 1.0 (50 ng/µL).

3.2. PCR Preparation (see Note 8)

1. Mix one part of each of the following 10× solutions with four parts water: DNA polymerase, PCR buffer, dNTPs, LCGreen Plus, and primer/probe mix.
2. Aliquot nine parts of the above master mix into each well or capillary.
3. Add one part of DNA to each well or capillary.

3.3. PCR Amplification in 96/384 Plates

1. Overlay each well with 10 µL (384-well) or 15 µL (96-well) of mineral oil.
2. Briefly spin the plate in a centrifuge.
3. Amplify with an initial denaturation of 94°C for 10 s followed by 50 cycles of 94°C for 5 s, 57°C for 2 s, and 72°C for 2 s.
4. After amplification, heat to 94°C for 1 s then cool to 10°C before melting.

3.4. PCR Amplification in Capillaries

1. Spin all samples down into the capillaries on a centrifuge.
2. Amplify on a LightCycler with an initial denaturation of 94°C for 10 s followed by 50 cycles of 94°C for 0 s, 55°C for 0 s, and 72°C for 2 s.
3. After amplification, heat to 94°C for 1 s then rapidly cool at 20°C/s to 40°C before melting.

3.5. Melting Acquisition on the LightScanner

1. Transfer the plate from the thermocycler to the LightScanner.
2. Heat the plate from 55°C to 88°C at 0.1°C/s, giving ~25 points/°C.

3.6. Melting Acquisition with HR1

1. Transfer each capillary to the HR-1 high-resolution melting instrument.
2. Perform melting from 55°C to 88°C with a slope of 0.3°C/s, giving 65 points/°C.

3.7. Melting Analysis

Software on most instruments allows visualization of probe and product melting transitions as derivative peaks, usually by Salvitsky—Golay polynomial estimation of the slope at each point *(10)*. Analysis of a 384-well run for factor V Leiden is shown in **Fig. 2**. Part A shows the data without background subtraction, both as the original melting curve (top) and its derivative (bottom). Part B shows the data after exponential background subtraction (*see* **Note 9**), both as a normalized melting curve (top) and a derivative plot (bottom).

4. Notes

1. Although a common commercial DNA preparation kit is referred to here, any DNA purification procedure can be used.
2. Various Taq polymerases can be used, including Taq DNA Polymerase (Roche Applied Science), KlenTaq1™ (AB Peptides) with TaqStart™ antibody (Clontech), and FastStart Taq DNA polymerase (Roche). A chemically modified hot start Taq polymerase or the addition of TaqStart antibody is not necessary, but makes the PCR more robust. However, whether the polymerase has 5′-exonuclease activity will impact the design of the probes. If a 5′-*exo*-negative polymerase is used, the probe can block enzyme extension if it is bound to the template during enzyme extension. Therefore, the probe T_m should be lower than the PCR extension temperature (<70°C). If a 5′-*exo*-positive polymerase is used, the probe T_m can be as high as 85°C.
3. A dNTP mix that includes dUTP may also be used and will shift the probe T_m to lower temperatures. If the total dNTP concentration changes, adjust the $MgCl_2$ concentration accordingly. Uracil-*N*-glycosylase may also be included *(12)* although it is usually not considered necessary in closed-tube systems *(13)*.
4. Saturating dyes, LCGreen I or LCGreen Plus (Idaho Technology) can be used for unlabeled probe genotyping. These dyes differ from the commonly used SYBR Green I in their ability to be used at high concentrations, allowing all available double-stranded binding sites to be saturated while not inhibiting PCR amplification. The excitation maximum for the LCGreen dyes is 450 nm with emission at 470 nm, allowing use on the SYBR Green I channel of most real-time instruments. However, the LCGreen dyes are not excited by argon-ion laser-based instruments *(1)*. The use of LCGreen dyes increases the melting temperature of DNA probes by about 1–3°C, and adjustment of cycling parameters may be required. LCGreen Plus has higher fluorescence intensity than LCGreen I and is best used on plate-based instruments.

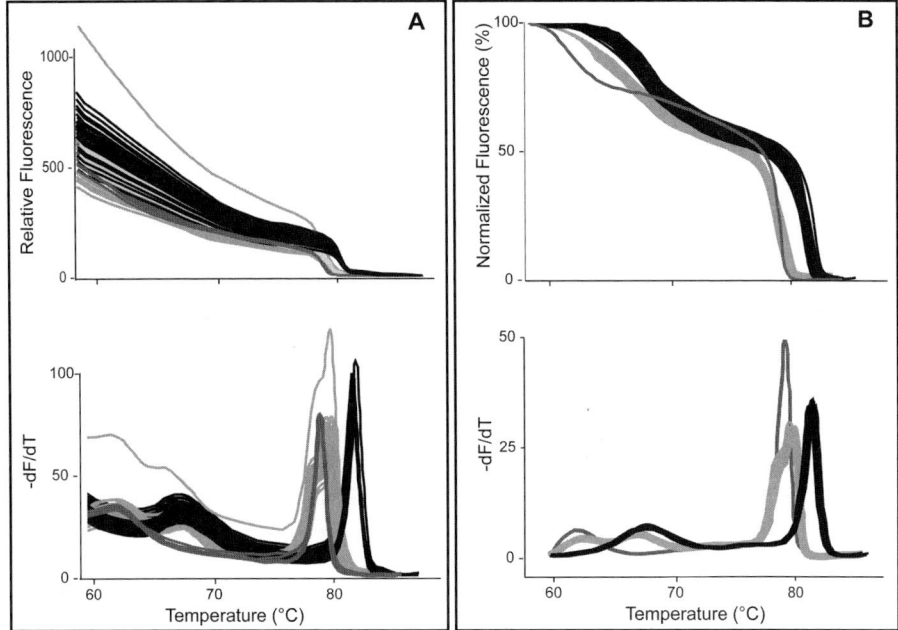

Fig. 2. Unlabeled probe genotyping of factor V Leiden in 384-well format on the LightScanner. Data are displayed as melting curves (*top*) and derivative plots (*bottom*) either without (**A**) or with (**B**) exponential background subtraction. It is much easier to cluster the curves by genotype after exponential background subtraction.

5. Primers are prepared with standard desalting. Unlabeled probe signals are stronger with shorter PCR products. Amplicons <100 bp are optimal, while 200 bp amplicons are usually fine and many probes within 400 bp amplicons can still be used. Asymmetric PCR (1:5 to 1:10) produces both double-stranded amplicon and the appropriate single-stranded DNA for probe binding (**Fig. 1**).

6. Probes are blocked on the 3′-end to prevent extension. The most common blocker is a phosphate. Incomplete phosphate blocking (from either incomplete synthesis or hydrolysis on storage) produces longer probes that appear as extra peaks on derivative plots between the expected probe and product melting temperatures. When this occurs, a C3 blocker may be more effective in preventing extension *(14)*. The length of an unlabeled probe is usually 20–40 bases depending on the GC content and the desired T_m *(7)*. Longer probes give stronger signals. T_ms between 50°C and 85°C have been successfully used. Probe T_m can be lowered by using dUTP instead of dTTP during probe synthesis *(15)*, or increased by substituting locked nucleic acids into the probe *(16)*. As with any probe-based system, mismatches destabilize the probe and lower the T_m. Irrelevant polymorphisms can be masked by incorporating mismatches, deletions, or universal bases into the probes *(17)*.

7. Unlabeled probes can be analyzed on any fluorescent melting instrument compatible with LCGreen dyes. Different alleles are best discriminated on instruments specifically designed for high-resolution melting analysis (HR-1 and the LightScanner). The HR-1 instrument performs melting analysis on samples contained in capillaries that were previously amplified on a LightCycler. The HR-1 processes one sample at a time with a throughput of ~40 samples per hour. The LightScanner analyzes 96- or 384-well plates that have been amplified in any 96- or 384-well thermocycler. The time for analysis in the LightScanner is 5–15 min for a full plate, depending on ramp times and temperature ranges. An extensive comparison of 9 instruments capable of melting analysis has been recently reported (1). In addition to the HR-1 instrument and the LightScanner, this study evaluated the melting capabilities of the ABI PRISM® 7000, ABI PRISM® 7900HT, BioRad iCycler, Cepheid SmartCycler®, Corbett Rotor-Gene™ 3000, and the Roche LightCycler 1.2 and 2.0. All but the 7900HT were compatible with unlabeled probe analysis with LCGreen Plus. Roche has recently introduced the LightCycler 480 System, which is a real-time 96-/384-well PCR instrument that should also be useful for melting analysis of unlabeled probes.

8. The usual total reaction volume is 10 µL. However, reactions can be scaled down to 5 µL if desired in either capillaries or plates.

9. Exponential background subtraction was implemented as follows: The fluorescence melting curve $F(T)$ is the sum of the melting signal $M(T)$ and the fluorescence background $B(T)$. Choose two temperatures, T_L and T_H, below and above the melting transition temperatures, so that the slopes, $M'(T_L) = M'(T_H) = 0$ and therefore $F'(T_L) = B'(T_L)$ and $F'(T_H) = B'(T_H)$. In order to fit $B(T)$ to the exponential model $Ce^{a(T-T_L)}$, fit the measured slopes at T_H and T_L to $B'(T) = aCe^{a(T-T_L)}$. At $T = T_L$, this gives $B'(T_L) = aC$ and at $T=T_H$ this gives $B'(T_H) = aCe^{a(T_H-T_L)}$. Dividing the second equation by the first gives $B'(T_H)/B'(T_L) = e^{a(T_H-T_L)}$ so that $a = \ln(B'(T_H)/B'(T_L))/(T_H-T_L)$ and the first equation gives $C = B'(T_L)/a$. Finally, obtain the signal with the background removed by subtraction: $M(T) = F(T) - Ce^{a(T-T_L)}$ with the parameters C and a determined as above. The signal $M(T)$ may optionally be normalized to the range 0–100 by applying the linear shift and rescaling: $M(T) = 100(M(T) - m)/(M - m)$, where $m = \min[M(T)]$ and $M = \max[M(T)]$ on the interval of interest.

References

1. Herrmann, M. G., Durtschi, J. D., Bromley, L. K., Wittwer, C. T., and Voelkerding, K. V. (2006) DNA melting analysis for mutation scanning and genotyping: a cross platform comparison. *Clin. Chem.* **52**, 494–503.

2. Ririe, K. M., Rasmussen, R. P., and Wittwer, C. T. (1997). Product differentiation by analysis of DNA melting curves during the polymerase chain reaction. *Anal. Biochem.* **245**, 154–160.

3. Wittwer, C. T., Herrmann, M. G., Moss, A. A., and Rasmussen, R. P. (1997) Continuous fluorescence monitoring of rapid cycle DNA amplification. *Biotechniques* **22**, 130–131, 134–138.

4. Jobs, M., Howell, W. M., Stromqvist, L., Mayr, T., and Brookes, A. J. (2003) DASH-2: flexible, low-cost, and high-throughput SNP genotyping by dynamic allele-specific hybridization on membrane arrays. *Genome Res.* **13**, 916–924.

5. Prince, J. A., Feuk, L., Howell, W. M., Jobs, M., Emahazion, T., Blennow, K., and Brookes, A. J. (2001) Robust and accurate single nucleotide polymorphism genotyping by dynamic allele-specific hybridization (DASH): design criteria and assay validation. *Genome Res.* **11,** 152–162.

6. Wittwer, C. T., Reed, G. H., Gundry, C. N., Vandersteen, J. G., and Pryor, R. J. (2003) High-resolution genotyping by amplicon melting analysis using LCGreen. *Clin. Chem.* **49,** 853–860.

7. Zhou, L., Myers, A. N., Vandersteen, J. G., Wang, L., and Wittwer, C. T. (2004) Closed-tube genotyping with unlabeled oligonucleotide probes and a saturating DNA dye. *Clin. Chem.* **50,** 1328–1335.

8. Wittwer, C. T., Dujols, V. E., Reed, G., and Zhou, L. (2004) Amplicon melting analysis with saturation dyes. US Patent Application 2006-0019253 A1.

9. Zhou, L., Wang, L., Palais, R., Pryor, R., and Wittwer, C. T. (2005) High-resolution DNA melting analysis for simultaneous mutation scanning and genotyping in solution. *Clin. Chem.* **51,** 1770–1777.

10. Wittwer, C. T. and Kusukawa, N. (2004) Real-time PCR. Pages 71-84 in D. H. Persing, Tenover, F. C., Versalovic, J., Tang, Y. W., Unger, E. R., Relman, D. A., and White, T. J., eds. *Diagnostic molecular microbiology; principles and applications.* ASM Press, Washington DC.

11. Liew, M., Nelson, L., Margraf, R. L., et al (2006) Genotyping of human platelet antigens 1-6 &15 by high-resolution amplicon melting and conventional hybridization probes. *J. Molec. Diagn.* in press **8,** 97–104.

12. Liew, M., Pryor, R., Palais, R., et al. (2004) Genotyping of single-nucleotide polymorphisms by high-resolution melting of small amplicons. *Clin. Chem.* **50,** 1156–1164.

13. Wang, J., Chuang, K., Ahluwalia, M., et al. (2005) High-throughput SNP genotyping by single-tube PCR with T_m-shift primers. *Biotechniques* **39,** 885–893.

14. Cradic, K. W., Wells, J. E., Allen, L., Kruckeberg, K. E., Singh, R. J., and Grebe, S. K. (2004) Substitution of 3′-phosphate cap with a carbon-based blocker reduces the possibility of fluorescence resonance energy transfer probe failure in real-time PCR assays. *Clin. Chem.* **50,** 1080–1082.

15. Zhou, L., Vandersteen, J., Wang, L., et al. (2004) High-resolution DNA melting curve analysis to establish HLA genotypic identity. *Tissue Antigens* **64,** 156–164.

16. Chou, L.-S., Meadows, C., Wittwer, C. T., and Lyon, E. (2005) Unlabeled oligonucleotide probes midified with locked nucleic acids for improved mismatch discrimination in genotyping by melting analysis. *BioTechniques* **39,** 644–650.

17. Margraf, R. L., Mao, R., and Wittwer, C. T. (2006) Masking selected sequence variation by incorporating mismatches into melting analysis probes. *Human Mutation* **27,** 269–278.

V

NON-NUCLEIC ACID TARGETS

15

Molecular Beacons for Protein—DNA Interaction Studies

Jun Li, Zehui Charles Cao, Zhiwen Tang, Kemin Wang and Weihong Tan

Abstract

Real-time monitoring of DNA—protein interactions involving molecular beacon (MB) and molecular beacon aptamer (MBA) was discussed in this chapter. MBs are single-stranded oligonucleotide probes with a hairpin structure. MBs have been designed for oligonucleotide recognition and protein—DNA interaction studies. Real-time monitoring of enzymatic reactions, such as cleavage, ligation, and phosphorylation of single-stranded DNA by specific enzyme, has been studied using MBs. Meanwhile, a new generation of molecular probes, MBA, was designed by combining the excellent signal transduction properties of MBs with the specificity of aptamers for protein recognition. Two different aptamers, the one for thrombin and that for platelet-derived growth factor, have been successfully used to construct MBA probes. The interaction between the proteins and the MBA probes was investigated by fluorescence resonance energy transfer, fluorescence anisotropy, and time-resolved fluorescence. This chapter has reviewed our recent progress in this area.

Key Words: DNA—protein interaction; molecular beacon; aptamer; molecular beacon aptamer; fluorescence resonance energy transfer; fluorescence anisotropy.

1. Introduction

The protein analysis and the study of DNA—protein interactions are of great interest in understanding many important biological processes because the functions of living cells are mostly executed and regulated by proteins and protein—DNA interaction *(1)*. Traditional methods for protein detection and DNA— protein interaction investigation include gel electrophoresis and autoradiography, etc. *(2,3)*. These methods have many complex treatments, which are time consuming, discontinuous, and offline. Meanwhile, the development of non-isotopic and sensitive methods for real-time monitoring of DNA—protein interactions in

From: *Methods in Molecular Biology, Vol. 429: Molecular Beacons: Signalling Nucleic Acid Probes, Methods and Protocols* Edited by: A. Marx and O. Seitz © Humana Press Inc., Totowa, NJ

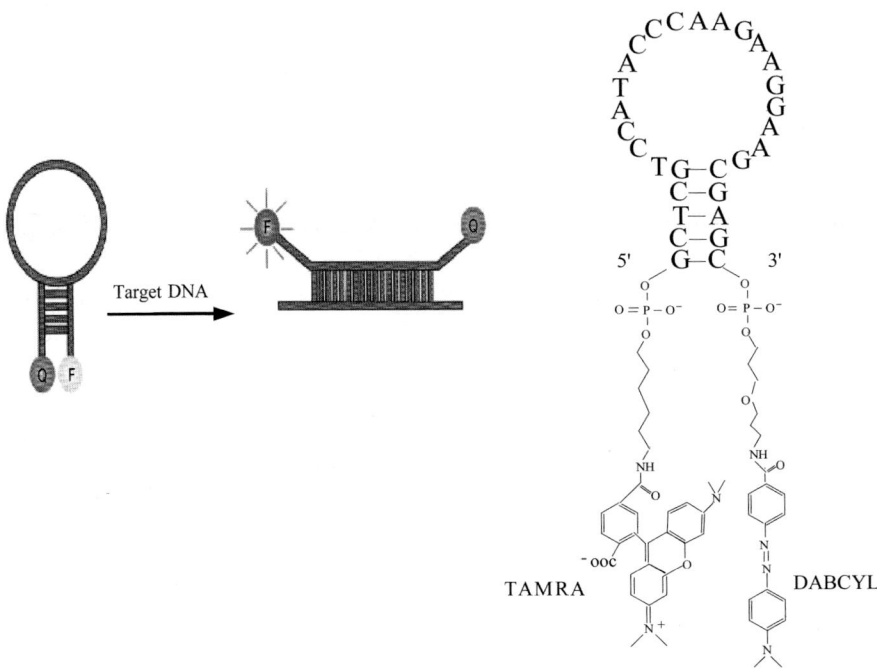

Fig. 1. (Left) The conformation change of the hybridization of MB and cDNA, F represents the fluorophore, Q represents the quencher. (Right) The structure of a typical MB.

homogeneous solutions is still a big challenge. Therefore, in the post-genome era, biomolecule recognition probes with high sensitivity and selectivity would have a significant potential for simple and precise protein analysis and the study of DNA—protein interactions.

Fluorescence probes offer high sensitivity, selectivity, and applicability in separation free detection and *in situ* monitoring. Molecular beacon (MB), one of the most interesting DNA probe, is a promising probe for quantitative genomic studies *(4,5)*. MB is hairpin-shaped oligonucleotide that is labeled with a fluorophore and a quencher at two termini (as shown in **Fig. 1**). MBs act like switches that are normally closed to bring the fluorophore/quencher pair together to turn fluorescence "off." When MB hybridizes with its complementary DNA (cDNA), it is prompted to undergo a conformational change that open the hairpin structure. The fluorophore and the quencher are separated, and fluorescence is turned "on." Based on these characters, MB was used as enzyme detection and real-time monitoring of DNA—protein detection except for the DNA analysis *(6,7)*.

Although MBs have been demonstrated as reporters for a few DNA-binding proteins, the lack of selectivity for different proteins has limited their use. To meet this challenge, molecular beacon aptamer (MBA) *(10)*, a new class of protein probe, has been developed by combining the excellent signal transduction capability of MB and the protein-binding specificity of aptamer. Aptamer is a novel class of short DNA/RNA sequence that rivals antibody in protein recognition *(8,9)*. Now, aptamers possess many advantages over antibodies: easier synthesis, more flexible labeling, better reproducibility, easier storage, faster tissue penetration, and shorter blood residence, etc. Unfortunately, an aptamer itself cannot be used as a fluorescent probe as it lacks signal transduction capability to report the recognition with target. The MBAs have shown great potential in protein analysis and DNA—protein interaction by taking advantage of MB and aptamer. In this chapter, we will discuss the protein—DNA interaction studies by using MB and MBA probes, respectively.

2. MB for Protein Recognition and Protein—DNA Interaction Study

2.1. The Nonspecific Protein Detection by MB

While MB probes are originally designed for nucleic acid studies, their hairpin structure could also be disturbed to restore fluorescence upon binding to some proteins. This protein recognition ability was first realized by the use of *E. coli* single-stranded DNA-binding protein (SSB) *(11)*. The fluorescence enhancements caused by SSB and cDNA were very comparable. Using MB-SSB binding, the SSB at a concentration as low as 2×10^{-10} mol/L could be detected by a conventional spectrophotometer. The interaction between SSB and MB was found to be much faster than that between the cDNA and the MB. This fast speed of the protein—DNA-binding reaction will provide the basis for rapid protein assays. In addition, there are significant differences in MB-binding affinity with different proteins, such as albumin and histone. This will lead to the exploration of potential for selective-binding studies of a variety of proteins using designed MBs.

The MB-based DNA probe was also used for detail binding studies of an enzyme, lactate dehydrogenase (LDH) *(12)*. The fluorescence signal of the MB was increased through interaction with LDH. This was then used for the elucidation of the binding properties and the study of the binding process. Different LDH isoenzymes were found to have different ssDNA-binding affinities. The results showed that the stoichiometry of LDH-5/MB binding was 1:1, and the binding constant was 1.9×10^{-7} mol/L. Detailed studies of LDH/MB binding, such as salt effects, temperature effects, pH effects, binding sites, binding specificity for different isoenzymes, and competitive binding with different substrates, were carried out by means of a simple fluorescent method using the MB probe.

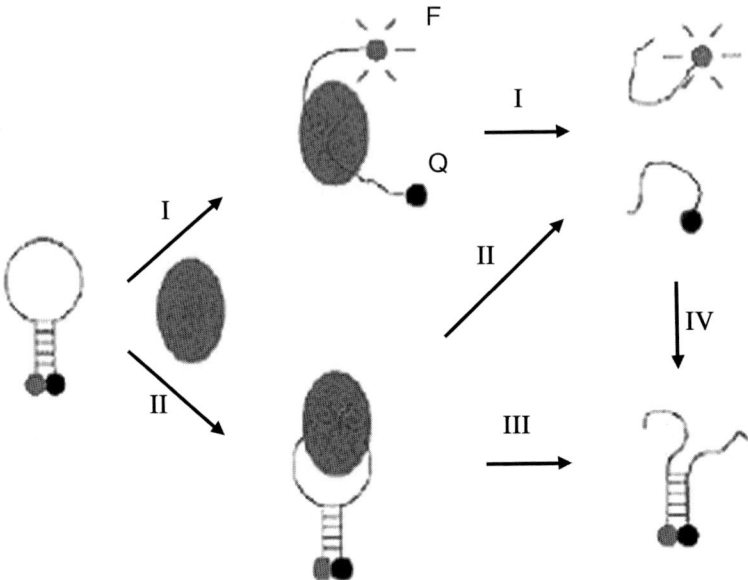

Fig. 2. Schematic of the fluorescence mechanism of the MB during cleavage by single-strand-specific DNA nuclease. The ball represents the nuclease. MB, F, and Q represent molecular beacon, fluorophore, and quencher, respectively. Here, the fluorophore and quencher are tetramethylrhodamine (TAMRA) and 4-(4′-dimethylaminophenylazo)benzoic acid (DABCYL), respectively.

2.2. Real-Time Enzymatic Cleavage Assay Using MBs

Traditional methods to assay enzymatic cleavage of single-stranded DNA (ssDNA) are discontinuous and time consuming. The lack of suitable fluorescent probes is an obstacle for the development of fluorescence methods which are continuous and convenient. Based on MB probes, a new method has been proposed to assay the ssDNA cleavage reaction by single-strand specific nuclease *(13)*. The single-stranded nuclease binds to and cleaves the single-stranded loop portion of the MB. The cleavage results in the dissociation of the stem since the five to seven bp in the stem are unstable at the cleavage temperature (37°C) when the loop is broken. Consequently, the fluorophore and quencher are completely separated from each other, giving rise to an irreversible fluorescence enhancement, which is higher than that caused by the cDNA during hybridization. **Fig. 2** shows the two possible mechanisms of the nuclease analysis by MBs. The solid arrows indicate two paths (I and II) leading to fluorescence enhancement during digestion. The dashed arrows represent two possible processes (III and IV) in which no fluorescence enhancement is produced.

Even though the nuclease may keep on cutting one single strand many times, only the first cut contributes to the fluorescence signal increase. Therefore, only the first cut is shown here.

It is shown that the fluorescence method permits the real-time monitoring of the enzymatic cleavage reaction process, easy characterization of the activity of DNA nucleases, and the study of steady-state cleavage reaction kinetics. Due to its high sensitivity, reproducibility, and convenience, MBs have been used to observe the study of single-stranded DNA cleavage reactions. Because MB carries an appropriate cleavage site within the stem, it has been applied to develop a continuous assay for cleavage of DNA by enediynes. The generality of this approach is demonstrated by using the assay to directly compare the DNA cleavage of by naturally occurring enediynes, non-enediyne small molecule agents, as well as the restriction endonuclease BamHI *(14)*. Meanwhile, MBs were used to quantify low levels of type I endonuclease activity *(15)*. Given the simplicity, speed, and sensitivity of this approach, the described methodology could easily be extended to a high-throughput format and become a new method of choice in modern drug discovery to screen for novel protein-based or small molecule-derived DNA cleavage agents.

2.3. Real-Time Monitoring of Enzymatic Ligation and phosphorylation by MB

Traditional methods for the analysis of the oligonucleotide ligation process include gel electrophoresis and autoradiography. For example, the oligos should be labeled with ^{32}P at 5′-terminus before ligation, and then the ligation reaction must be stopped with urea or EDTA. After that, a denaturing polyacrylamide gel electrophoresis has to be done, and a succeeding autoradiography visualizing process is used to assay the ligation product quantitatively *(16)*. A nucleotide fragment sensing (NTFS) technique has been developed for protein monitoring based on MB *(17–19)*. The mechanism is described in **Fig. 3**.

The reaction system is composed of ligase, two oligos to be ligated and a MB, in which two oligos match the 3′ and 5′ half part of MB's loop. At the beginning of ligation reaction, each oligo hybridizes to half of MB's loop to form a complex with a nick. Instead of opening the stem completely, this can only make the stem take apart slightly. Whereas the ligation reaction goes on, the ligation product can open the stem of MB completely, and then the fluorescence intensity rises simultaneously. Therefore, the mechanism of this novel approach illustrated in **Fig. 3** can be characterized as a "ligate and light" process. According to this mechanism, the MB is not only an indicator but also a template for two oligos in the ligation process. The fluorescence signal gives information of ligation proceeding continually, sensitively and precisely. Based on this principle, novel assay methods were developed to real-time monitoring of DNA—ligase interaction *(17,18)*.

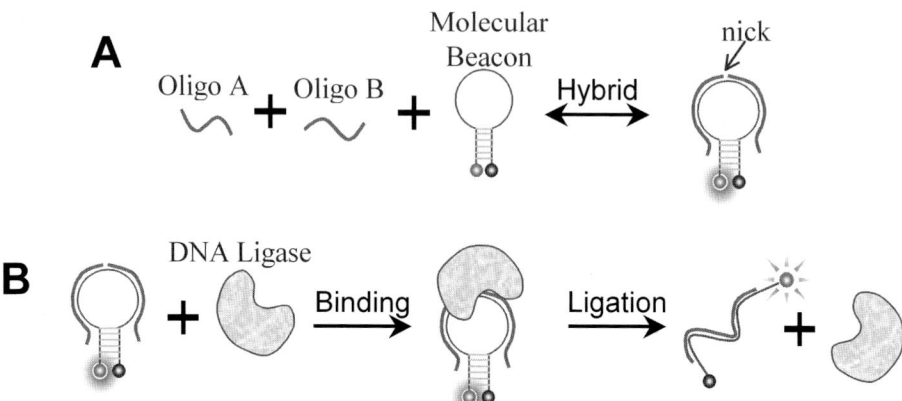

Fig. 3. 'Ligate and light'—Schematics diagram of real-time monitoring of the nucleic acid ligation process by employing MB. (**A**) The two half-matching oligos hybridize with MB to form a nick and open the stem slightly. (**B**) The ligase bands to the nick and catalyzes the ligation of two short oligos to form a long oligo, after which the MB will would be opened and the fluorescence will be restored.

Another application of this principle is developed for real-time monitoring of the phosphorylation of nucleic acids. As illustrated in **Fig. 4**, the oligo with 5′-hydroxyl can not be ligated to open the MB restoring fluorescence, unless the oligo phosphorylated with 5′-phosphate. In these procedures, the DNA ligase plays a role of a converter that transforms the information of "oligos have been phosphorylated" into fluorescence enhancement. Thus phosphorylation of nucleotide can be monitored in real time using this "phosphorylation and ligation" enzyme coupled reaction. Utilizing the high selectivity and excellent sensitivity of MB and rapid ligation feature of DNA ligase, this approach offers selective, sensitive, and real-time information of DNA—polynucleotide kinase interaction (*19*).

3. Molecular Beacon Aptamer for Protein Detection

While the study of the nonspecific DNA-binding proteins is important, the specific interactions between the MBs and proteins will be more interesting and useful. Through the investigation of nonspecific protein—DNA binding, it opens the possibility for further development of easily obtainable and modified DNA molecules for real-time specific protein detection. The most interesting example is MBA.

As described above, aptamer has a high specificity for protein binding. However, an aptamer itself cannot be used as a fluorescent probe as it lacks

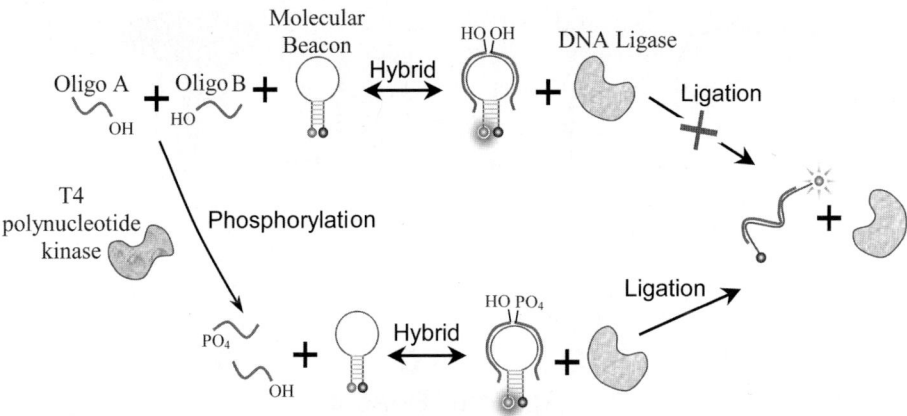

Fig. 4. Schematic of monitoring the nucleic acids phosphorylation. (Up) The nick constructed by oligos with hydroxyl linking to 5′-end cannot be ligated by DNA ligase. (Down) While T4 polynucleotide kinase introduced, the phosphorylation would progress that the hydroxyl at 5′-end of oligo be replaced by phosphoryl. Thus the nick formed by phosphorylated oligos can be sealed by DNA ligase and enhance the fluorescence.

signal transduction capability to report the binding event. One could combine the binding specificity of aptamers and the signal transduction capability of MB to form a new molecular probe, MBA, to analyze proteins and study DNA—protein interactions. For example, an aptamer-derived MB was used to analyze the Tat of HIV *(20)*. The molecule included a non-structured oligomer and a hairpin structure MB, which contained a fluorophore (fluorescein) and a quencher (DABSYL). Unlike other conventional MBs, this new MB had a RNA sequence and the Tat aptamer was an RNA. In the presence of Tat or its peptides, the two oligomers underwent a conformational change to form a duplex followed by the stabilization of the ternary complex. This change led to the restoration of the fluorescence of the fluorophore, and thus a significant enhancement of the fluorescence signal was observed.

MBA have been successfully designed in the last few years. The sequence of an aptamer is usually designed as the loop sequence of a MB. In the presence of protein, there will be a fluorescence signal change which depends on the protein concentration. Up to now, it has shown a great potential in monitoring protein production in standard and actual biological sample. However, it is shown that the successful design of DNA probe is still limited by two reasons. The first one is that there are limited number of aptamers for specific protein with high affinity. The other is that the conformation change upon protein binding is not universal and special design of each probe is needed to construct a useful MBA. Up to now,

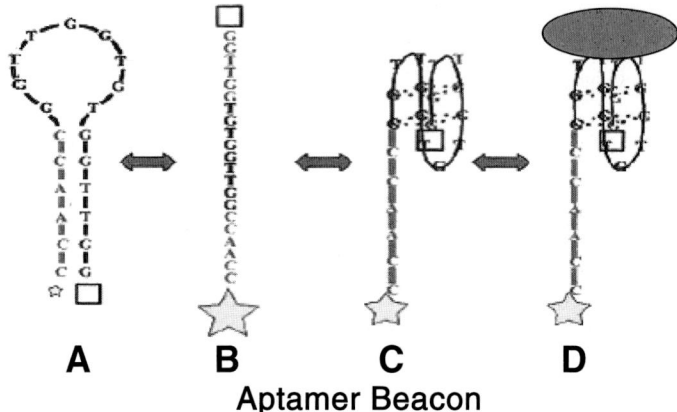

Aptamer Beacon

Fig. 5. Schematic of the mechanism of signaling by thrombin MBA probe. **(A)** Thrombin bound to MBA probe. **(B)** Aptamer in G-quartet conformation, which allows thrombin binding. **(C)** Unfolded conformation. **(D)** MBA probe in quenched stem-loop conformation. Yellow stars represent the fluorophore, and the white square represents the quencher. The emission intensity of the fluorophore is represented by the size of the star.

only a few aptamers such as thrombin and platelet-derived growth factor (PDGF) aptamers have been successfully used to construct MBA probes. We will focus on these two MBA probes in this discussion.

3.1. MBA Probes for Thrombin

Thrombin was a coagulation protein that had many effects in the coagulation cascade with two exosites for suitable ligand *(21)*. In 1992, Bock et al. screened a 15mer ssDNA (5′-GGT TGG TGT GGT TGG-3′) that bound thrombin in its exosite-1 with K_d around 25–200 nM *(22)*. In 1997, Tasset et al. found another 27 or 29mer ssDNA to thrombin in the exosite-2 with Kd around 5 nM *(23)*. The sequence for the 29 mer aptamer was 5′-AGT CCG TGG TAG GGC AGG TTG GGG TGA CT-3′, and that for the 27mer aptamer was 5′-ACC CGT GGT AGG GTA GGA TGG GGT GGT-3′. The 15mer aptamer has a simpler structure to be used as a MBA probe for thrombin binding.

Stanton et al. have designed a MBA probe for detecting a wide range of thrombin and the mechanism was described in **Fig. 5** *(24)*. An thrombin aptamer was engineered into a MBA probe by adding several nucleotides to the 5′-end, which are complementary to nucleotides at the 3′-end of the aptamer. A fluorescence-quenching pair was used to report changes in conformation induced by thrombin binding. In the absence of thrombin, the added nucleotides will form a duplex with the 3′-end, forcing the probe into a stem-loop structure.

The fluorescence will be turn "off" because the distance is close between the fluorophore attached to the 5′-end and the quencher attached to the 3′-end. On the other hand, the MBA probe forms the ligand-binding structure when thrombin is in the presence. This conformational change causes a long distance separation between the fluorophore and the quencher. The fluorescence turns "on". The fluorescence intensity change is used as the thrombin−DNA interaction study or the analysis of thrombin.

Li et al. introduced another MBA probe for real-time thrombin recognition and quantitative analysis *(25)*. This thrombin-binding aptamer-based MAB probe was prepared as a model to demonstrate the feasibility as shown in **Fig. 6**. The MAB probe exists at equilibrium between a non-structured random coil and a compact intra-molecular quadruplex. When no thrombin is presence, the probe is in a random coil structure. The addition of thrombin (gray ellipse) shifts the equilibrium in favor of the quadruplex structure, which draws the fluorophore (F) and quencher (Q) closer, leading to fluorescence quenching. This MAB recognizes its target protein with high specificity and high sensitivity (112 p*M* thrombin concentration) in homogeneous solutions.

Another MBA probe was developed to real-time monitor of protein−protein interactions. In this strategy, protein is not labeled to pose minimum effects on the binding properties of the proteins *(26)*. Two kinds of signal transduction strategies, fluorescence resonance energy transfer (FRET) and fluorescence anisotropy, have been used to study the interactions of thrombin with different proteins. As shown in **Fig. 7A**, aptamer is dual-labeled with a fluorophore and a quencher. The folded form of the aptamer when it binds to the bait protein results in a quenched fluorescence. The bait-prey protein interaction causes the release of aptamer from the bait protein, leading to a restored fluorescence. In the case of **Fig. 7B**, aptamer is labeled with only one dye. When bound to the much larger bait protein, the aptamer displays slow rotational diffusion. The interaction between bait and prey proteins displaces the aptamer. The unbound aptamer has much faster rotational diffusion. The change in the rotation rate is reported by fluorescence anisotropy of the dye molecule. The FRET and the fluorescent anisotropy approach complement each other in providing insight into the kinetics, mechanisms, binding sites, and binding dynamics of the interacting proteins. Based on these two kinds of design, the interaction of protein−DNA is used to obtain detailed protein−protein interaction information by simplicity and affectivity, does not require labeling of proteins.

3.2. MBA probe for PDGF

PDGF is an oncoprotein. The different isomers of PDGF are important in life science researches because they are largely responsible for cancer growth and are often found overexpressed or mutated in malignant tumor. PDGF can be

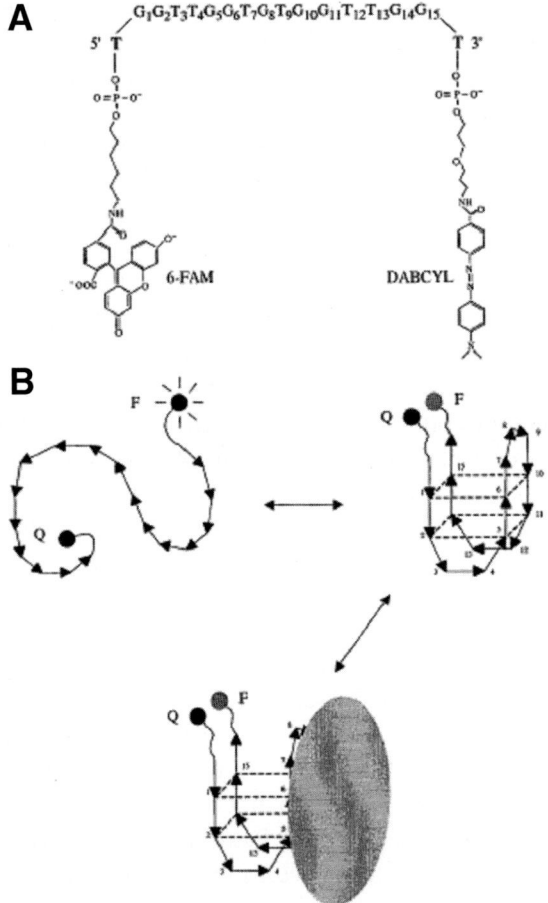

Fig. 6. (A) Molecular structure of the fluorophore—quencher-labeled MBA for throm-bin binding. Letters G1 to G15 represent the sequence of the 15-mer aptamer. **(B)** The mechanism of the fluorophore—quencher-labeled MBA. F, fluorophore; Q, quencher.

assembled in at least three isoforms: heterodimers PDGF-AB, homodimers PDGF-BB, and PDGF-AA. Green et al. *(27)* selected a DNA aptamer molecule with high-binding affinity for PDGF BB homodimers, and showed by using a binding assay that the affinities of these DNA molecules for the three different homodimeric forms of PDGF are distinguishable. The development of MBA probe based on PDGF-BB aptamer was shown as the following scheme (**Fig. 8**). F represented the fluorophore group on the 5′-terminus. Q was the quencher on the 3′-end. The present structure was the stable conformation of the aptamer under physiological conditions in the presence of PDGF. Upon PDGF protein

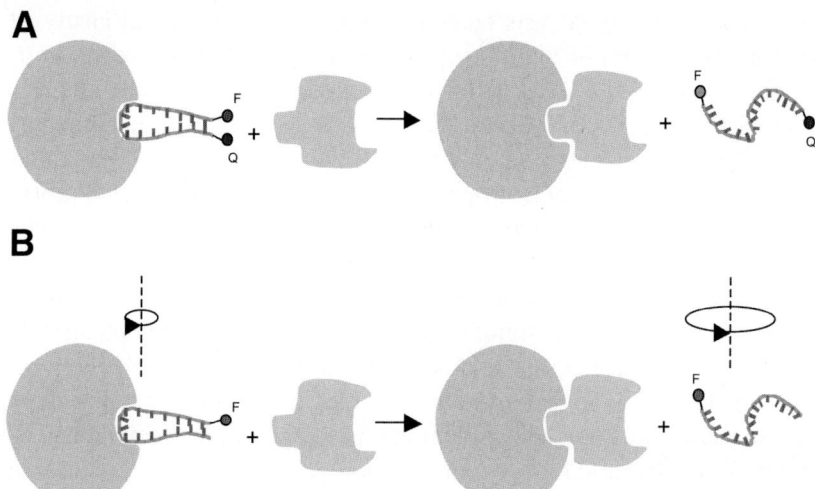

Fig. 7. Dye-labeled protein-binding aptamers reporting protein—protein interactions F, fluorophore; Q, quencher.

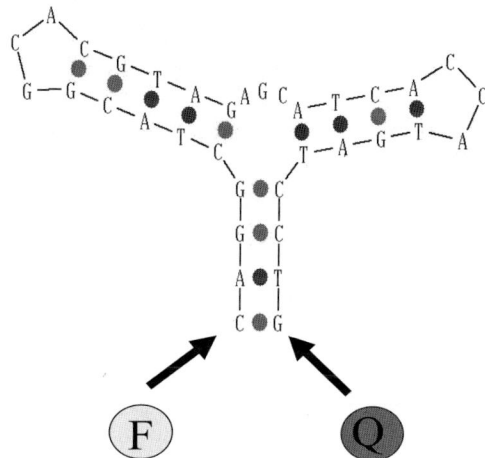

Fig. 8. Structure of the PDGF probe based on its aptamer.

binding, the aptamer formed a close-packed tight structure, which reduced the distance between the two termini of the aptamer and caused fluorescence quenching. Therefore, the DNA—protein interaction could be easily described by the FRET between the fluorophore (F) and the quencher (Q). For further development, the quencher could be replaced by another fluorophore or just no label. Then, the signal was detected by the fluorescence intensity of the second fluorophore or just by fluorescent anisotropy of the labeled fluorophore.

The first successful PDGF MBA probe made use of a single fluorophore labeled molecular probe for the ultrasensitive detection of PDGF in homogeneous solution *(28)*. The aptamer was only labeled with fluorescein and no quencher was added. Fluorescence anisotropy was used for the real-time monitoring of the binding between the aptamer and the protein. When the labeled aptamer was bound with its target protein, the rotational motion of the fluorophore attached to the complex become much slower because of an increased molecular weight after binding, resulting in a significant fluorescence anisotropy change. Using the anisotropy change, it was available to detect the binding events between the aptamer and the protein in real time and in homogeneous solutions. The detection limit of this assay was down to subnanomolar range and the method had a high selectivity. When fivefold higher concentration of different extracellular proteins (moles) were added to a 20 nM aptamer solution in physiological buffer at 25°C, these proteins almost had no anisotropy change. As for the comparison of the binding capability of the PDGF-B to other growth factors, such as PDGF-AA and PDGF-AB, epidermal growth factor, and insulin-like growth factor-I (IGF-I), PDGF-BB had the highest anisotropy change. This DNA—protein study had the potential to find wide applications in protein monitoring, in cancer protein detection as well as other studies in which protein analysis was important.

Another PDGF MBA was employed with a synthetic DNA aptamers labeled by a fluorophore and a quencher at the two termini, measuring fluorescence quenching *(29)*. The specific quencher, DABCYL, could be used to detect PDGF at sub-nanomolar concentrations even in the presence of serum and other cell-derived proteins in cell culture media. Similarly, the three highly related molecular variants of PDGF (AA, AB, and BB dimers) could be distinguished from one another with a single-step assay. The use of fluorescence quenching as a measure of binding between the DNA probe and the target protein eliminated the potential false signals that may arise in traditional fluorescence enhancement assays because of degradation of the DNA aptamer by contaminating nucleases in biological specimens.

Since FRET is a distance-dependent phenomenon, it should be possible to improve the performance of FRET by changing the fluorophore—quencher pairs. Some other FRET pairs were investigated to compare with the standard MBA probe linked with fluorescein-Dabcyl pair: Texas Red—Black Hole Quencher 2 (BHQ2), Cy5—Black Hole Quencher 2 and fluorescein—tetramethylrhodamine (TMR) *(30)*. MBA probes with these FRET pairs were tested for the decrease in fluorescence intensity upon addition of a fourfold molar excess of PDGF-BB in the standard fluorescence-quenching assay. All of the selected FRET pairs

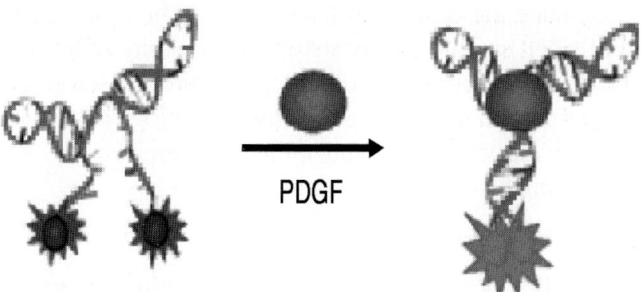

Fig. 9. The use of the pyrene excimer to probe PDGF.

showed specific quenching of the fluorescence signal when PDGF was added, albeit with different degrees. The availability of choices in the selection of fluorophore and quencher was an important part of MBA-based bioassay development because various biological specimens might present potential interference at certain excitation—emission spectra that might be effectively addressed by selecting the proper FRET pair. In addition, the development of a multiplex assay for the simultaneous detection of multiple proteins in a single homogeneous solution might be possible by using more MBAs that are selective for different biomarkers and that are labeled with different fluorophore—quencher pairs.

The analysis of protein by MBA probe meets the challenges in a complex biological fluid because of the background signals from both the probe and the biological fluids where the proteins reside. To solve this problem, a molecular engineered light-switching excimer aptamer probe was designed for rapid and sensitive detection of PDGF in complex biological fluids by time-resolved fluorescence *(31)*. Labeled with one pyrene at each end, the aptamer switched its fluorescence emission from 400 nm (pyrene monomer) to 485 nm (pyrene excimer) upon PDGF binding. As shown in **Fig. 9**, PDGF aptamer (red) is end-labeled with pyrene molecules (dark) that are separated from each other because of the open structure of the aptamer. The pyrene molecule had monomer emission peaks at ≈ 378 and 398 nm. After binding to PDGF, the aptamer adapted a close conformation, bringing two pyrene molecules close to each other. Consequently, pyrene excimer (grey) formed and light (≈ 485 nm) was emitted after photo-excitation. This fluorescence wavelength changing from monomer to excimer emission was a result of aptamer conformation rearrangement induced by target binding. The excimer probe was able to effectively detect picomolar PDGF in homogeneous solutions and with the naked eye. Moreover, because the excimer had a much longer fluorescence lifetime (≈ 40 ns) than that of the background (≈ 5 ns), time-resolved measurements were used to eliminate the biological background. Therefore, PDGF was able to be detected in a cell sample quantitatively without any sample pretreatment. This DNA—protein interaction-based strategy could be used to develop other aptamer probes for protein monitoring in complex biological system. Combined with lifetime-based measurements and molecular engineering, it holds great potential in protein analysis for biomedical studies.

4. Outlook and Conclusion

MBs have shown a highly efficient signal transduction ability based on FRET and possessed high sensitivity for separation-free analysis. Its capability of binding with proteins has also induced great interest in protein analysis and the study of DNA—protein interactions. Although MB could be straightway used for the real-time monitoring of the DNA—DNase, DNA—ligase or polynucleotide kinase interaction, the specificity is a problem in DNA—protein study if the sequence of MB was not properly chosen. The combination of MB's excellent signal transduction capability with the binding specificity of aptamer resulted in a novel MBA probe for protein analysis and DNA—protein study. These probes took advantage of many fluorescence signaling schemes, such as fluorescence enhancement, fluorescence quenching, fluorescent anisotropy, for real-time monitoring of protein activity in homogenous solution or for monitoring of DNA—protein interactions with a high sensitivity. The specificity was dependent on the binding ability of the aptamers and the corresponding proteins. The application in actual biological complex fluid was also investigated via different signal transduction approaches in order to overcome high-cellular background signal. The development of MB and MBA probes for protein analysis or DNA—protein interactions in a homogeneous solution or even in living cells should be feasible, providing a great potential in protein monitoring, biomarker discovery, clinic diagnosis, and therapeutic drug screening.

References

1. Phizicky, E., Bastiaens, P. I. H., Zhu, H., et al. (2003) Protein analysis on a proteomic scale. *Nature* **422**, 208–215.
2. Barnouin, K. (2004) Two-dimensional gel electrophoresis for analysis of protein complexes. *Methods Mol. Biol.* **261**, 479–498.
3. Gullberg, M., Fredriksson, S., Taussig, M., et al. (2003) A sense of closeness: protein detection by proximity ligation. *Curr. Opin. Biotechnol.* **14**, 82–86.
4. Tyagi, S. and Kramer, F. R. (1996) Molecular beacons: probes that fluoresce upon hybridization. *Nat. Biotechnol.* **14**, 303–308.
5. Tyagi, S., Bratu, D. P., and Kramer, F. R. (1998) Multicolor molecular beacons for allele discrimination. *Nat. Biotechnol.* **16**, 49–53.
6. Fang, X., Li, J. J., Perlette, J., et al. (2000) Molecular beacons: novel fluorescent probes. *Anal. Chem.*, **72**, 747A–753A.

7. Goel, G., Kumar, A., Puniya, A. K., et al. (2005) Molecular beacon: a multitask probe. *J. Appl. Microbiol.* **99,** 435–442.

8. Ellington, A. D. and Szostak, J. W. (1990) In vitro selection of RNA molecules that bind specific ligands. *Nature* **346,** 818–822.

9. Robertson, D. L. and Joyce, G. F. (1990) Selection in vitro of an RNA enzyme that specifically cleaves single-stranded DNA. *Nature* **344,** 467–468.

10. Tan, W. H., Wang, K., and Drake, T. J. (2004) Molecular beacons. *Curr. Opin. Chem. Biol.* **8,** 547–553.

11. Li, J. J., Fang, X. H., Schuster, S., and Tan, W. H. (2000) Molecular Beacons: A novel approach to detect protein - DNA Interactions. *Angew. Chem. Int. Ed. Engl.* **39,** 1049–1052.

12. Fang, X. H., Li, J. J., and Tan, W. H. (2000) Using molecular beacons to probe molecular interactions between lactate dehydrogenase and single-stranded DNA. *Anal. Chem.* **72,** 3280–3285.

13. Li, J. J., Geyer, R., and Tan, W. (2000) Using molecular beacons as a sensitive fluorescence assay for enzymatic cleavage of single-stranded DNA. *Nucleic Acids Res.* **28,** e52.

14. Biggins, J., Prudent, J., Marshall, D., et al. (2000) A continuous assay for DNA cleavage: the application of 'break lights' to enediynes, iron-dependent agents, and nucleases. *Proc. Natl Acad. Sci. USA* **97,** 13,537–13,542.

15. Biggins, J., Onwueme, K., and Thorson, J. (2003) Resistance to enediyne antitumor antibiotics by CalC self-sacrifice. *Science* **301,** 1537–1541.

16. Landegren, U., Kaiser, R., Sanders, J., et al. (1988) A ligase-mediated gene detection technique *Science* **241,** 1077–1080.

17. Tang, Z. W., Wang, K. M., Tan, W. H., et al. (2003) Real-time monitoring of nucleic acid ligation in homogenous solutions using molecular beacons *Nucleic Acids Res.* **31,** e148.

18. Liu, L. F., Tang, Z. W., Wang, K. M., et al. (2005) Using molecular beacon to monitor activity of *E. coli* DNA ligase. *Analyst* **130,** 350–357.

19. Tang, Z., Wang, K., Tan, W., et al. (2005) Real-time investigation of nucleic acids phosphorylation process using molecular beacons. *Nucleic Acids Res.* **33,** e97.

20. Yamamoto, R. and Kumar, P. K. R. (2000) Molecular beacon aptamer fluoresces in the presence of Tat protein of HIV-1. *Genes Cells* **5,** 389–396.

21. Bode, W., Turk, D., and Karshikov, A. (1992) The refined 1.9-A X-ray crystal structure of D-Phe-Pro-Arg chloromethylketone-inhibited human α-thrombin: Structure analysis, overall structure, electrostatic properties, detailed active-site geometry, and structure-function relationships. *Protein Sci.* **1,** 426–471.

22. Bock, L. C., Griffin, L. C., Latham, J. A., et al. (1992) Selection of single-stranded DNA molecules that bind and inhibit human thrombin. *Nature*, **355,** 564–567.

23. Tasset, D. M., Kublik, M. F., and Steiner, W. (1997) Oligonucleotide inhibitors of human thrombin that bind distinct epitopes. *J. Mol. Biol.* **272,** 688–698.

24. Hamaguchi, N., Ellington, A., and Stanton, M. (2001) Aptamer beacons for the direct detection of proteins. *Anal. Biochem.* **294,** 126–131.

25. Li, J. J., Fang, X. H., and Tan, W. H. (2002) Molecular aptamer beacons for real-time protein recognition. *Biochem. Biophys. Res. Commun.* **292,** 31–40.
26. Cao, Z. H. and Tan, W. H. (2005) Molecular aptamers for real-time protein-protein interaction study. *Chem. Eur. J.* **11,** 4502–4508.
27. Green, L. S., Jellinek, D., Jenison, R., et al. (1996) Inhibitory DNA ligands to platelet-derived growth factor B-chain. *Biochem.* **35,** 14,413–14,424.
28. Fang, X. H., Cao, Z. H., Beck, T. et al. (2001) Molecular aptamer for real-time oncoprotein platelet-derived growth factor monitoring by fluorescence anisotropy. *Anal. Chem.* **73,** 5752–5757.
29. Fang, X. H., Sen, A., Vicens, M. et al. (2003) Synthetic DNA Aptamers to detect protein molecular variants in a high-throughput fluorescence quenching assay. *Chembiochem,* **4,** 829–834.
30. Vicens, M. C., Sen, A., Vanderlaan, A., et al. (2005) Investigation of molecular beacon aptamer-based bioassay for platelet-derived growth factor detection. *Chembiochem* **6,** 900–907.
31. Yang, C. J., Jockusch, S., Vicens, M., et al. (2005) Light-switching excimer probes for rapid protein monitoring in complex biological fluids. *Proc. Natl. Acad. Sci. USA* **102,** 17,278–17,283.

16

DNA Polymerase Profiling

Daniel Summerer

Abstract

We report a simple homogeneous fluorescence assay for quantification of DNA polymerase function in high throughput. The fluorescence signal is generated by the DNA polymerase triggering opening of a molecular beacon extension of the template strand. A resulting distance alteration is reported by fluorescence resonance energy transfer between two dyes introduced into the molecular beacon stem. We describe real-time reaction profiling of two model DNA polymerases. We demonstrate kinetic characterization, rapid optimization of reaction conditions, and inhibitor profiling using the presented assay. Furthermore, to supersede purification steps in screening procedures of DNA polymerase mutant libraries, detection of enzymatic activity in bacterial expression lysates is described.

Key Words: High throughput screening; FRET; DNA polymerase; directed evolution; inhibitor screening; fluorescent probes.

1. Introduction

DNA polymerases catalyze the template-directed DNA synthesis in all living organisms occurring in replication, repair, and recombination *(1)*. This central role has rendered these enzymes to key components of numerous biotechnological applications like polymerase chain reaction (PCR), cDNA synthesis for expression profiling, mutagenesis techniques, and genotyping of single nucleotide polymorphisms (SNPs) *(2,3)*. In addition, DNA polymerases represent important drug targets, particularly in the therapy of antiretroviral diseases like acquired immunodeficiency syndrome (AIDS) *(4–6)*. Today, progress in drug discovery is tightly linked to the massive parallel evaluation of large compound libraries like natural product collections or libraries generated by combinatorial synthesis for inhibitor potency *(7)*. Similarly, to tailor DNA polymerases for artificial reaction conditions used in biotechnological applications, directed

From: *Methods in Molecular Biology, Vol. 429: Molecular Beacons: Signalling Nucleic Acid Probes,*
Methods and Protocols Edited by: A. Marx and O. Seitz © Humana Press Inc., Totowa, NJ

evolution has emerged as a powerful tool to isolate new or improved enzyme functions from genetically engineered protein libraries *(8)*.

Current demands on assay formats for high-throughput screening are high sensitivity to allow minimal assay volumes, ease of automation, and the avoidance of hazardous components like radioisotopes *(7,8)*. Additionally, online profiling of enzymatic activity is much preferred over endpoint measurements of the generated product, since reaction rates can be deduced from slopes of the obtained reaction graphs. This enhances the assays dynamic range for a given substrate concentration especially if a broad range of activity is present in the screened library and adjustment of reaction time and substrate consumption is difficult to achieve for all variants.

We have developed a homogenous, fluorescent assay format for rapid evaluation of DNA polymerase function that allows real-time monitoring of reaction graphs *(9)*. The assay can be used for inhibitor screening and evaluation, rapid optimization of reaction conditions, assessment of kinetic constants, and for the measurement of DNA polymerase activity in crude lysates of bacterial expression cultures. The design of the setup should allow screening of variant libraries for properties like altered fidelity, stability, and processing of artificial substrates. Employment of crude expression lysates simplifies the screening procedure since laborious purification steps can be omitted.

The assay relies on a molecular beacon equipped with fluorescence resonance energy transfer (FRET)-donor and -acceptor dye within an 8mer stem. This stem bears a 3′-extension that allows binding of a primer and acts as template for the DNA polymerase (**Fig. 1**).

Proceeding DNA synthesis results in opening of the stem region, separation of the dyes and increase of donor emission. We took advantage of the widely used FAM/TAMRA pair (carboxyfluorescein, *N,N*′-tetramethylrhodamine) for FRET though other dyes might be used that can be bypassed by the DNA polymerase employed *(10,11)*. The assay setup allows investigation of a wide variety of DNA polymerases. We tested phage T7 DNA polymerase and its 3′–5′-exonuclease deficient mutant Sequenase™, Human immunodeficiency virus type 1 reverse transcriptase (HIV-1 RT), human DNA polymerase β, bacterial *Bacillus stearothermophilus* (Bst) DNA Polymerase I (large fragment) and several archaeal thermostable enzymes (Vent™, 3′-5′-exonuclease-deficient Vent (Vent⁻)™), and *Thermus thermophilus* (Tth) DNA polymerase.

In the following, we describe the evaluation of this assay principle with the Klenow Fragment of *E. coli* Pol I (3′–5′ e-xo⁻, KF⁻), i.e., the assessment of Michaelis–Menten kinetics, optimum concentration of magnesium ions, and detection of DNA polymerase activity in crude bacterial expression lysates. Additionally, we describe the profiling of inhibitor potency for HIV-1 RT with two well-characterized inhibitors of this target enzyme.

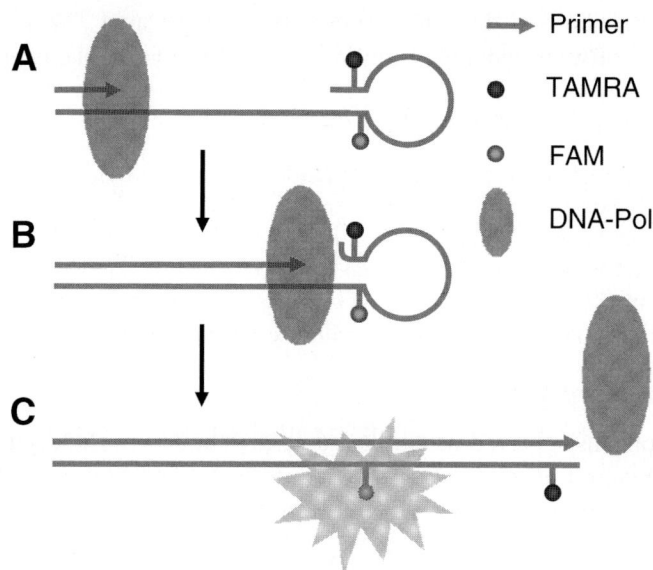

Fig. 1. Assay design for real-time observation of the DNA polymerase reaction. **(A)** The molecular beacon template labeled with fluorophor (carboxyfluorescein, FAM) and acceptor (*N,N'*-tetramethylrhodamine, TAMRA) bears a loop structure in closed conformation before start of reaction. **(B)** While extension proceeds, the DNA polymerase opens the stem and prevents reannealing by DNA duplex formation. **(C)** The increase in the distance between the two labels is reported by restoration of FAM emission. Figure was taken from Ref. *(9)*.

2. Materials

2.1. Design and Synthesis of a Molecular Beacon Template

The molecular beacon template can be synthesized by the researcher or purchased by a commercial supplier. Several oligonucleotide synthesis companies provide synthesis and purification of long, double internally labeled oligos. These include IBA GmbH (Germany), MWG-Biotech (Germany), Eurogentech (Belgium), and Integrated DNA Technologies (USA).

The construct we used consisted of a 23mer loop connected by a GC-rich 8mer stem region and a 43mer extension with the sequence 5′-*GGC CCG* U^{TAMRA}*AG*-GAG GAA AGG ACA TCT TCT AGC A U^{FAM}*A CGG GCC* GTC AAG TTC ATG-GCC AGT CAA GTC GTC AGA AAT TTC GCA CCA C-3′ where italicizes identify the stem sequences. TAMRA and FAM were connected to C5 of 2′-deoxy-uridines as 5,6-regioisomeric mixtures. The employed primer had the sequence 5′-(GTG GTG CGA ATT TTC TGA C)-3′. If changing part of the sequence is desired, evaluation of the folding properties using David Zukers web-based DNA-folding software *mfold* is recommended (www.bioinfo.rpi.edu/applications/mfold/old/dna/form1.cgi).

Following materials were used for synthesis:

1. ABI 392 DNA/RNA synthesizer (Applied Biosystems, USA).
2. Ultramild phosphoramidites (Glen Research, USA).
3. Fluorescein dT phosphoramidite (cat. no. 10-1056; Glen Research).
4. TAMRA dT phosphoramidite (cat. no. 10-1057; Glen Research).
5. DNA synthesis reagents, controlled pore glass solid supports (1000 Å pore size) and deprotection solution of the Ultramild series (Glen Research).

2.2. Expression of KF⁻

Expression was achieved using plasmid pQKlenowExo⁻ (kindly provided by Dr. S. Brakmann, Leipzig) *(11)*. The plasmid is derived from pQE30 (Qiagen, Germany) and allows tightly regulated transcription from the T5 promoter by two lac operator sequences.

Following materials were used for expression:

1. Vector pQKlenowExo⁻ (Dr. S. Brakmann).
2. Vector pQE30Xa (Qiagen).
3. *E. coli* strains M15 [pREP4] (Qiagen) or XL1blue (Stratagene, USA) (*see* **Note 1**).
4. Luria Bertani medium (LB).
5. Super broth.
6. Ampicillin (Sigma, USA).
7. Kanamycin (Sigma).
8. Isopropyl-β-D-thiogalactopyranoside, 50 mM stock solution (IPTG; Sigma).
9. 1.2 mL 96-deepwell plates (cat. no. 0030 127.544; Eppendorf, Germany).
10. 2.2 mL 96-deepwell plates (cat. no. 0030 127.560; Eppendorf).
11. Airpore tape sheets (cat. no. 19571; Qiagen).
12. Lysis buffer: 50 mM NaH$_2$PO$_4$, 300 mM NaCl (pH 8.0) containing 1.5 mg/mL lysozyme (Sigma).

2.3. Automated Online Profiling of DNA Polymerase Activity by FRET

1. Fluorescence plate reader with at least one dispenser, shaking option, and filters for fluorescein excitation and emission like Fluostar, Polarstar (BMG Labtechnologies, Germany) or Fluoroskan Ascent (Thermolabsystems, USA).
2. Black 96-well fluorescence assay plates, low protein binding (cat. No. 3694, Corning, USA) (*see* **Note 2**).
3. Klenow fragment (3′–5′-exonuclease-deficient, NEB, USA).
4. HIV-1 reverse transcriptase (Roche, Germany).
5. KF⁻ reaction buffer: 50 mM Tris–HCl (pH 7.3) and 1 mM dithiothreitol (DTT).
6. HIV-1 RT reaction buffer: 50 mM Tris–HCl (pH 8.3) and 50 mM KCl.
7. 3′-Azido-2′3′-dideoxythymidine-5′-*O*-triphosphate (AZT TP, IBA GmbH, Germany).

8. Nevirapine (AIDS Research and Reference Reagent Program, NIAID, NIH).
9. Deoxynucleoside triphosphates (Sigma).
10. 0.1 M MgCl$_2$.

3. Methods

3.1. Synthesis of a Molecular Beacon Template

In principle, double labeling can be done postsynthetically via aminolinkers or sulfhydryl functions, by direct coupling of labeled phosphoramidites or both (mixed synthesis). If possible, direct coupling of both dyes is recommended in terms of coupling efficiencies and in particular to obtain clean 1:1 ratios of fluorophor and dye. We experienced strong variations in signal-to-noise ratio when DNA was prepared by mixed synthesis. In our case, FAM dT and TAMRA dT phophoramidites were used and Ultramild chemistry was applied since TAMRA is labile under standard deprotection conditions. Synthesis should be performed using 1 µmol solid support with 1000 Å pore size using standard DNA-coupling protocols, DMT off mode and increased coupling time of 10 min for the labeled phosphoramidites. Purification by 12% denaturing PAGE (8 M urea) using a 1.5 mm gel and a 50 cm chamber is strongly recommended to get optimal results (*see* **Note 3**). DNA can be stored in ddH$_2$O at–20°C protected from light.

3.2. Expression of KF⁻

The following protocol is optimized for automated parallel expression in 96-well plates using a liquid handling station like the Microlab Star (Hamilton Robotics, Switzerland) or the Genesis (Tecan, Switzerland). In principle, all steps except for centrifugation steps can be automated.

1. A volume of 500 µL LB medium containing 100 µg/mL ampicillin (for XL1blue and additionally 25 µg/mL kanamycin when M15 [pREP4] is used) in a well of a 1.2 mL 96-deepwell plate were inoculated with a single colony of the respective strain harboring pQKlenowExo⁻.
2. Plates were sealed with airpore tape sheets and grown overnight at 37°C and 200 rpm shaking (*see* **Note 4**).
3. Fifty microliters of overnight cultures were added to 930 µL super broth in 2.2 mL 96-deepwell plates.
4. Cultures were sealed with airpore tape sheets, grown at 37°C to an OD600 of 0.6–0.7 under shaking at 200 rpm and induced by addition of 20 µL IPTG (50 mM resulting in 1 mM final concentration).
5. After 4 h shaking at 200 rpm at 37°C cells were harvested by centrifugation.
6. For lysis, pellets were subjected to one freeze–thawing step, resuspended in 500 µL lysis buffer and shaked for 45 min at 4°C.
7. Lysates were cleared by centrifugation and directly used for primer extension reactions.

8. Control lysates of same strain harboring vector pQE30Xa without KF⁻ coding insert were prepared identically.

3.3. Automated Online Profiling of DNA Polymerase Activity by FRET

3.3.1. General Procedure for Fluorescent Primer Extension Reactions with a Molecular Beacon Template

1. Fifty microliter reactions contained 200 nM molecular beacon template, 300 nM primer 10 mM MgCl$_2$, and varying amounts of equimolar dATP, dGTP, dCTP, and dTTP in reaction buffer specific for the respective DNA polymerase.
2. Molecular beacon template, primer and 5 μL 10 × reaction buffer were mixed in 40 μL total volume/well.
3. Mixture was heated to 95°C for 5 min and was put on room temperature for 1 h.
4. dNTP and enzyme were added to a total volume of 45 μL.
5. Mixture was transferred to 96-well assay plate.
6. Plate was preincubated in fluorescence plate reader at 37°C for 5 min.
7. Reaction was initiated by automated dispensing of 0.1 M MgCl$_2$ solution followed by a short shaking step (3 s, 2–3 mm shaking diameter, 150 rpm).
8. FAM fluorescence increase was measured in appropriate time intervals by exciting at 485 nm and emission at 518 nm as arbitrary units.

3.3.2. Kinetic Measurements with KF⁻

The kinetics of enzymatic DNA synthesis exhibit Michaelis–Menten dependence of substrate concentration, thus an apparent Michaelis–Menten constant (K_M) can be derived from experiments conducted with a given amount of enzyme and varied substrate concentrations *(12)*.

Reactions were conducted using 1.4 U KF⁻ and a range of 1.25–37.5 μM of each dNTP. A measurement of an equivalent reaction mix without KF⁻ was conducted in parallel and subtracted from the data of the experiments including KF⁻ (*see* **Note 6**). Reaction graphs were obtained by fluorescence measurements in time intervals of 20 s for 6.5 min (**Fig. 2a**).

Initial apparent reaction rates (v_i (F_{518}/min⁻¹)) were obtained from data by determination of the slopes of the initial linear portion of individual reactions graphs and plotted versus dNTP concentration (**Fig. 2b**). A K_M–value was obtained by nonlinear curve fitting using the Michaelis–Menten equation. We obtained a K_M value (3.6 ± 0.2 μM) that compares well with that measured independently using a conventional radiometric assay (3.9 ± 1.1 μM) *(9)*.

3.3.3. Rapid Assessment of Optimum Magnesium Ion Concentration

It is often desirable to rapidly determine optimum conditions of a given enzyme or variant. We employed the assay setup to determine the optimal magnesium chloride concentration in terms of activity. Reactions were performed using 1.4 U KF⁻ and different MgCl$_2$ concentrations as depicted in **Fig. 3**.

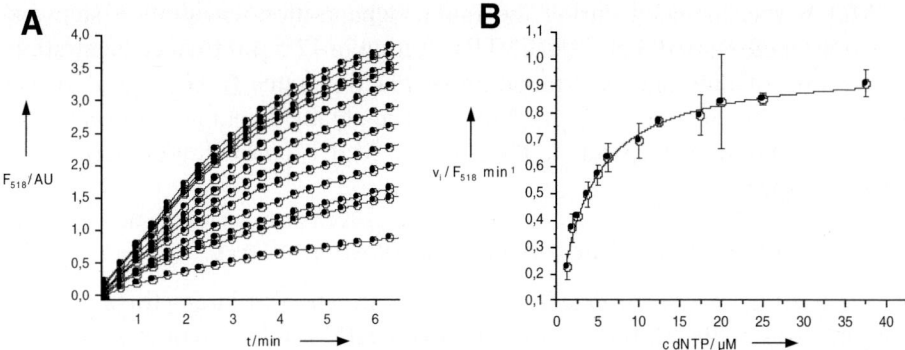

Fig. 2. Fluorescent real-time determination of the K_M constant using KF⁻. **(A)** Time courses of reactions employing different dNTP concentrations. Reactions contained 1.4 *U* KF⁻ and dNTP concentrations ranging from 1.25 µ*M* (lowest curve) to 37.5 µ*M* (highest curve). Data are results of multiple experiments. For clarity, error bars are not shown. **(B)** Michaelis–Menten plot of reaction velocities obtained from **Fig. 2A**. Initial apparent reaction rates (v_i (F_{518}/min^{-1})) were obtained from data by determination of the slopes of the initial linear portion of individual reaction graphs and plotted versus dNTP concentration. Line graph corresponds to nonlinear curve fit using the Michaelis–Menten equation. Figure was taken from Ref. *(9)*.

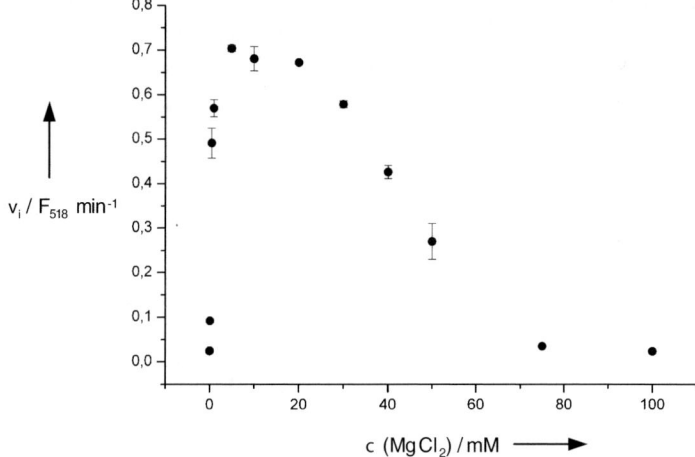

Fig. 3. Rapid assessment of optimal magnesium chloride concentration for KF⁻ using the molecular beacon template. Reactions contained 1.4 U KF⁻ and different $MgCl_2$ concentrations as depicted. Reactions were started by addition of dNTP to a final concentration of 12.5 µ*M* of each. Initial apparent reaction rates (v_i (F_{518}/min^{-1})) were obtained from data by determination of the slopes of the initial linear portion of individual reaction graphs and plotted against dNTP concentration.

$MgCl_2$ was included during the initial denaturation/renaturation step and reactions were started with 5 μL dNTP resulting in 12.5 μM final concentration. Correction of data and determination of reaction rates (v_i (F_{518}/min^{-1})) was conducted as in Section 3.3.2 (*see* **Note 6**). The obtained optimum concentration of about 10 mM corresponds to the standard $MgCl_2$ concentration used in the literature for KF$^-$.

3.3.4. Profiling of HIV-1 Reverse Transcriptase Inhibitors

We used the assay for detection and quantification of interactions of DNA polymerases with inhibitors. We employed HIV-1 RT as model system. In current HIV drug therapy, two classes of RT inhibitors are in use that differ in their mode of action. Non-nucleoside RT inhibitors (NNRTIs) allosterically inhibit RT function, whereas nucleoside RT inhibitors (NRTIs) are first transformed to 5′-*O*-triphosphates by cellular processes and then incorporated into the nascent DNA strand by the RT to cause chain termination *(4–6)*. To validate whether the assay format rapidly identifies inhibitors of HIV-1 RT acting by both mechanisms, we studied nevirapine and AZT TP representing well characterized NNRTIs and NRTIs, respectively. Reactions were performed using 2.1 U HIV-1 RT and inhibitor in different concentrations. Inhibitors were added to reactions along with 50 μM dNTP (final concentration) and enzyme before preincubation in the plate reader and reactions were started with $MgCl_2$. Correction of data and determination of reaction rates (v_i (F_{518}/min^{-1})) was conducted as in Section 3.3.2 (*see* **Note 6**). Inhibition profiles from which IC$_{50}$–values can be deduced were obtained by plotting reaction rates against inhibitor concentration (**Fig. 4**).

The resulting IC$_{50}$ values of nevirapine (0.13 ± 0.03 μM) and AZT TP (2.10 ± 0.22 μM) correspond well with reported data indicating the suitability of the assay for faithful inhibitor characterization *(13,14)*.

3.3.5. Profiling of KF$^-$ Activity in Crude Bacterial Lysates

Fluorescent reactions were conducted under the same conditions as used with purified KF$^-$. $MgCl_2$ was added before the denaturation/renaturation step and reactions were started with dNTP (200 M final concentration). Instead of purified enzyme, 1 μL of cleared lysate of 1 mL culture was employed (*see* Section 3.2). As control, same amount of lysate of the control culture (harboring pQE30Xa) was used. Data were not corrected as in previously described measurements (*see* **Note 6**). The obtained reaction graphs indicate activity only when KF$^-$ expressing cultures were employed and exhibit no background activity that might be caused by host factors like DNA polymerases or nucleases (**Fig. 5**).

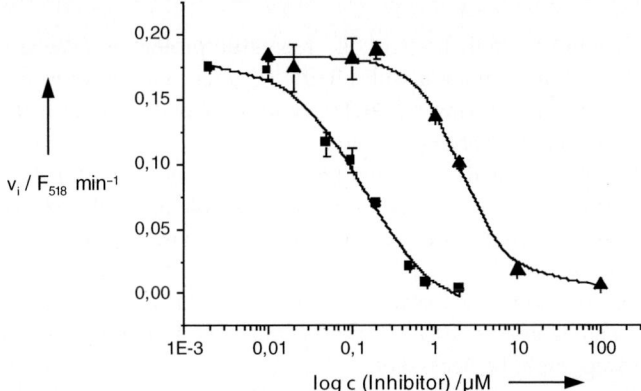

Fig. 4. HIV-1 RT inhibition profiles obtained from measurements with the molecular beacon template. Reactions were conducted using 2.1 U HIV-1 RT and 50 μM of each dNTP in the presence of inhibitors at different concentrations as depicted. (Triangles) AZT TP and (squares) nevirapine. Initial apparent reaction rates (v_i) were obtained from data by determination of the slopes of the initial linear portion of individual reaction graphs. Figure was taken from Ref. *(9)*.

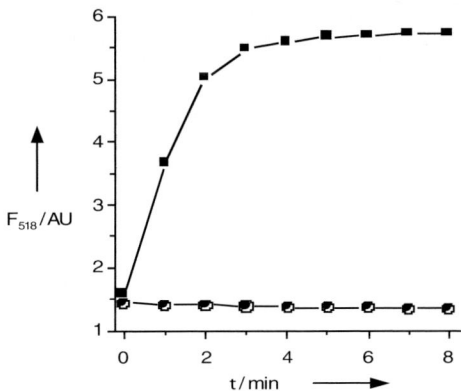

Fig. 5. Primer extension reactions promoted by crude lysates of KF⁻ overexpressing *E. coli* cultures (strain M15 [pREP4]) using the molecular beacon template. Reactions contained 10 mM MgCl$_2$ and lysate of expression-induced M15 [pREP4] cultures and were started by addition of dNTP to a final concentration of 200 μM. (Squares) Reactions employing lysate of M15 [pREP4] cells harboring the KF⁻ coding vector pQKlenowExo 4 h after induction of expression. (Circles) Reactions employing lysate of M15 [pREP4] cells harboring the noncoding vector pQE30Xa 4 h after induction of expression. Data are results of multiple experiments. Presented graphs correspond to raw data without correction (*see* **Note 6**). Figure was taken from Ref. *(9)*.

4. Notes

1. For tight control, a high level of lac repressor protein is needed. This can be achieved by cotransformation with pREP4 (Qiagen) that encodes lac repressor or by usage of strains carrying the lacI^q genotype (like XL1blue) that increases lac repressor concentration about 10-fold.

2. Plates have unusual small wells and protocols are optimized for these plates. The described 50 μL reaction volumes are not applicable to regular 96-well plates.

3. Gel temperature should be constantly at 45–50°C to avoid formation of secondary structures by the GC-rich stem. If the crush and soak method is used for gel extraction, sample should be incubated at −80°C thoroughly for precipitation. Solubility of the oligonucleotide in 70% ethanol is enhanced by the dyes and a second precipitation step might be necessary.

4. Shaking diameter seemed not to be critical. We shaked with a diameter of several centimeters with good expression yield so that a common incubator can be used.

5. For generation of mutant libraries, long-term storage of the lysates is desirable. This can be done by addition of three volumes storage buffer (50 mM Tris–HCl (pH 7.3), 1 mM DTT, 0.1 mM EDTA, 1 mM PMSF, 1 mM benzamidine, 1 μg/mL leupeptine, and 1 μg/mL aprotinine) after lysis. Expression culture and lysis volume should be scaled by a factor of 0.6 in this case to allow proper handling in 96-deepwell plates. Samples can be stored at −80°C and tolerate at least three freeze–thawing steps without detectable loss in activity.

6. We observed slight photobleaching of FAM that can be corrected by this procedure. This effect can be observed in the negative control of **Fig. 5** that is not corrected. In case of the optimization experiment for MgCl$_2$, a quenching effect is observable and initial fluorescence decreases with increasing magnesium concentration. This effect is eliminated by analyzing reaction rates as described rather than absolute fluorescence data.

Acknowledgments

The author thank Prof. Dr. Andreas Marx at the University of Konstanz for continuous support and helpful discussions.

References

1. Kornberg, D. G. and Baker, T. A. (1991) *DNA Replication.* W. H. Freeman, New York, NY.
2. Sambrook, J. and Russell, D. W. (2001) *Molecular cloning: A Laboratory Manual,* Cold Spring Harbor Laboratory Press, Cold Spring Harbor, NY.
3. Shi, M. M. (2001) Technologies for individual genotyping: detection of genetic polymorphisms in drug targets and disease genes. *Clin. Chem.* **47,** 164–172.
4. Naeger, L. K. and Miller, M. D. (2001) Mechanisms of HIV-1 nucleoside reverse transcriptase inhibitor resistance: is it all figured out? *Curr. Opin. Invest. Drugs* **2,** 335–339.
5. Tozser, J. T. (2001) HIV inhibitors: problems and reality. *Ann. N. Y. Acad. Sci.* **946,** 145–159.

6. Pillay, D., Taylor, S., and Richman, D. D. (2000) Incidence and impact of resistance against approved antiretroviral drugs. *Rev. Med. Virol.* **10**, 231–253.

7. Hertzberg, R. P. and Pope, A. J. (2000) High-throughput screening: new technology for the 21st century. *Curr. Opin. Chem. Biol.* **4**, 445–451.

8. Olsen, M., Iverson, B., and Georgiou, G. (2000) High-throughput screening of enzyme libraries. *Curr. Opin. Biotechnol.* **11**, 331–337.

9. Summerer, D. and Marx, A. (2002) A molecular beacon for quantitative monitoring of the DNA polymerase reaction in real time. *Angew. Chem. Int. Ed.* **41** (19), 3620–3622.

10. Brakmann, S. and Lobermann, S. (2001) High-Density Labeling of DNA: Preparation and characterization of the target material for single-molecule sequencing. *Angew. Chem. Int. Ed.* **40**, 1427–1429.

11. Brakmann, S. and Nieckchen, P. (2001) The large fragment of Escherichia coli DNA polymerase I can synthesize DNA exclusively from fluorescently labeled nucleotides. *Chem. Bio. Chem* **2**, 773–777.

12. Fersht, A. (2000) *Structure and Mechanism in Protein Science*, Freeman, New York, NY.

13. Reardon, J. E. and Miller, W. H. (1990) Human immunodeficiency virus reverse transcriptase. Substrate and inhibitor kinetics with thymidine 5′-triphosphate and 3′-azido-3′-deoxythymidine 5′-triphosphate. *J. Biol. Chem.* **265**, 20,302–20,307.

14. Merluzzi, V. J., Hargrave, K. D., Labadia, M., et al. (1990) Inhibition of HIV-1 replication by a nonnucleoside reverse transcriptase inhibitor. *Science*, **250**, 1411–1413.

17

Modular Reporter Hairpin Ribozymes for Analyzing Molecular Interactions

S. Hani Najafi-Shoushtari and Michael Famulok

Abstract

Methods for the detection of biologically relevant interactions by highly precise catalytic control elements based on hairpin ribozymes, and their subsequent analysis are described. These include ribozyme design, catalytic performance in real time as a function of fluorescence signal amplification, and applications for sensing protein and nucleic acid interactions in high-throughput formats. Detailed instructions for two of our main reporter ribozyme formats that either follow repressible or inducible regulatory mechanisms are provided. We have shown that these techniques can be applied for detecting diverse target molecules including microRNAs, or protein—protein interactions. These reporter systems thus represent a general way to obtain signal-amplifying sensors for diverse applications in molecular profiling.

Key Words: Ribozyme engineering; allostery; regulation mechanism; hairpin ribozyme; microRNAs; molecular profiling; FRET; signaling probes; real-time detection.

1. Introduction

Novel methods which facilitate the rapid and reliable identification of molecular interactions in biological systems, compatible with high-throughput screening formats are required in biomedical and biotechnological research. Knowledge about the interaction partners, binding affinity, and interacting domains represent the basis for identification of the biological function and discovery of novel inhibitors, modulators and drug leads of a given protein or nucleic acid. Although there are several powerful methods available for detection and quantification of molecular interactions, they are often not generally applicable or compatible with high-throughput screening methodology, in real time. Hence, the development of novel, broadly applicable methods, independent of target protein function, is of fundamental interest. We have pioneered the use of ribozymes for developing

From: *Methods in Molecular Biology, Vol. 429: Molecular Beacons: Signalling Nucleic Acid Probes, Methods and Protocols* Edited by: A. Marx and O. Seitz © Humana Press Inc., Totowa, NJ

Table 1
Two Classes of Hairpin Ribozyme Probes that can either Respond Positively (iHPs) or Negatively (rHPs) to their Cognate Target Molecules

Ribozyme probe	Traget molecule	Reference
inducible (iHP)		
iHP-TRAP	*trp*-mRNA(5'-UTR)	*(9)*
iHp-let7	let-7a microRNA	*(10)*
iHP-miR1	dme-miR-1	*(10)*
iHP-miR2	dme-miR-2	*(10)*
iHP-miR4	dme-miR-4	*(10)*
iHP-miR5	dme-miR-5	*(10)*
iHP-miR7	dme-miR-7	*(10)*
iHP-miR10	dme-miR-10	*(10)*
iHP-miR34	dme-miR-34	*(10)*
iHP-miR79	dme-miR-79	*(10)*
repressible (rHP)		
rHP-Thrombin	α-thrombin DNA aptamer	*(11)*
rHP-TRAP	*trp*-mRNA (5'-UTR)	*(9)*
rHP-let7	let-7a microRNA	

functional assays allowing the analysis of interactions of biologically relevant molecules in real-time. Here, we describe methods of general applicability for the detection of biologically relevant interactions by highly precise catalytic control elements based on hairpin ribozymes, and their subsequent analysis.

Site-specific ribonuclease activity of the hairpin ribozyme requires a physical interaction (docking) of two independently folded internal loop domains A and B *(1–6)*. These are connected via a bendable linkage, often referred to as the hinge region, with structural features that confer appropriate conformational flexibility and stability in hairpin ribozyme mediated catalysis *(7,8)*. Previously, we have shown that the conversion of this region into an oligonucleotide effector-binding site (loop domain C) allows regulation of the hairpin ribozyme catalytic performance by precise conformational control that involves effector-triggered formation of distinct structural motifs within the interface between the internal loops *(9–12)*. In this context, two design strategies have been explored and employed to create either allosterically inducible (iHPs) or repressible (rHP) hairpin ribozyme constructs for various effector target molecules (**Table 1**). The generation and analysis of two of them, rHP-let7 and iHP-let7, bearing a binding site for the let-7 microRNA will be described here to illustrate these methods. In addition, a general framework is provided for molecular profiling using such regulatory hairpin ribozymes as convenient signaling probes in FRET- or fluorescence-dequenching assays.

2. Materials

2.1. Oligonucleotide Amplification and Workup

1. Phenol/chloroform/isoamyl alcohol for DNA or RNA precipitation, Store at 4°C.
2. Large volumes of TBE buffer (1×): 89 mM Tris, 89 mM boric acid, and 2.0 mM EDTA (Merck, Roth). Store at room temperature.
3. Materials and equipments for polyacrylamide and agarose gel-electrophoresis.
4. Polyacrylamide gel loading buffer: 97% (v/v) formamide, 15 mM EDTA, 0.25% (w/v) xylene cyanol, and 0.25% (w/v) bromophenol blue.
5. Ammonium persulfate (APS): prepare 10% working solution in water and store at 4°C.
6. 25% acrylamide/bis solution containing urea (stored at room temperature to avoid urea precipitation) and N,N,N,N'-tetramethyl-ethylenediamine (TEMED, Fluka, Germany; stored at 4°C).
7. Agarose gel DNA-loading buffer (pH 8): 50 mM Tris—HCl, 50 mM EDTA, 50% (v/v) glycerol, 0.05% (w/v) bromophenol blue sodium salt, and 0.05% (w/v) xylene cyanol. Store in aliquots at 4°C.
8. Conventional programmable polymerase chain reaction (PCR) heating block (Biometra).
9. 1 U/µL RNase free DNase I (Roche/Boehringer, Mannheim).
10. 5 U/µL DAp-DNA-polymerase (Goldstar, Eurogentec, Belgium).
11. UV-Transilluminator (Biorad) for agarose gel imaging.
12. PCR buffer (pH 8.3): 100 mM Tris—HCl and 750 mM KCl.
13. Ribonuclease inhibitor RNasin (Promega).
14. 50 U/µL T7-RNA polymerase (Stratagene).
15. 3.0 M sodium acetate (pH 5.4).
14. Transcription buffer (pH 7.9): 40 mM Tris—HCl, 6.8 mM spermidine, 22 mM MgCl$_2$, and 0.01% (v/v) Triton X-100.
15. 2 U/µL inositol-pyrophosphatase.

2.2. Ribozyme Reactions

1. Ribozyme reaction buffer (pH 7.5): 50 mM Tris—HCl and 20 mM MgCl$_2$. The buffer is sterile filtrated and stored at 4°C. Working solutions can be kept at room temperature.
2. UV-absorbance spectroscope for quantification of nucleic acid concentration (Perkin Elmer lambda 2).
3. Fluorescence scanner with appropriate optical equipments for real-time detection of FAM emission light (Fluoroscan Ascent FL, Labsystems, in this example).
4. 96-well plates (Corning, Inc.).

2.3. Oligonucleotides

All ribozyme constructs described here were synthesized using following DNA templates and primers. (The T7 promotor sequence is italicized and the transcription initiating sequence is shown in bold letters.)

iHP-let7: 5′-AAA TAG AGA AGC GAG CAC GAC TAT ACA ACC TAC TAC CTC TTC TCT ACG TGC ACC AGA GAA ACA CAC GAC GTA AGT CGT GGT ACA-3′

rHP-let7: 5′-AAA TAG AGA AGC GAA GAG TAT ACA ACC TAC TAC CTC UAC CAG AGA AAC ACA CGA CGT AAG TCG TGG TAC A-3′

5′-primer: 5′-*TCT AAT ACG ACT CAC TAT A*GG AAA TAG AGA AGC GA-3′

3′-primer: 5′-TAC CAG GTA ATG TAC CAC GAC TTA CGT-3′

RNA-oligonucleotides (FRET-labeled substrate and let-7 microRNA) were purchased from Dharmacon (Lafayette, CO, USA) and IBA (Göttingen, Germany), respectively and worked up according to manufacturer's guidelines:

FRET-labeled RNA substrate: 5′-TAMRA-UCGCAGUCCUAUUU-FAM-3′, stored at −20°C at stock concentrations of 10 mM in ddH$_2$O.

let-7: 5′-UGA GGU AGU AGG UUG UAU AGU U-3′

3. Methods

The methods described below outline (i) the construction and generation of an inducible and a repressible hairpin ribozyme for the detection of let-7 microRNA, (ii) the kinetic analysis of the resulting constructs by real-time fluorescence signal measurements, and (iii) the application and utilization of such regulatory hairpin ribozymes for molecular profiling.

3.1. Construction of Regulatory Hairpin Ribozymes: rHP-let7 and iHP-let7

3.1.1. Design Strategies

As previously described, two distinct design strategies were applied to generate two different classes of regulatory hairpin ribozymes, referred to as repressible (rHP) and inducible (iHP) hairpin ribozymes *(9)*. This was achieved by incorporating a new domain as an oligonucleotide binding site (domain C) into the CH1 ribozyme construct (*see* **Note 1**). Briefly, in the first strategy, domain-specific hybridization of a single-stranded oligonucleotide to the cognate hairpin loop domain C enforces a conformational switch towards an extended, and therefore inactive, ribozyme structure (*see* **Note 2**). However, in the second strategy, oligonucleotide annealing restores ribozyme activity from an originally inactive ribozyme construct by triggering the formation of an alternate conformer that adopts a stable pseudo-half-knot structure (*see* **Note 3**). **Fig. 1** shows the predicted mechanisms that are expected to occur in the absence and presence of effector molecule following the appropriate design strategies.

3.1.2. Design Procedures

The ribozyme constructs described here were designed at secondary structure level consulting RNA *mfold* software program (version 3.1) provided by Zuker and Turner; *see*: http://www.bioinfo.rpi.edu/applications/mfold/old/rna/ *(13,14)*.

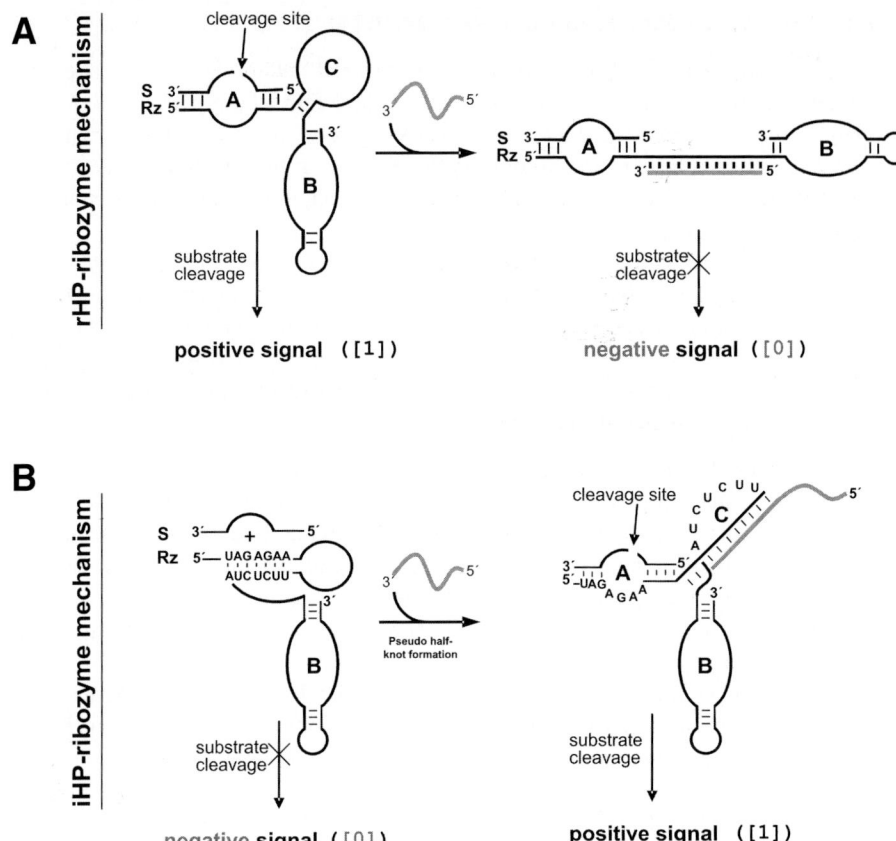

Fig. 1. Schematic of the allosteric mechanism investigated in the first and second design strategy. **(A)** Domain C is attached to the ribozyme sequence via a pseudo-three-way junction that is expected to favor the bent conformation by inhibiting the coaxial stacking of the helical stem regions H2 and H3 (**Note 4**). Oligonucleotide (grey line) annealing leads to formation of an inactive ribozyme conformer by inhibiting the appropriate docking of the A and B domains. **(B)** To produce an inducible ribozyme variant domain C is extended by seven consecutive nucleotides (5′-UUCUCUA-3′), complementary to the conserved sequences in the substrate binding arm (amino acid residues 4–10) that mediate intramolecular interaction between domains A and B. This can hybridize to the ribozyme-part of domain A, preventing substrate binding to the ribozyme to create an inactive ribozyme construct. However, when the oligonucleotide effector binds to the loop the terminal helix of the domain C is regained following a conformational switch in ribozyme which is expected to create a pseudo-half-knot structure, thereby inducing cleavage activity.

The design procedure occurs in two main steps:

1. The respective hairpin ribozyme sequences (either rHP-let7 or iHP-let7) were probed in absence of substrate molecule to check for destructive base pairing and folding within the ribozyme construct (*see* **Note 4**). Thereby, the following prerequisites were considered for a proper folding in the absence of the substrate sequence:

   a. The substrate-binding region of domain A should display fully unpaired sequence if the first design strategy is applied.

   b. The substrate-binding region of domain A display a proper formations of the AC-stem if the second design strategy is applied. To promote preferential hybrid formation, this structural element is introduced downstream of the let-7-binding region.

   c. Domain B is properly folded as observed for CH1 secondary structure independent from the applied design strategy.

2. Once the predicted structure matches all the requirements, extend the ribozyme sequence by the substrate sequence with a closing tetraloop (ribozyme-5'-GUAA-3'-substrate) on H1 connected to the 5'-end of the ribozyme (**Fig. 2**). This configuration allows probing the domain C folding in the presence of an intact A-domain. Thereby, the following points should be considered: to be prerequisites:

   a. Both domains A and B are correctly folded according to the CH1 structure prediction.

   b. Domain C should form an unpaired loop by avoiding self-complementary and nonspecific interference with other domains (in some cases, this can be achieved by composing the C domain binding region which preferably comprises only A, U, and C bases). Furthermore, this will enable a thermodynamically more preferred hybridization of the oligonucleotide effector.

   c. The terminal connecting domain (TCD) is properly folded according to an active ribozyme conformer. Basically, the base identity chosen in the second design strategy should contain a symmetric sequence with C2 geometry comprising a 5'-SSWSS-3' sequence order (whereby S = G or C and W = A or U). In comparison, the thermodynamic energy barrier (d[ΔG^0] as a difference of ΔG^0 determined in the absence and presence of the substrate sequence) should be minimal, whereby the substrate containing sequence exhibits a lower ΔG^0 value.

In some cases, all these sequences were designed in such a way that only one structure is formed and predicted by the RNA *mfold* program.

3.2. Preparation and Real-Time Analysis of iHP-let7 and rHP-let7 Ribozymes

3.2.1. Generation of iHP-let7 and rHP-let7 DNA Templates

The dsDNA constructs encoding ribozymes is generated and amplified by PCR *(15)* using Goldstar *DAp* DNA polymerase.

1. Prepare a PCR mixture (50 µL) consisting of 2 µL of a 100 µ*M* solution µg iHP-let7 or rHP-let7 ssDNA template (this amount corresponds to 1–2 µg) dissolved in 48 µL

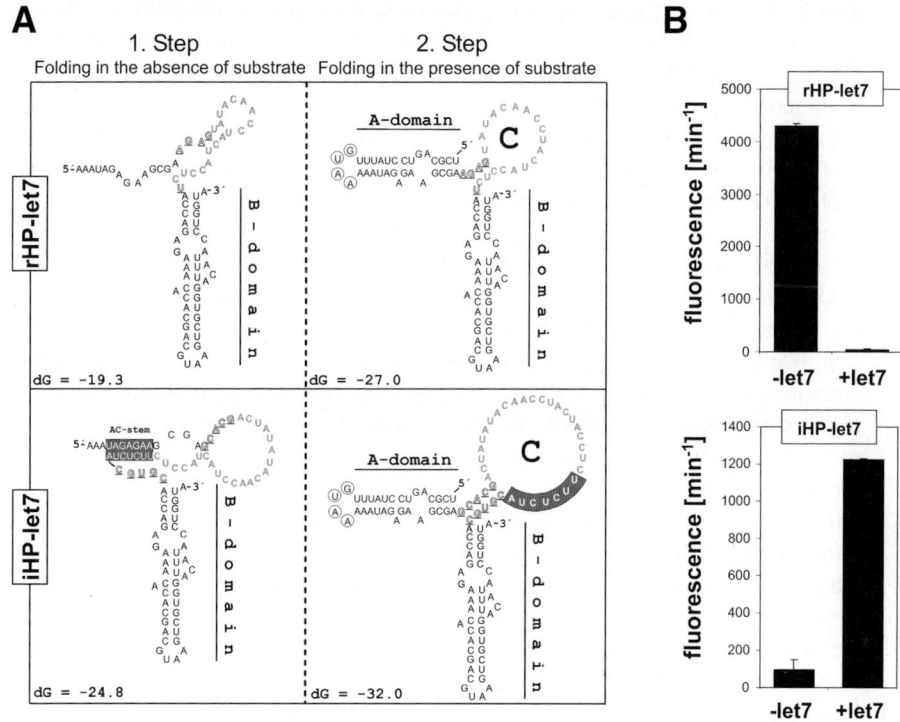

Fig. 2. Design and analysis of rHP-let7and iHP-let7 ribozyme constructs. (**A**) Computational design strategy for rHP-let7 and iHP-let7 in two main steps based on *mfold* software program. Both constructs shows proper folding of B-domain in the absence of substrate. The let-7 binding region is shown in grey letters and the TCD sequences are in bordered letters. While the substrate-binding region in rHP-let7 shows no significant intramolecular base pairing, same region in iHP-let7 is blocked by the formation of AC-stem (gray box) as expected. However, to evaluate appropriate formation of C-domain RNA folding is validated in the presence of substrate (Step 2). Thereby, the 5′-inserted tetraloop thermodynamically contributes to the right folding of domain A as a prerequisite for catalytic activity. Domain B is retained in both constructs as required. The dG values depicted in each box refer to the respective predicted secondary structures. (**B**) Analysis of ribozymes activity by fluorescence measurements. As expected, let-7 microRNA is specifically detected as a function of either increasing (iHP-let7) or decreasing (rHP-let7) fluorescent intensity as a result of substrate cleavage (*see* **Note 11**).

10× PCR mix containing 5 μL 10× PCR buffer, 2 μL of a 25 m*M* MgCl$_2$, 0.5 μL of a 5 U/μL DAp polymerase solution, 1 μL of a 100 μ*M* the 3′-primer, 1 μL of the 100 μ*M* 5′-primer, 0.7 μL of a 25 m*M* dNTP mixture (Sigma, Deisenhofen), and 37.8 μL ddH$_2$O.

2. Run the PCR in 14 cycles beginning with 94°C for 1 min (denaturing) followed by 53°C for 1 min (hybridization) and ending with 72°C for 1 min (primer extension), thereby set the lid temperature at 110°C. After 14 cycles centrifuge the tubes at room temperature for 3 min.

3. For quality control add 4 μL agarose loading buffer to a reaction aliquot of 7 μL and perform electrophoresis on a 2.5% (w/v) agarose gel. Identify full-length dsDNA (rHP-let7 and iHP-let7 with 101 and 115 bp, respectively) on an UV-*trans*-illuminator by using Peqgold 100 bp DNA-ladder containing 11 bands ranging from 80 to 1031 bp (PeqLab Biotechnology, Erlangen, Germany).

4. Add an equal volume of phenol:chloroform:isoamyl alcohol (25:24:1, v/v/v) to the aqueous nucleic acid solution, vortex for 1 min, and centrifuge at room temperature at $20,000 \times g$ for 5 min to separate the phases. Transfer the upper aqueous phase into to a new tube by pipetting (be careful to avoid transfer of precipitated material from the interface). Add double volume of chloroform to remove residual phenol from the nucleic acid containing aqueous solution using the same procedure.

5. After extraction, precipitate DNA by adding 0.1 volumes 3.0 M sodium acetate (NaOAc) pH 5.4, 0.01 volumes glycogen and 3.0 volumes of ice-cold absolute ethanol. After shaking thoroughly, the samples were incubated for 10 min at −80°C. Finally, the DNA was recovered by centrifugation 15 min at 4°C and the resulting pellet washed with 70 μL ethanol 70% (v/v), and resuspended in 20 μL water. Store at −25°C until needed.

6. Determine the total DNA amount by UV-absorption spectroscopy at 260 nm. Thereby, an optical density of 1 corresponds to a concentration of approx. 50 μg/mL double-stranded DNA (*see* **Note 5**).

3.2.2. In vitro Transcription of Ribozyme Constructs

1. Use a 10 μL aliquot of amplified dsDNA template (approx. 300 pmol) for each transcription in a final volume of 10 μL containing 20 μL 5× transcription buffer, 10 μL of a 25 mM NTP mix, 2.5 μL of a 0.1 M DTT, 1.25 μL of a 40 U/μL RNasin, 1 μL of a 2 U/μL IPP, 5 μL of a 50 U/μL T7 polymerase, and 50.25 μL water. Place tubes in a water bath and allow transcription to occur overnight (~16 h) at 37°C.

2. Add 0.5 μL of a 1 U/μL DNAse I solution to the reaction mix to digest the DNA template by further incubation at 37°C for at least 30 min.

3. Stop reaction by adding 10 μL of a 0, 5 M EDTA and collect the RNA after phenol/chloroform extraction and ethanol-precipitation as described above (*see* points 5 and 6 in Section 3.1.2., and **Note 6**).

3.2.3. Electrophoretic Purification

1. Prepare a 8% PAA-gel solution (70 mL) consisting of 36 mL acrylamide/bis-acrylamide (25% w/v), 27 mL of 8.3 M urea solution and 7.0 mL 10× TBE containing 8.3 M urea. Mix thoroughly and start polymerization by adding 400 μL 10% ammonium peroxydisulfate and 20 μL TEMED.

2. Pour immediately between two glass plates held apart by spacers of 1.5 m*M* and sealed with clamps. Thereby, avoid air bubbles as these greatly affect the current flow. Insert the comb and place plate horizontally at room temperature for 30–40 min.

3. Prior to loading, pre-run gel at 10 V/cm^2 for 20 min. Resuspend RNA samples (20 µL) in PAA-loading buffer at a 2:1 ratio, denature 3 min at 95°C and load into the gel wells. Run gel vertically with 1× TBE as electrophoresis buffer for 1–2 h until the bromophenol blue band runs out of the gel.

   a. After electrophoretic separation, check the right size of the bands for rHP-let7 (82 bp) and iHP-let7 (96 bp) visualized by UV-shadowing at 254 nm. Excise the bands using a sterile scalpel, crush the gel slices, and soak in 1.0 vol. of 0.3 *M* sodium acetate solution.

   b. Incubate the suspension at 65°C for 2 h or overnight by room temperature to allow passive elution.

   c. Filter the sample through a syringe which is pre-packed with silanized glass wool (Pierce) to separate the nucleic acid containing liquid phase from the gel fragments.

   d. Finally, precipitate the RNA with ethanol and collect by centrifugation as described in previous sections. Dissolve the RNA pellet in 20 µL of water and store at −20°C until needed.

3.2.4. Ribozyme Assay Procedure

Ribozymes reactions were analyzed under multiple turnover conditions in real time using excess fluorescent (FRET)-labeled substrate over ribozyme at a 200:1 ratio.

1. Thaw ribozyme samples on ice and centrifuge at high speed for 20 s.

2. Prepare a ribozyme reaction mix (50 µL) in hairpin ribozyme reaction buffer containing 50 n*M* ribozyme in 96-well plates (*see* **Note 7**).

3. Add the let-7 miRNA (50 n*M*) and preincubate with the ribozyme mix in the Ascent Fluoroscan apparatus at 32°C for 15 min.

4. Start cleavage reaction upon addition of 5 µL of a 10 µ*M* FRET-labeled substrate stock solution. Insert plate immediately in the apparatus and start scanning (*see* **Note 8**).

5. Allow the cleavage reaction to proceed for at least 20 min at 30 s intervals until it has reached a constantly increasing linear phase by measuring the FAM-dye fluorescence (*see* **Note 9**).

6. Determine the initial velocity (fluorescence/min) by plotting the fluorescence intensity (*y*), multiplied by factor 10,000, versus time (*x*) by linear regression analysis using the Microsoft EXEL software and fitting to the following equation:

$$k\,(\text{fluorescence}/\text{min}) = \frac{n\sum xy - \left(\sum x\right)\left(\sum y\right)}{n\sum x^2 - \left(\sum x\right)^2} \times 10.000$$

where *n* is the number of data points (*see* **Note 10**).

oligonucleotide detection:

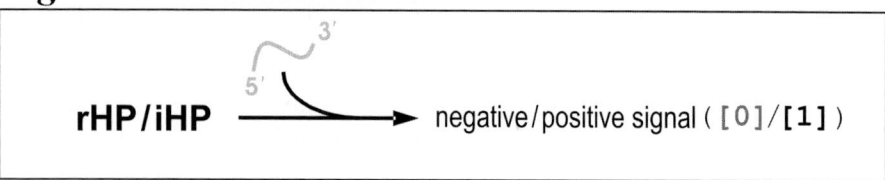

nucleic acids interactions:

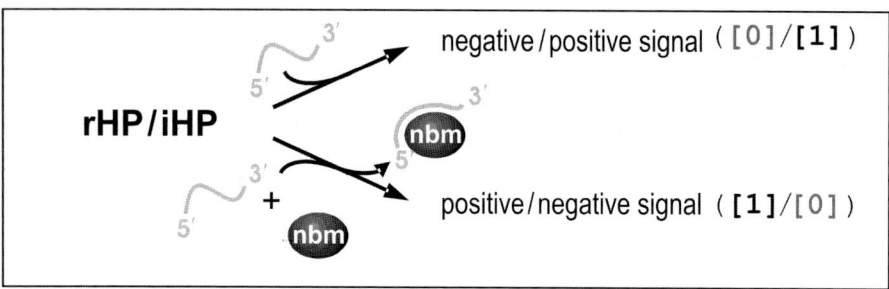

protein interactions:

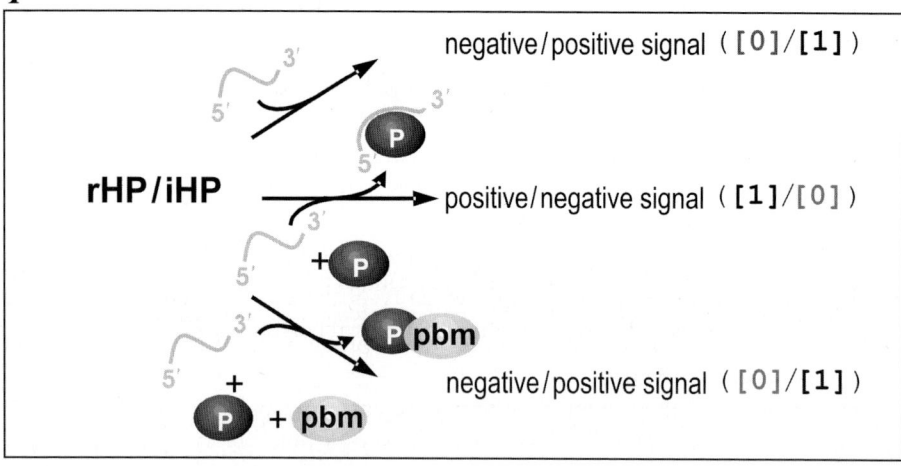

Fig. 3. Different application approaches of rHP and iHP ribozymes in molecular profiling of biological relevance. Both formats are suitable to generate sequence specific signals is response to their molecular environment. These can be utilized for oligonucleotide (grey line) detection, evaluation of nucleic acids interaction or protein (p) interactions. A schematic approach for each application is provided. All assay

Fig. 2B shows the plots (depicted in bars) of the corresponding ribozyme activity in presence and absence of the let-7 microRNA after data analysis.

3.3. A General Framework for Molecular Profiling

As described in previous sections, regulatory ribozymes can be designed in a rational and straightforward fashion and simply tuned to produce sequence-specifically a negative or a positive fluorescence signal. These can be generated for the direct detection of a wide array of functional oligonucleotides containing mRNAs, DNA/RNA aptamers, and microRNAs independently of their sequence. In contrast to the existing detection methods for molecular profiling this approach exhibits several advantages: (i) detection is achieved without extra labeling of the target molecules or their cognate detection probes. (ii) Higher sensitivity due to the advantage of signal amplification as a result of substrate cleavage reaction. (iii) In every detection set-up, the substrate molecule acts as the fluorescence tagged-probe. These features can be set as a basis to define a more general approach to gauge molecular interactions in real time. For example, in a high-throughput assay different microRNAs could be profiled in a higher accuracy when compared with standard molecular beacons *(10)*. Moreover, in a more complex set-up, the presence of distinct proteins and their target molecules could be easily sensed and evaluated. **Fig. 3** outlines the extent of applications in molecular profiling using these new catalytic probes.

4. Notes

1. As the basis for the design of regulatory hairpin ribozyme variants, a modified version of the wild-type hairpin ribozyme was used (**Fig. 1A**). Briefly, helix H4 in this construct is extended by 3 bp and a stable GUAA tetraloop is used to stabilize the loop B domain. Furthermore, this construct contains a rate-enhancing U47C mutation.

   Three base pair changes, one in H1 and two in H2, further contribute to a well-behaved kinetics by minimizing self-complementarity of the substrate. For more detailed descriptions consult the corresponding literature reports *(4,16,17)*.

---

formats relies on binding competition between interacting species. In case of RNA or DNA nucleic acids-binding molecules (nbm) can be detected as a result of positive or negative fluorescence read out. This is performed in two reaction setups that either contain or lack the cognate nbm. To evaluate protein interactions one further reaction setup is required which contains all three components for competition. The binding of the protein to the protein-binding molecule (pbm) results in release of the protein-binding oligonucleotide effector which affects ribozyme activity accordingly. Consequently, the fluorescence readout of each reaction can also be rendered into binary numbers.

2. As a criterion to allosteric behavior, the oligonucleotide complementary sequence is positioned in a manner that is distal from the catalytic core of the ribozyme. In this arrangement, this new site serves as a structurally independent-binding site. This is linked via a pseudo-three-way junction to enable catalysis to occur in a *trans*-cleaving mode, as well as to selectively favor the active bent conformation. Thereby, the base identities of the TCD are chosen in order to minimize the potential of domain C for disruptive interactions during catalysis by stabilizing the stem-loop formation. In contrast to reported "communication modules", TCD module can be rationally designed without being necessarily a subject for in vitro selection strategies.

3. If a complementary oligonucleotide hybridizes asymmetrically to the loop of a hairpin the topology of the resulting complex resembles one-half of a pseudo-knot, referred as a pseudo-half-knot *(18)*. This contains two stems stacked at the junction to form a connective double-stranded helix, which is similar to the reported pseudo-knot structure *(19)*.

4. In case of different predicted structures, only the thermodynamically more stable folding was considered as a basis for further investigations. However, sequences showing different foldings (including the appropriate one) but much closed energy values can only be tolerated if the difference in respective dG values does not exceed 0.5–0.7.

5. For a more accurate concentration calculation consult a free software program (Biopolymer Calculator) available on the World Wide Web site at http://paris.chem.yale.edu/extinct.html.

6. It is also recommended to use 6.0 *M* ammonium acetate instead of the sodium acetate for a more "protective" precipitation of RNA.

7. Reaction should be run in at least duplicates. As negative control experiments run additional reactions in the absence of ribozyme, substrate, and oligonucleotide effector molecules, respectively.

8. Importantly, begin scanning with vigorous shaking of the plate upon insertion for 5–10 s.

9. Changes in FRET decay occurs through a variable emission scan (500–650 nm) at a fixed excitation wavelength (492 nm). This is caused by dissociation of the 5′-and 3′-cleavage products from the ribozyme resulting in a loss of FRET and a shift in dominant emission spectrum from TAMRA (585 nm) to FAM fluorescein (520 nm). In this context, shorter excitation intervals should be avoided as this can result in bleaching of the FAM fluorescent molecule and a subsequent manipulation of the ribozyme activity during the reaction course.

10. The initial fluorescence/min values were corrected by subtracting values derived from reactions lacking ribozyme.

11. The catalytic activity in iHP constructs is highly Mg^{2+} dependent. At higher Mg^{2+} concentrations ribozyme activity is increased as this presumably contributes to the intermolecular stabilization of half-pseudo-knot structure. In this context, iHP-let7 achieves highest activity at 100 nm Mg^{2+} (4000–5000 fluorescence/min) which is comparable to the activity of rHP-let7 at basic conditions (data not shown).

Acknowledgments

This work was supported by the Deutsche Forschungsgemeinschaft and the Volkswagen-Stiftung (to M.F.). We thank Günter Mayer and Jörg S. Hartig for helpful comments.

References

1. Ferre-D'Amare, A. R. (2004) The hairpin ribozyme *Biopolymers* **73,** 71–78.
2. Zhuang, X., Kim, H., Pereira, M. J., Babcock, H. P., Walter, N. G., and Chu, S. (2002) Correlating structural dynamics and function in single ribozyme molecules *Science* **296,** 1473–1476.
3. Wilson, T. J., Nahas, M., Ha, T., and Lilley, D. M. (2005) Folding and catalysis of the hairpin ribozyme *Biochem. Soc. Trans.* **33,** 461–465.
4. Esteban, J. A., Banerjee, A. R., and Burke, J. M. (1997) Kinetic mechanism of the hairpin ribozyme. Identification and characterization of two nonexchangeable conformations *J. Biol. Chem.* **272,** 13,629–13,639.
5. Ferre-D'amare, A. R. and Rupert, P. B. (2002) The hairpin ribozyme: from crystal structure to function *Biochem. Soc. Trans.* **30,** 1105–1109.
6. Fedor, M. J. (2002) The catalytic mechanism of the hairpin ribozyme *Biochem. Soc. Trans.* **30,** 1109–1115.
7. Esteban, J. A., Walter, N. G., Kotzorek, G., Heckman, J. E., and Burke, J. M. (1998) Structural basis for heterogeneous kinetics: reengineering the hairpin ribozyme *Proc. Natl. Acad. Sci. USA* **95,** 6091–6096.
8. Walter, N. G., Burke, J. M., and Millar, D. P. (1999) Stability of hairpin ribozyme tertiary structure is governed by the interdomain junction *Nat. Struct. Biol.* **6,** 544–549.
9. Najafi-Shoushtari, S. H., Mayer, G., and Famulok, M. (2004) Sensing complex regulatory networks by conformationally controlled hairpin ribozymes *Nucleic Acids Res.* **32,** 3212–3219.
10. Hartig, J. S., Grüne, I., Najafi-Shoushtari, S. H., and Famulok, M. (2004) Sequence-Specific Detection of MicroRNAs by Signal-Amplifying Ribozymes *J. Am. Chem. Soc.* **126,** 722–723.
11. Hartig, J. S., Najafi-Shoushtari, S. H., Grüne, I., Yan, A., Ellington, A. D., and Famulok, M. (2002) Protein-dependent ribozymes report molecular interactions in real time *Nat. Biotechnol.* **20,** 717–722.
12. Najafi-Shoushtari, S. H. and Famulok, M. (2005) Competitive regulation of modular allosteric aptazymes by a small molecule and oligonucleotide effector *RNA* **11,** 1514–1520.
13. Mathews, D. H., Sabina, J., Zuker, M., and Turner, D. H. (1999) Expanded Sequence Dependence of Thermodynamic Parameters Improves Prediction of RNA Secondary Structure *J. Mol. Biol.* **288,** 911–940.
14. Zuker, M. (2003) Mfold web server for nucleic acid folding and hybridization prediction *Nucleic Acids Res.* **31,** 3406–3415.
15. Mullis, K., Faloona, F., Scharf, S., Saiki, R., Horn, G., and Erlich, H. (1986) Specific enzymatic amplification of DNA in vitro: the polymerase chain reaction *Cold Spring Harb. Symp. Quant. Biol.* **51 Pt 1,** 263–273.

16. Berzal-Herranz, A., Joseph, S., Chowrira, B. M., Butcher, S. E., and Burke, J. M. (1993) Essential nucleotide sequences and secondary structure elements of the hairpin ribozyme *Embo. J.* **12,** 2567–2573.

17. Romero-Lopez, C., Barroso-delJesus, A., Puerta-Fernandez, E., and Berzal-Herranz, A. (2004) Design and optimization of sequence-specific hairpin ribozymes *Methods Mol. Biol.* **252,** 327–338.

18. Ecker, D. J., Vickers, T. A., Bruice, T. W., et al. (1992) Pseudo–half-knot formation with RNA *Science* **257,** 958–961.

19. Tinoco, I., Jr. (1997) From RNA hairpins to kisses to pseudoknots *Nucleic Acids Symp. Ser.* **36,** 49–51.

18

Screening of Molecular Interactions Using Reporter Hammerhead Ribozymes

Jörg S. Hartig and Michael Famulok

Abstract

The characterization of molecular interactions is a central task in modern life sciences. Applications such as drug screening in pharmaceutics or the elucidation of biomolecular interactions in molecular biology rely on efficient methods to search for interacting partners. Here, we describe a novel technique that utilizes hammerhead ribozymes to signal molecular interactions. The ribozyme is modified by a domain that specifically binds to a target molecule such as a protein. Upon binding of the target, the catalytic activity of the ribozyme is changed, allowing for detection of the presence as well as the occurrence of interactions of the targeted ligand. The assay can be performed in high-throughput format by employing double-labeled ribozyme substrates, hence being well suited for drug-screening applications. The detection proceeds rapidly and in real-time. Moreover, the technique neither requires labeling of the target molecule nor the potential interaction partners or analytes since an indirect readout is facilitated by switching the catalytic activity of a reporter ribozyme. The assay can be utilized to sense a broad variety of biomolecular interactions, and is very sensitive due to signal amplification by the ribozyme reaction.

Key Words: Hammerhead ribozyme; aptamer; RNA reporter; drug-screening; protein-interactions; FRET-assay; signal amplification.

1. Introduction

Catalytic RNAs, also known as ribozymes, have been employed in the engineering of biosensors for the detection of target molecules *(1)*. In these studies, phosphodiester-cleaving ribozymes are modified by aptamers. By attaching small molecule-binding aptamers to ribozymes, biosensors were generated that signal the presence of the ligand by changing the catalytic activity of the ribozyme *(1,2)*. Further work on such fusions of aptamers and ribozymes (so-called aptazymes, or reporter-ribozymes) resulted in the development of

From: *Methods in Molecular Biology, Vol. 429: Molecular Beacons: Signalling Nucleic Acid Probes, Methods and Protocols* Edited by: A. Marx and O. Seitz © Humana Press Inc., Totowa, NJ

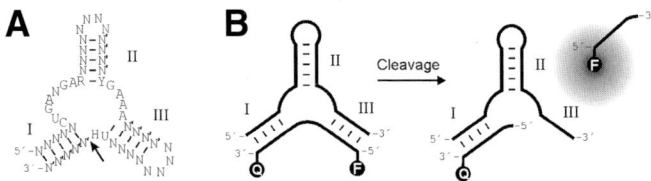

Fig. 1. The hammerhead ribozyme. **(A)** Secondary structure and conserved nucleotides. The ribozyme is shown in the intramolecular cleavage (in *cis*) format. The cleavage site is indicated by the arrow. *N*, any nucleotide; *N′*, any nucleotide complementary to *N*; *H*, any nucleotide except G; *R*, purine nucleotide; *Y*, pyrimidine nucleotide complementary to R. **(B)** Schematic of a hammerhead ribozyme-cleaving intermolecular (*in trans*). A pair of FRET dyes is attached to the RNA substrate (bottom strand), enabling readout of cleavage activity via fluorescence detection. *F*, fluorescent dye; *Q*, fluorescence quencher.

biosensors capable of detecting interactions between proteins and other biomolecules. These assays are based on ribozymes that change their catalytic activity in the presence of the target protein. If the protein-bound ribozyme is incubated with an interaction partner of the protein, the ribozyme is displaced and again undergoes a change of catalytic activity (**Fig. 2**). Detection of the interaction is greatly facilitated by monitoring the ribozyme-catalyzed cleavage reaction of fluorescently labeled ribozyme substrates *(3,4)*. Upon cleavage of the substrate, a fluorescent signal is generated by separating a pair of fluorescence resonance energy transfer (FRET) dyes attached to the ribozyme substrate. Hence, the readout of fluorescence allows the detection of biomolecular interactions, rendering the method suitable for high-throughput analysis. Here, we introduce different strategies for the construction of protein-dependent hammerhead ribozymes (HHRs), capable of detecting protein interactions. The method can be used to identify interactions of the target protein with varying classes of molecules, such as nucleic acids, smaller molecules, as well as other proteins. For example, we have employed reporter ribozymes in screens detecting protein interactions with small molecules for drug-screening purposes *(5)*.

The method described in this chapter exploits protein-regulated HHRs for the identification of biomolecular interactions. The HHR (*see* **Fig. 1A**) represents one of the best characterized catalytic RNAs. The HHR has relatively few requirements for catalytic activity: Its catalytic core, which is comprised of the conserved nucleotides shown in **Fig. 1A**, its three flanking variable helices, termed stems I, II, and III, and the presence of divalent metal ions such as Mg^{2+}. A detailed discussion of HHR kinetics can be found in the literature *(6,7)*. Hammerhead cleavage can be performed to occur in *trans* (intermolecularly) by deleting the terminal loop of one of the stems (*see* **Fig. 1B**, generated by deletion of stem III loop in **Fig. 1A**). Carrying out the cleavage reaction in *trans* allows the convenient attachment of a pair of fluorescent reporter dyes to the RNA substrate. In this system, the fluorescence of one reporter dye in the intact RNA substrate is quenched by FRET to the second dye (*see* **Fig. 1B**, left). In addition to easy readout of the cleavage reaction, multiple turnover of the substrate results in signal amplification, which is helpful when investigating low-affinity interactions or samples with low concentrations of target molecules *(8)*. A single interaction can trigger the catalytic activity of one ribozyme molecule, which then goes on to cleave multiple substrates, thereby amplifying the detection signal.

Ribozymes that change their catalytic activity in the presence of a target protein can be used to screen for interaction partners of the protein (*see* **Fig. 2**). To generate ribozymes that switch cleavage activity upon binding of the targeted molecule, protein-binding RNA sequences are connected to the ribozyme. So far, three distinct strategies for the construction of a protein-regulated aptazyme have been reported: direct selection, starting from partly randomized ribozyme pools *(1,9–12)*; selection of the connection site that links a preexisting aptamer and a ribozyme *(10,13)*; and rational design of aptazymes, lacking the selection step. Since the direct selection of an RNA enzyme responding to a protein has not yet been demonstrated and selection procedures for generation of aptazymes are still relatively complex and time-consuming, we will focus on the rational design of protein-dependent ribozymes for reporting molecular interactions.

1.1. Rational Design of Reporter Hammerhead Ribozymes

The technique of constructing an aptazyme by means of rational design is made possible by the often predictable and easily programmable structural features of nucleic acids, based on simple rules, such as Watson—Crick base pairing. Aptazymes have been constructed by simply fusing an existing aptamer to a ribozyme. For example, aptamers can be linked to a ribozyme sequence in a way that the aptamer hybridizes to the substrate recognition site of the ribozyme (*see* **Fig. 3A**). In order to facilitate the intramolecular interaction of the ribozyme's substrate-binding site with the aptamer sequence, the sequences of the substrate-binding site of the ribozyme, and also the substrate itself, can be altered. The HHR requires a substrate with the conserved sequence UH, where H is any nucleobase but G (*see* **Fig. 1A**). In addition, the cleavage site must be flanked by sequences that allow hybridization to the ribozyme so that the helical stem structures I and III are formed.

1.1.1. Detecting Interactions of HIV-1 Rev

For example, an aptamer sequence for the human-immunodeficiency virus type 1 (HIV-1) Rev protein was attached to the 5′-end of a HHR *(5)*. In the absence of Rev protein, the aptamer sequence hybridizes to the 5′-end of the ribozyme,

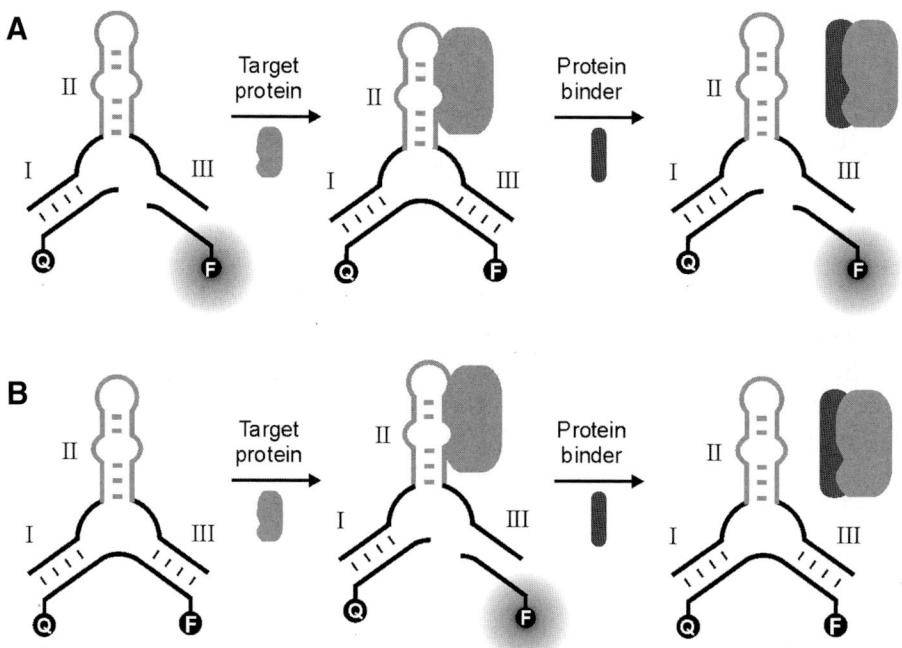

Fig. 2. General outline for the detection of protein interactions using reporter ribozymes. **(A)** Starting from catalytically active ribozymes. The ribozyme is switched off by the target protein via interaction with a protein-binding RNA domain (light grey). In the protein-bound state (middle), the complex can be used as sensor for interaction partners of the protein. In the presence of a protein binder (dark grey) that competes with the ribozyme binding of the protein, the ribozyme is liberated and catalytic activity is switched on. **(B)** Starting from inactive ribozymes: cleavage activity is induced by the target protein, addition of the interaction partner of the protein leads to inactivation of the ribozyme. When aiming at the detection of interaction partners of the target protein, the first approach is preferred since it generates a positive readout if an interaction takes place. If solely the detection of the presence or concentration of the target protein alone is desired, the second setup should be used since a signal is generated upon presence of the target protein.

thereby blocking substrate binding and hence cleavage activity (*see* **Fig. 3A**). Upon addition of HIV-1 Rev, the attached aptamer sequence changes its conformation driven by protein binding. Hence, the protein interaction allows the substrate to access the 5′-end of the ribozyme, thereby facilitating its cleavage. In a second design, the naturally occurring HIV-1 Rev-binding RNA sequence (Rev-binding element) was attached to the ribozyme in place of helix II (*5*). Upon addition of Rev protein, this ribozyme was switched off, displaying catalytic

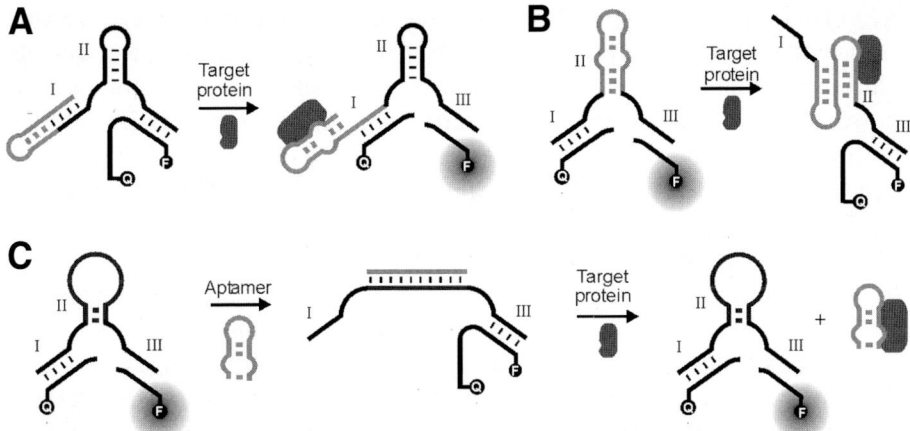

Fig. 3. Examples of protein-responsive ribozymes. Aptamer sequences are indicated in light grey. **(A)** Reporter ribozymes constructed by fusion of an aptamer to a ribozyme. The fused aptamer sequence prevents binding of the ribozyme to its substrate. Addition of the target protein results in folding of the aptamer and liberation of the substrate-binding site enabling cleavage activity. **(B)** In a second approach, aptamers undergoing structural reorganization upon binding of the target protein are fused to the hammerhead ribozyme to build stem II. For example, protein binding induces a pseudo-knotted structure of the aptamer, thereby disrupting the formation of the catalytic core (right panel). **(C)** This approach differs significantly from the others since a sequence complementary to the aptamer, instead of the aptamer itself, is introduced into the ribozyme (dark grey). Addition of the free aptamer (light grey) results in its hybridization to stem II of the ribozyme, resulting in disruption of the catalytically active ribozyme fold. If the target protein is added, the aptamer is decoyed from the ribozyme when complexing its target, thereby releasing the catalytically active ribozyme.

behavior opposite to that of the first Rev-responsive ribozyme. These ribozymes were subsequently used to screen a structurally diverse library of antibiotics for Rev-binding activity. If the Rev protein is bound by the small molecule, the ribozyme gets displaced from the complex and the catalytic activity changes. The experiment demonstrated the suitability of aptazymes for high-throughput assays and identified the first small molecules that bind directly to HIV-1 Rev. One of the identified antibiotics was subsequently shown to repress HIV-1 replication in cell culture *(5)*.

1.1.2. Detecting Interactions of HIV-1 Reverse Transcriptase

A different strategy for constructing protein-responsive ribozymes was exploited when we generated an aptazyme controlled by reverse transcriptase

(RT) of HIV-1. The aptamer for HIV-1 RT folds into a pseudoknot structure when complexed with the protein (*see* Refs. *[14–16]*). A competing, non-binding fold of the aptamer sequence is represented by a hairpin-looped structure. As shown in **Fig. 3B**, the aptamer was inserted into stem II of the HHR, resulting in a fusion construct FK-1 with competing folds of the ribozyme and the aptamer pseudo-knotted structure *(17)*. The simultaneous folding of both domains is impossible in this design, since pseudoknot formation results in disruption of the catalytic core of the ribozyme. In the absence of HIV-1 RT, the ribozyme folds and displays cleavage activity, with the inserted aptamer sequence folded as a hairpin loop in stem II (*see* **Fig. 3B**, left). Upon binding of HIV-1 RT, the pseudo-knotted fold of the aptamer is induced, disrupting stem II, and thus inhibiting catalytic activity of ribozyme FK-1. The ribozyme senses the presence of the protein at concentrations as low as 1 nM with a half-maximum inhibition of cleavage activity occurring at 10 nM HIV-1 RT. In addition to high sensitivity, the ribozyme displays high specificity since the reporter is unaffected by the presence of the homologue RT of HIV-2.

By generating variants of the reporter ribozyme for HIV-1 RT, we were able to demonstrate that rational design can be used to fine-tune nucleic acid function. Deletion of a single GC base pair in stem II yielded a construct with a destabilized stem II, which showed no cleavage activity in the absence of protein. In contrast, insertion of an additional GC base pair stabilizes stem II. This mutant is catalytically active, with the cleavage activity being unaffected even by the presence of HIV-1 RT. Owing to the increased stability of stem II, the protein is no longer able to induce folding of the pseudoknot. The generation of such reporter variants demonstrates that ribozyme properties can be designed by rational means *(17)*.

Interaction partners of HIV-1 RT were detected using the aptazyme FK-1. A complex of double-stranded DNA, which competes with the aptazyme for binding of HIV-1 RT, was detected in a concentration-dependent manner. In addition, drugs binding to the primer-template domain of the polymerase were detected as well. Interestingly, a RT inhibitor that also binds to the protein, but at a different site from the primer-template site was not detected, suggesting a domain-specific mode of detection *(17)*.

1.1.3. Detecting Interactions of Human α-Thrombin

A different approach was taken when we generated RNA reporters capable of detecting human α-thrombin and interaction partners *(5)*. The strategies for rational design discussed so far are best suited for generating reporters composed of natural RNA. Modified RNA or DNA aptamers are difficult to employ in such strategies since they cannot be introduced by standard in vitro transcription. In **Fig. 3C**, a novel strategy is introduced, allowing the use of aptamers composed of nucleic acids other than natural RNA. A ribozyme detecting human α-thrombin was generated by incorporating the RNA sequence complementary to the anti-thrombin DNA aptamer in stem II of the HHR (*see* **Fig. 3C**). The ribozyme itself is catalytically active. Upon addition of the DNA aptamer, hybridization to stem-loop II of the ribozyme prevents the formation of the catalytic core, thus switching the ribozyme off. Cleavage activity is restored by addition of α-thrombin, which complexes the DNA aptamer, thus releasing the active conformation of the ribozyme (*see* **Fig. 3C**, right). This protein-dependent release of the ribozyme turned out to be highly specific for human α-thrombin. Related clotting factors as well as less closely related proteins were unable to restore cleavage activity (**5**).

The thrombin-dependent ribozyme was used to detect the interaction between α-thrombin and a naturally occurring inhibitor peptide (hirudin) isolated from the leech *Hirudo medicinales*. Even small alterations of the analyte could be detected. For example, modification of hirudin by a single sulfate group was sensed (**5**). Moreover, the interaction between thrombin and a second interaction partner (anti-thrombin III) was not sensed. Since anti-thrombin III binds to thrombin at a different site than both hirudin and the DNA aptamer, interference with the aptamer binding did not occur. This result again shows that detection is likely based on competition of binding, enabling the detection of interactions in a domain-specific fashion.

2. Materials

2.1. Reagents and Solutions

Since the methodology utilizes RNA as reporters, extreme care should be dedicated to work with ribonuclease-free solutions and reagents. Buffers and solutions should exceptionally be prepared with DEPC-treated water. Additional measures should be taken in order to prevent contamination, *see* also (**18**).

HHR catalysis is effective under a wide range of conditions and will tolerate variations in buffer composition, temperature, and salt concentrations. The buffer can therefore be chosen to meet the specific requirements of the target protein. The buffer of choice should not contain divalent metal ions such as Mg^{2+} or Ca^{2+}, since these are added separately in order to initiate ribozyme cleavage.

2.2. Synthesis of Aptazymes and Substrate

Aptazymes are best prepared by in vitro transcription starting from a dsDNA template. Large quantities of the template can be obtained by PCR-amplification of the corresponding DNA sequence, containing a T7 promoter for in vitro transcription. The ribozyme itself is then purified from the subsequent in vitro transcription reaction by standard denaturing polyacrylamide gel electrophoresis (PAGE). For details of methodology, *see* Chapter 12.

Ribozyme sequences shorter that 35 nucleotides are best prepared using solid phase phosphoramidite chemistry since transcription yields tend to be low *(19)*. This technique also allows the synthesis of the RNA substrate modified with the two reporter dyes, which subsequently enable fluorescence readout of the cleavage reaction *(20)*. Many companies offer customized synthesis of modified or natural RNA (*see* http://www.glenres.com/ExtraPages/oligohouses.html).

3. Methods

3.1. Aptamer Selection

The first step of constructing a protein-responsive ribozyme is the selection of a nucleic acid sequence that binds to the target protein. This sequence might have a natural origin, as discussed above in the case of the ribozyme inhibited by the presence of HIV Rev *(5)*. In this example, a sequence that specifically binds to an arginine-rich domain of the Rev protein was adapted from the RNA genome of HIV in order to render the ribozyme Rev-dependent. If a naturally occurring nucleic acid is not available for the protein of interest, then the systematic evolution of ligands by exponential enrichment procedure can be used to generate protein-binding nucleic acid sequences *(21–23)*.

3.2. Design of Potential Aptazymes

If an RNA sequence that binds to the protein of interest has been identified, fusions of this sequence and the HHR can be designed based on the principles discussed above. Fusion sequences have to be examined for the desired secondary structures using *m*fold or similar folding algorithms prior to aptazyme synthesis *(24)*. Evaluating the secondary structure of potential aptazymes by such algorithms represents a powerful tool for predicting the catalytic behavior of a designed sequence. Although these algorithms do not take into account binding energies resulting from protein interactions or possible tertiary structure motifs, they often enable reliable prediction of the cleavage activity of the ribozyme in the absence of the target protein. For example, to generate a ribozyme that possesses catalytic activity, the secondary structure of the catalytic core, shown in **Fig. 1A**, should be among the lowest folding energy predictions generated by *m*fold or other algorithms (*see* **Note 1**). If promising candidates are found, synthesis of substrate and ribozyme RNA should be carried out as described in the Materials section.

3.3. Characterization of Aptazyme Sequences

In order to test whether the constructed ribozyme sequences respond to the presence of the target protein with a change in catalytic activity, cleavage reactions are carried out in the presence and absence of the target protein (*see* **Note 2**).

If the examined ribozymes show pronounced (10-fold or more) differences in catalytic activity in the presence and absence of the target protein, they are suited for screening for interactions of the target protein. The identification of a ribozyme sequence that is unaffected by the presence of the target protein is not without its uses. Such non-responsive ribozymes can be used in control reactions during the final screening experiment, as described in Section 3.4.

Reactions for the characterization of ribozymes can be carried out in multi-well plate fluorescence readers, using 96- or 384-well plates, in volumes as low as 10 µL *(8)*. Nevertheless, for easy handling, a total reaction volume of 50 µL is appropriate. The ribozyme reaction is initiated by adding $MgCl_2$ (*see* **Note 3**) to a final concentration of 10 mM (*see* **Note 4**). During the course of the reaction, fluorescence is measured as a function of time. Especially in cases where the ribozyme reaction appears to be relatively slow ($\kappa_{obs} \ll 1$ min), data acquisition should be carried out at reasonable intervals, e.g., every 2–5 min since excessive excitation of the fluorescent markers results in bleaching of the dyes (*see* **Note 5**).

Typical starting reaction conditions are as follows:

Final volume: 50 µL

Combine	Final concentration
Buffer	As required
Ribozyme	50 nM
Substrate	500 nM
Target protein	50 nM–5 µM
$MgCl_2$	10 mM (reaction start)

Prior to starting the ribozyme reactions with $MgCl_2$, the ribozyme, RNA substrate, target protein, and analyte should be pre-incubated for an appropriate time to allow formation of the desired complexes (*see* **Note 6**).

3.4. Screening and Evaluation

If a protein-responsive ribozyme is identified, analytes can be screened for interactions with the target protein. The individual reactions should be performed in a multi-well plate format, including duplicate standard reactions of the ribozyme alone, as well as ribozyme with target protein on the same well plate. Ribozyme-only reactions (lacking protein) are used to normalize the data obtained from reactions containing potential interaction partners.

Two independent reactions should be carried out for each analyte tested: one containing the reporter ribozyme and the target protein, and one containing target protein and a ribozyme that is not regulated by the protein (*see* Section 3.3.). The latter reaction serves as a control to identify analytes that influence ribozyme activity by interacting directly with the ribozyme rather than with the target protein. When this control is factored in, all influences affecting solely the ribozyme activity can be excluded. The target-insensitive control ribozyme should display cleavage activity independent of the target protein. For this purpose, variants of the aptazyme-containing scrambled sequences or mutations in the aptamer part of the molecule generally work well since such changes usually prevent interactions with the target protein. If such a mutant is not easily found, unmodified ribozymes such as shown in **Fig. 1A** may serve as control (*7,25*).

Strong ribonuclease activity in the analyte preparations represents a challenge and should be avoided. Ribonuclease activity results in digestion of the RNA substrate, which could be misinterpreted as cleavage activity originating from the ribozyme. Nevertheless, weak-to-moderate nuclease activity is tolerable if a control reaction for each analyte, containing only the RNA substrate, is included. By doing so, correction of the ribozyme cleavage data is possible by subtracting the "substrate only" control.

The following equation can be used to evaluate the raw cleavage data:

$$A_{rel} = \frac{A_{analyte}^{reporter} \Big/ A_{standard}^{reporter}}{A_{analyte}^{control} \Big/ A_{standard}^{control}}$$

With

$A_{analyte}^{reporter}$: Activity of reporter ribozyme in presence of target protein and analyte;

$A_{standard}^{reporter}$: Activity of reporter ribozyme in presence of target protein only;

$A_{analyte}^{control}$: Activity of control ribozyme in presence of target protein and analyte;

$A_{standard}^{control}$: Activity of control ribozyme in presence of target protein only.

The value that is described by this equation represents a relative activity, corrected for nonspecific influences affecting the ribozyme directly, and thus revealing true interaction partners of the target protein. The interaction partners identified by the aptazyme-based screening protocol should be validated using alternate methods for characterizing biomolecular interactions.

4. Notes

1. The aptazyme is designed to function as an in *trans*-cleaving ribozyme. The substrate sequence should be included when calculating the secondary structure of a potential ribozyme. Since folding of a single sequence is achieved more easily, we connect the substrate and ribozyme strand via a short loop (e.g., a GNRA tetraloop, with N: any nucleotide; R: G or A), as shown in **Fig. 1A** when calculating the lowest energy fold.

2. The optimal ratio of ribozyme to substrate will vary with the catalytic fitness of the ribozyme, but ribozyme concentrations ranging from 5 to 100 nM with substrate concentrations between 200 and 1000 nM are a good starting point. An excess of substrate enables signal amplification, since the *in trans* versions of the ribozymes used here are able to perform multiple substrate turnovers.

3. The reactions can be performed in a highly parallel fashion by simultaneously adding MgCl$_2$ to all reaction wells. The reaction can be started manually by adding MgCl$_2$ using a multi-channel pipette since the ribozyme reactions are slow when compared with most protein kinetics. The microtiter plate should then be assayed immediately using the fluorescence reader. Nevertheless, we usually use an automated dispensing device of the fluorescent reader to start the ribozyme reactions.

4. The concentration of Mg^{2+} required for optimum cleavage activity could be influenced by the modification of the ribozyme by fusing the aptamer sequence. Ten millimolar of MgCl$_2$ is usually appropriate for unmodified HHRs and works well for most modified ribozymes.

5. Alternatively, ribozyme catalysis can be monitored by assessing cleavage of radioactively labeled RNA substrates. This might be helpful if the analytes are fluorescent and interfere with the fluorescent readout. Nevertheless, the high-throughput character of the method is lost when using radioactively labeled substrates.

   Aliquots are removed at certain time points during the reaction and are quenched in formamide loading buffer. Cleavage products are separated from the uncleaved substrate by 12% PAGE, and subsequently analyzed by autoradiography. The fraction of cleaved substrate is divided by the sum of cleaved and intact substrate, and plotted versus the time, yielding a time course of the cleavage reaction.

6. The order of addition of the components usually has little influence on the resulting ribozyme activity, although comparison of different orders of addition, as well as a range of incubation times can be carried out for optimizing the reaction.

Acknowledgments

This work was supported by the Deutsche Forschungsgemeinschaft and the Volkswagen-Stiftung (to M.F.).

References

1. Breaker, R. R. (2002) Engineered allosteric ribozymes as biosensor components. *Curr. Opin. Biotechnol.* **13**, 31–39.

2. Seetharaman, S., Zivarts, M., Sudarsan, N., and Breaker, R. R. (2001) Immobilized RNA switches for the analysis of complex chemical and biological mixtures. *Nat. Biotechnol.* **19**, 336–341.

3. Jenne, A., Gmelin, W., Raffler, N., and Famulok, M. (1999) Real-Time Characterization of Ribozymes by Fluorecence Resonance Energy Transfer (FRET). *Angew. Chem. Int. Ed.* **38**, 1300–1303.

4. Jenne, A., Hartig, J. S., Piganeau, N., et al. (2001) Rapid identification and characterization of hammerhead-ribozyme inhibitors using fluorescence-based technology. *Nat. Biotechnol.* **19**, 56–61.

5. Hartig, J. S., Najafi-Shoushtari, S. H., Grune, I., Yan, A., Ellington, A. D., and Famulok, M. (2002) Protein-dependent ribozymes report molecular interactions in real time. *Nature Biotechnology* **20**, 717–722.

6. Hertel, K. J., Stage-Zimmermann, T. K., Ammons, G., and Uhlenbeck, O. C. (1998) Thermodynamic dissection of the substrate-ribozyme interaction in the hammerhead ribozyme. *Biochemistry* **37**, 16,983–16,988.

7. Stage-Zimmermann, T. K. and Uhlenbeck, O. C. (1998) Hammerhead ribozyme kinetics. *Rna* **4**, 875–889.

8. Hartig, J. S., Grune, I., Najafi-Shoushtari, S. H., and Famulok, M. (2004) Sequence-specific detection of MicroRNAs by signal-amplifying ribozymes. *J. Am. Chem. Soc.* **126**, 722–723.

9. Koizumi, M., Soukup, G. A., Kerr, J. N., and Breaker, R. R. (1999) Allosteric selection of ribozymes that respond to the second messengers cGMP and cAMP. *Nat. Struct. Biol.* **6**, 1062–1071.

10. Soukup, G. A. and Breaker, R. R. (1999) Engineering precision RNA molecular switches. *Proc. Natl Acad. Sci. USA* **96**, 3584–3589.

11. Piganeau, N., Thuillier, V., and Famulok, M. (2001) In vitro selection of allosteric ribozymes: theory and experimental validation. *J. Mol. Biol.* **312**, 1177–1190.

12. Piganeau, N., Jenne, A., Thuiller, V., and Famulok, M. (2001) An allosteric ribozyme regulated by doxycycline. *Angew. Chem. Int. Ed.* **39**, 4369–4373.

13. Kertsburg, A. and Soukup, G. A. (2002) A versatile communication module for controlling RNA folding and catalysis. *Nucleic Acids Res.* **30**, 4599–4606.

14. Tuerk, C., MacDougal, S., and Gold, L. (1992) RNA pseudoknots that inhibit human immunodeficiency virus type 1 reverse transcriptase. *Proc. Natl Acad. Sci. USA* **89**, 6988–6992.

15. Jaeger, J., Restle, T., and Steitz, T. A. (1998) The structure of HIV-1 reverse transcriptase complexed with an RNA pseudoknot inhibitor. *Embo. J.* **17**, 4535–4542.

16. Kensch, O., Connolly, B. A., Steinhoff, H. J., McGregor, A., Goody, R. S., and Restle, T. (2000) HIV-1 reverse transcriptase-pseudoknot RNA aptamer interaction has a binding affinity in the low picomolar range coupled with high specificity. *J. Biol. Chem.* **275**, 18,271–18,278.

17. Hartig, J. S. and Famulok, M. (2002) Reporter ribozymes for real-time analysis of domain-specific interactions in biomolecules: HIV-1 reverse transcriptase and the primer-template complex. *Angew. Chem. Int. Ed.* **41**, 4263–4266.

18. Sambrook, J. and Russell, D. W. (2001) In Molecular Cloning, Vol. 1, p. 7.82, Cold Spring Harbor Laboratory Press, Cold Spring Harbor, New York.

19. Beaucage, S. L. (2002) Synthesis of Unmodified Oligonucleotides, In Current Protocols in Nucleic Acid Chemistry (Beaucage, S. L., Bergstrom, D. E., Glick, G. D., Jones, R. A., eds.), John Wiley & Sons, Chapter 3.

20. Beaucage, S. L. (2003) Synthesis of Modified Oligonucleotides and Conjugates. In Current Protocols in Nucleic Acid Chemistry (Beaucage, S. L., Bergstrom, D. E., Glick., G. D., Jones, R. A, eds.), John Wiley & Sons, Chapter 4.

21. Sun, S. (2000) Technology evaluation: SELEX, Gilead Sciences Inc. *Curr. Opin. Mol. Ther.* **2**, 100–105.

22. Joyce, G. F. (1994) In vitro evolution of nucleic acids. *Curr. Opin. Struct. Biol.* **4,** 331–336.
23. Klug, S. J. and Famulok, M. (1994) All you wanted to know about SELEX. *Mol. Biol. Rep.* **20,** 97–107.
24. Zuker, M. (2003) Mfold web server for nucleic acid folding and hybridization prediction. *Nucleic Acids Res.* **31,** 3406–3415.
25. Hertel, K. J., Herschlag, D., and Uhlenbeck, O. C. (1994) A kinetic and thermo-dynamic framework for the hammerhead ribozyme reaction. *Biochemistry* **33,** 3374–3385.

Index

Printed in the United States of America